FRONTIERS IN PLURIPOTENT STEM CELLS RESEARCH AND THERAPEUTIC POTENTIALS

Bench-to-Bedside

Edited By

Kuldip S. Sidhu

Director Stem Cell Lab, Faculty of Medicine
The University of New South Wales
Australia

FIRST EDITION

Bentham Science Publishers
Executive Suite Y - 2
PO Box 7917, Saif Zone
Sharjah, U.A.E.
subscriptions@benthamscience.org

Bentham Science Publishers
P.O. Box 446
Oak Park, IL 60301-0446
USA
subscriptions@benthamscience.org

Bentham Science Publishers
P.O. Box 294
1400 AG Bussum
THE NETHERLANDS
subscriptions@benthamscience.org

DEDICATION

To my beautiful and conscientious wife, Dr Kiran Sidhu and our two most adorable sons, Dr Ankur Sidhu and Mr Robin Sidhu who are my inspiration and their inquisitive minds always put me on the spot and make me think to explore further.

CONTENTS

FOREWORD

Scientific research continues to add new dimensions to the life of mankind. In the recent years there has been renewed interest in the field of stem cell particularly the embryonic stem cell research at the cutting edge molecular level, and this has given rise to the whole new field of stem cell biotechnology – harnessing their full potentials for alleviating the sufferings of human with diseases for which there is no cure. The rapid expanding knowledge in this area has open up new vistas and products which will affect many aspects of human health at large.

The collection of scholarly written chapters in this eBook focuses on the impact of rapid development in the field of pluripotent stem cells research on human health and many debilitating human diseases such as diabetes, Alzheimer's, macula degeneration and many more where the traditional medical approaches are less than ideal. Because this field of stem cell is becoming so broad, other areas of stem cell biology like pluripotency and disease modelling, tissue engineering, toxicological perspectives, ethics, animal models are also discussed in length to move this field forward from laboratory to clinics.

The future impact of these advances will allow a much greater control over people's health, and to enhance both the length and quality of their lives. Understanding of the disease process, disease modelling and tissue engineering will be enhanced considerable by using pluripitent stem cells in the future and that will pave the development of cell therapy more quickly.

However, it would be entirely misplaced optimism to believe that such advances will usher in an era of stem cell panacea for all kind of human ailments because with the increase of human life expectancy recently, unexpected diseases may also crop up in the population. At best stem cell field will provide a definitive further strong support towards human health and can offer answers more quickly.

This excellent compilation by the Professor Kuldip Sidhu is very timely when pluripotent stem cells field once again is emerging as an important area of human biotechnology and regenerative medicine with number of more clinical trials in horizon.

Professor Glyn Stacey

Director, UK Stem Cell Bank
EN6 3QG Hertfordshire
UK

PREFACE

Stem cells in general and pluripotent stem cells in particular have gained tremendous interest in the recent years, primarily driven by the hope of finding cures for several debilitating human diseases through cell transplantation (regenerating medicine). Pluripotent stem cells have the inherent ability to reproduce indefinitely and have the capability to produce all the 220 different types of cells constituting the human body and thus offer tremendous therapeutic potentials. The isolation of human embryonic stem cells (hESCs) from embryos and their successful culture in the petrii dish in 1998 has been considered the biggest breakthrough of the 21st century. This has been followed by another remarkable breakthrough in 2006 when scientists demonstrated for the first time that such pluripotent stem cells could be produced from adult somatic tissues by reprogramming without having to use human embryos. These pluripotent stem cells are called the induced pluripotent stem (iPS) cells. Production of iPS cells has been considered as the biggest discovery of this decade. Both hESCs and iPS cells are pluripotent and are highly versatile and offer tremendous therapeutic potential for finding cures for many incurable diseases such as diabetes, Parkinson's, Alzheimer's, and many other diseases *via* stem cell therapeutic in the next decade or so.

Both ES and iPS cells are pluripotent and capable of differentiating into three primary germ layer derivatives—an important characteristic for producing healthy cells for therapeutic purposes. While there are some proofs of principle with embryonic stem cells for therapeutic approaches in animal models, two major roadblocks to their use in humans must be overcome: post-transplantation immune rejection and ethical issues. iPS cell technology addresses both concerns: it is possible to use a patient's own somatic cells to generate therapeutic iPS cells (thus eliminating the potential for immune rejection), and they represent an acceptable alternative to the use of human embryos for stem cell production. An added benefit is that patient-specific iPS cells can be used for drug and regenerative medical research. However, there are some significant issues that need further investigations : (i) the low efficiency of cell reprogramming, (ii) the integration of viral transgenes into the somatic genome, especially oncogenes such as c-MYC and KLF4 and (iii) iPS cell can lead to tumor formation. There has been a significant progress during the last 4-5 years to address to these issues and the efficiency of reprogramming has been increasing and the non-integrating systems developed to eliminate the possibility of mutagenesis with integrating approaches.

The use of stem cells in the therapeutic field and in drug development has been of considerable interest in contemporary bioscience. In recent years our knowledge and understanding of stem cell biology and regenerative medicine has increased substantially. ESCs not only continuously divide, they are also able to differentiate into all cell types of the human body. The isolation of tissue stem cells could offer a less ethically contentious and more practical source of replacement tissue for organs that are susceptible to age-related diseases or traumatic injury. These diseases include Alzheimer's and Parkinson's disease, but also stroke, myocardial infarction and diabetes to mention a few. They have become a serious health problem in our societies as people now live longer. A large part of stem cell research aims to identify the ideal cell type and time point of cell transplantation. Future research will also lead to improved protocols that generate more pure populations of transplantable cells. With the continuous progress in stem cell research, modern clinical medicine is at the threshold of transformation.

Cellular therapies with products from pluripotent stem cell have the potential to treat many conditions where present conventional treatments are inadequate. As such, public expectation remains high that these novel therapies will be ''panacea,'' even though there have been few preclinical animal trials. However, recent FDA approval for two clinical trials with cells derived from hESC for spinal cord injury patients and those with macula degeneration will give a big boost once some preliminary data are made available. Therefore, there is still a significant gap between promising laboratory-based research and approved final products in this emerging field. the information generated during the last one decade on these pluripotent stem cells is compartmentalized in different journals and reports. This eBook brings together the state-of-the-art on these stem cells, compiled by reputed scientists from different part of the world who are experts in the relevant fields. The major focus of this eBook has been to bring together all aspects of pluripotent stem cells from basic biology to their use in understanding disease process, toxicology, drug discovery and towards developing therapeutics.

The major credit goes to all those who contributed scholarly written chapters on fields of their expertise to this eBook and have laid the foundation for continuous progression of innovative investigations for inquisitive minds to explore further the areas of stem cell biology that can offer respite to many debilitating human diseases for which there is no cure. Members of the stem cell lab contributed sincerely for completion of this eBook. The patience of my family, particularly my wife, Dr Kiran Sidhu greatly helped me in completing this task particularly Dr Sophia Dean who prepared the subject index and also carried out general editing. The cooperation of the Bentham Press and its staff to bring out the final form of this eBook that is more presentable and readable is appreciable.

Kuldip S. Sidhu
Director Stem Cell Lab, Faculty of Medicine
The University of New South Wales
Australia

List of Contributors

Susanne B.-Hoffman

In Vitro Methods Unit (ECVAM)
Institute for Health & Consumer
Protection Joint Research Centre
European Commission *via* E. Fermi
Ispra (VA) Italy
E-mail: susanne.bremer@jrc.ec.europa.eu

Romulo M. Brena

Epigenome Center
Norris Comprehensive Cancer Center
University of Southern California
1450 Biggy Street, NRT G511
Los Angeles, CA 90033-9601
E-mail: Brena@usc.edu

Sven Büttner

Alzheimer's & Parkinson's Disease Lab
Brain & Mind Research Institute
University of Sydney
Sydney, Australia
E-mail: sven.buettner@dzne.de

H. Cao

College of Veterinary Medicine
Shaanxi Center of Stem Cell Engineering
& Technology
Northwest A&F University
Yangling, Shaanxi, China
E-mail: 04204142caohui@163.com

Methichit Chayosumrit

Siriraj Center of Excellence for
Stem Cell Research
Faculty of Medicine, Siriraj Hospital
Mahidol University
Thailand
E-mail: methichit@gmail.com

Henry Chung

Stem Cell Lab
Faculty of Medicine
The University of New South Wales
Sydney, Australia
E-mail: henc218@gmail.com

Taraprasad Das

L V Prasad Eye Institute
Bhubaneswar Campus

Orissa, India
E-mail: tpdbei@gmail.com

Mark Denham

Centre for Neuroscience
Department of Pharmacology
The University of Melbourne
Australia
E-mail: mdenham@unimelb.edu.au

Mirella Dottori

Centre for Neuroscience
Department of Pharmacology
The University of Melbourne
Melbourne, Australia
E-mail: mdottori@unimelb.edu.au

Roberto E.-Waser

In Vitro Methods Unit (ECVAM)
Institute for Health & Consumer
Protection Joint Research Centre
European Commission *via* E. Fermi
Ispra (VA) Italy
E-mail: Roberto.ENSENAT-WASER@ec.europa.eu

Subhash Gaddipati

L.V.Prasad Eye Institute
Kallam Anji Reddy Campus
L.V.Prasad Marg, Banjara Hills
Hyderabad, India
E-mail: subhashgaddipati@gmail.com

Jürgen Götz

Alzheimer's & Parkinson's Disease Lab
Brain & Mind Research Institute
University of Sydney
Sydney, Australia
E-mail: jgoetz@med.usyd.edu.au

Anand Hardikar

Diabetes and Pancreas Biology Section
The O'Brien Institute, Department of Surgery
St Vincent's Hospital
The University of Melbourne
Australia
E-mail: aahardikar@gmail.com

Y. Hu

College of Veterinary Medicine
Shaanxi Center of Stem Cell Engineering & Technology
Northwest A&F University
Yangling, Shaanxi, China

E-mail: shelley4@163.com

Jinlian Hua

College of Veterinary Medicine
Shaanxi Center of Stem Cell Engineering
& Technology
Northwest A&F University
Yangling, Shaanxi, China
E-mail: jlhua2003@126.com

Brock James

Centre for Neuroscience
Department of Pharmacology
The University of Melbourne
Australia
E-mail: bjames@unimelb.edu.au

Martina Klaric

In Vitro Methods Unit (ECVAM)
Institute for Health & Consumer
Protection Joint Research Centre
European Commission *via* E. Fermi
Ispra (VA) Italy
E-mail: Martina.KLARIC@ec.europa.eu

Khun H. Lie

Stem Cell Lab
Faculty of Medicine
University of New South Wales
Australia
E-mail: kunokuniya@yahoo.com

Robert Lindeman

Department of Haematology
Prince of Wales Hospital
Barker Street
Randwick NSW 2031
Australia
E-mail: Robert.Lindeman@SESIAHS.HEALTH.NSW.GOV.AU

Zichuan Liu

State Key Lab of Reproductive Biology
Institute of Zoology
Chinese Academy of Sciences
Graduate School of Chinese Academy of Sciences
Beijing, P R. China
E-mail: liuchuanfeng_81@163.com

David Ma

Blood Stem Cells and Cancer Research
St Vincent's Centre for Applied Medical Research
St Vincent's Hospital

Darlinghurst NSW 2010 Australia
E-mail: dma@arrow.org.au

Indumathi Mariappan

L.V.Prasad Eye Institute
Kallam Anji Reddy Campus
L.V.Prasad Marg, Banjara Hills
Hyderabad, India
E-mail: indumathi@lvpei.org

Michael Morris

School of Medical Sciences
Bosch Institute
Royal North Shore Hospital
The University of Sydney
Sydney, Australia
E-mail: michaelmorris@med.usyd.edu.au

Catalina A. Palmer

Blood Stem Cells and Cancer Research
St Vincent's Centre for Applied Medical Research
St Vincent's Hospital
Darlinghurst NSW 2010
Australia
E-mail: catalina1001@gmail.com

Martin Pera

The Eli and Edythe Broad Center
Regenerative Medicine and Stem Cell Research
USC, USA
E-mail: mpera@unimelb.edu.au

Perminder Sachdev

Neuropsychiatry Institute
The University of New South Wales
Sydney, Australia
E-mail: p.sachdev@unsw.edu.au

Virender Sangwan

L.V.Prasad Eye Institute
Kallam Anji Reddy Campus
L.V.Prasad Marg, Banjara Hills
Hyderabad, India
E-mail: vsangwan@lvpei.org

Kuldip S. Sidhu

Stem Cell Lab
Faculty of Medicine
The University of New South Wales
Sydney, Australia
E-mail: k.sidhu@unsw.edu.au

Loane Skene

Melbourne Law School
University of Melbourne
Parkville, Melbourne
Australia 3010
E-mail: l.skene@unimelb.edu.au

J. Sun

College of Veterinary Medicine
Shaanxi Center of Stem Cell Engineering & Technology Northwest
A &F University Yangling
Shaanxi, China
E-mail: sunjw2005@163.com

Greg Sutherland

Discipline of Pathology
Sydney Medical School
University of Sydney
Sydney, Australia
E-mail: g.sutherland@sydney.edu.au

Geeta Vemuganti

L.V.Prasad Eye Institute
Kallam Anji Reddy Campus
L.V.Prasad Marg, Banjara Hills
Hyderabad, India
E-mail: gkvemuganti@gmail.com

Kinga Vojnits

In Vitro Methods Unit (ECVAM)
Institute for Health & Consumer
Protection Joint Research Centre
European Commission *via* E. Fermi, Ispra (VA) Italy
E-mail: Kinga.Vojnits@jrc.ec.europa.eu

L. Wang

College of Veterinary Medicine
Shaanxi Center of Stem Cell Engineering
& Technology Northwest A&F University
Yangling, Shaanxi, China
E-mail: wangjameslong@yahoo.com.cn

Baolong Xia

State Key Lab of Reproductive Biology
Institute of Zoology
Chinese Academy of Sciences
Graduate School of Chinese Academy of Sciences
Beijing, P R. China
E-mail: xiaxia-222@hotmail.com

Shun Zhang

State Key Lab of Reproductive Biology

Institute of Zoology
Chinese Academy of Sciences
Graduate School of Chinese Academy of Sciences
Beijing, P R. China
E-mail: zs87sky@126.com

S. Zhang

College of Veterinary Medicine
Shaanxi Center of Stem Cell Engineering & Technology
Northwest A&F University
Yangling, Shaanxi, China
E-mail: shanshan_331986@126.com

Qi Zhou

State Key Lab of Reproductive Biology
Institute of Zoology
Chinese Academy of Sciences
Beijing, P R. China
E-mail: zhouqi@ioz.ac.cn

Shuya Zhou

State Key Lab of Reproductive Biology
Institute of Zoology
Chinese Academy of Sciences
Graduate School of Chinese Academy of Sciences,
Beijing, P R. China
E-mail: zhousy@ioz.ac.cn

H. Zhu

College of Veterinary Medicine
Shaanxi Center of Stem Cell Engineering
& Technology
Northwest A&F University
Yangling, Shaanxi, China
E-mail: haijing12585@163.com

CHAPTER 1

Stem Cells, Definition, Classification and Sources

Kuldip S. Sidhu[1*], Methichit Chayosumrit[2] and Khun H. Lie[1]

[1]Stem Cell Laboratory, Faculty of Medicine, University of New South Wales, Australia and [2]Siriraj Center of Excellence for Stem Cell Research, Faculty of Medicine Siriraj Hospital, Mahidol University, Thailand

Abstract: The human body is made up of about 220 different kinds of specialized cells such as nerve cells, muscle cells, fat cells and skin cells. All specialized cells originate from stem cells. Stem cells are not specialized and the process of their specialization is called differentiation. Depending on the source, stem cells can be classified into two broad categories *i.e.* embryonic stem cells that are derived from embryos and non-embryonic stem cells that are derived from adult and fetal tissues. Induced pluripotent stem cells, another types of pluripotent stem cells derived from any tissue by reprogramming and are the homologous source of stem cells. Stem cells are emerging as an important source of material for diseases in regenerative medicine. The non-embryonic stem cells like adult stem cells are in clinical use for many years and embryonic stem cells are now emerging as an alternative source for the same purpose with huge potentials in drug discovery and toxicological studies. The study of biology of stem cells is the hallmark of the recent emerging field of regenerative medicine and medical biotechnology.

Keywords: Embryonic stem cells, adult stem cells, feeder fibroblast, differentiation, pluripotency, induced pluripotent stem cells.

INTRODUCTION

Stem cell research has attracted tremendous interest recently because of their potential applications in regenerative medicine and also their growing need in drug discovery and toxicological studies. Several types of stem cells have been isolated and identified and that all can be broadly classified into two main categories, the *embryonic-* that are derived from embryos and *non-embryonic*-type that are found in adult and fetal tissues. Induced pluripotent stem cells on the contrary can be derived from any tissue by reprogramming and are the recent addition to the repertoire of stem cells. The stemness of "stem cells" resides in their ability to self-renew and differentiate to other lineages. The degree of stemness determines whether stem cells are *totipotent, pluripotent, multipotent, monopotent* with increasing restrictions respectfully to differentiate into different cell types - totipotent being the least and monopotent the most restricted stem cells. While the non-embryonic stem cells such as bone marrow stem cells and fetal stem cells including fetal tissues have been used clinically for many years in such diseases like cancer and ailments of central nervous system, the embryonic stem cells are emerging only recently as an alternative sources of stem cells with great potentials in regenerative medicine for being pluripotent and almost all cell types can be derived from such cells. However, there are number of issues, both the ethical and technical that need to be resolved before their full potentials can be harnessed fully. Recently there has been a significant progress in that direction with FDA approval of the first clinical trials with these cells for diseases for spinal cord injury and macular degeneration. This chapter attempts to define, classify and describe the sources of the various types of stem cells that have been isolated to date.

1. STEM CELLS

Stem cells are endowed with the property of self-renewal and the ability to give rise to specialized cell types (potency). Stem cell potency is arranged in a hierarchical manner starting with the early *totipotent*

***Address correspondence to Kuldip S. Sidhu:** Director, Stem Cell Laboratory, Faculty of Medicine, The University of New South Wales, Level 2, Clinical Sciences Building, Prince of Wales Hospital, Randwick, NSW 2031, Australia; Tel: + 61 2 9382 2622; Fax: +61 2 9382 2622; E-mail: k.sidhu@unsw.edu.au *stem cells* that are able to form both embryo and placenta, followed by pluripotent stem cells that are able only to form the embryo, *multipotent stem cells* that are able to form limited number of cell lineages within the three embryonic germ layers (endoderm, mesoderm and ectoderm), and finally *monopotent stem cells* which are committed to form a single cell type [1]. The two distinctive types of stem cells are: adult stem cells that are specifically found from a limited number of cell types in the adult tissues; and embryonic stem cells that are isolated from the inner cell mass (ICM) of an expanding blastocyst. These embryonic stem cells are pluripotent, whereas the adult stem cells are typically multipotent or monopotent. In the year 2006, another category of pluripotent stem cells known as induced pluripotent stem cells that can be derived from any adult tissue by reprogramming has been added [2].

Stem cell potency at least in animals can be assessed using the blastocyst complementation assay, which involves the injection of the candidate cells into a developing mouse blastocyst and the test for the formation of tissues from all three germ layers of the developing embryo [3]. Cell potency can also be undertaken *in vitro* through clonogenic assay which characterize the ability of single cells to self renew and differentiate. Regardless of their different sources, both adult stem cells (multipotent) and embryonic stem cells (pluripotent) have been proposed as promising candidates for cell-based therapies. The ability of stem cells to regenerate and differentiate into a specific type of cell could potentially provide a renewable source of replacement cells for various degenerative diseases.

1.1 Adult Stem Cells

Precursor or stem cells are present and active in the body throughout life to repair damaged tissues [4]. Adult stem cells (ASCs) have the ability to differentiate into cells within a specific lineage and regenerate tissues. ASCs are thought to reside within a special microenvironment known as the stem cell niche in a specific area of each tissue. A stem cell niche is comprised of cellular components as well as signals emanating from support cells [5]. ASCs exist in a variety of tissues and organs such as adipose tissue [6], bone marrow [7], umbilical cord blood [8], liver [9], brain [10], intestine [11] and pancreas [12]. Precursor cells within each tissue normally differentiate to form the specialized cell types of the tissue in which they reside. However, certain adult stem cell types can differentiate into cell types of other lineages. The process is called transdifferentiation [13].

Unlike embryonic stem cells, which can be defined by their derivation, the specific origin of ASCs in mature tissues still remain debatable, hence it is uncertain whether these adult cell populations are truly stem cells. There is a possibility that ASCs are merely remnants from developmental embryogenesis that can maintain a quiescent state based on the environmental niches or exposure to stimulatory signals. The capability of ASCs to undergo both symmetric and asymmetric cell division could enable these cells to control the balance between self-renewal and differentiation into other cell fates [14]. A genetic study using microarray has discovered a common set of genes that are highly expressed in the adult mouse hematopoietic stem cells, mouse neural progenitor cells and possibly other adult stem cell types [15]. This selective set of transcripts might therefore represent a conserved group of genes which are involved in the self-renewal and trans-differentiation of ASCs.

ASCs can be isolated and cultured *in vitro* for an extended period without losing their plasticity and can also differentiate into specialized cells. Most ASCs are restricted to certain lineages (multipotent) but may also exhibit greater plasticity by trans-differentiating to other specialized cell types (pluripotent). Mesenchymal stem cells derived from bone marrow have the ability to trans-differentiate into different early lineages *in vitro* (visceral mesoderm, neuroectoderm and endoderm) and also to more specified cell types (hematopoietic and gastrointestinal tract) when transplanted in an animal model [16].

Another valuable potential of ASCs is their use in cell therapy. ASCs can be obtained directly from individual patients, thereby preventing complications associated with immune rejection of allogenic tissue. Currently, adult hematopoietic stem cells derived from sources such as peripheral blood, umbilical cord blood and bone marrow represent the best characterized ASCs and have been routinely used for the clinical transplantation of hematological related malignancies [17]. The clinical potential of *ASCs* has also been demonstrated in the treatment of T1DM patients which resulted in c-peptide and insulin independence [18]. C-peptide, which remains as an integral part of proinsulin continues to serve as a diagnostic marker of insulin biosynthesis and secretion [19]. Overall, the potency and function of ASCs are advancing rapidly and will undoubtedly remain a growing research field.

1.2 Embryonic Stem Cells

Embryonic stem cells (ESCs) are derived from embryos after fertilization. These cells are unique due to their coordinated capabilities to either self-renew through cell division or to differentiate into any tissues in the body.

1.2.1 Derivation of Embryonic Stem Cells

The study of stem cell pluripotency was initially based on the teratocarcinoma-derived embryonal carcinoma cells (EC), as demonstrated by their abilities to self-renew for an indefinite period and also their capability to differentiate into cells from multiple different lineages [20]. The resemblances between EC cells and cells from the early embryo have suggested that mouse EC cells may represent the *in vivo* counterpart of the pluripotent cells in the inner cell mass (ICM) [21]. This suggestion appears mistaken as most EC formation and growth were commonly accompanied by numerous genetic alterations which led to genetic instability and limited development potential. Nonetheless, EC studies laid the experimental baseline for the establishment of embryonic stem cell (ESC) cultures. Mouse embryonic stem cells (mESCs) were derived from the ICM of an expanding mouse blastocyst and were shown to be capable of both self-renewal and differentiation into cells of the three primary embryonic germ layers (pluripotency) [22, 23]. mESCs express some similarities to the cells of the early epiblast and maintain their pluripotency when grown on a fibroblast feeder layer with serum containing culture conditions. Unlike EC cells, ESCs could retain a normal karyotype and exhibit a high differentiative capability when incorporated into a chimaeric mouse model [24]. It was found to be necessary to keep ESCs in an undifferentiated state to maintain a normal karyotype and unlimited capacity for self renewal.

The isolation and culturing of ICM cells from human blastocyst were found to be challenging, mainly due to the suboptimal human embryo culture media [25]. In 1998, Thomson and colleagues were the first to report the successful derivation of human embryonic stem cells (hESCs) lines using *in vitro* fertilized (IVF) / pre-implantation human embryos. Surplus embryos produced by IVF for assisted reproduction are usually frozen for transfer at a later stage or discarded. It is those cultured zygotes intended for destruction that have been donated after informed consent and used for the derivation of hESC lines [26]. Embryos are cultured to the 5-day old blastocyst stage (Fig. **1**), which contains three distinct structures: the trophoblast, which later becomes the placenta; the blastocoel, which is a fluid-filled cavity within the blastocyst; and the ICM, which later becomes the fetus [27]. Following fertilization and blastocyst formation, the ICM was isolated through immunosurgery/dissection and plated onto an irradiated mouse fibroblast feeder layer in the presence of high serum-containing growth medium [28, 29].

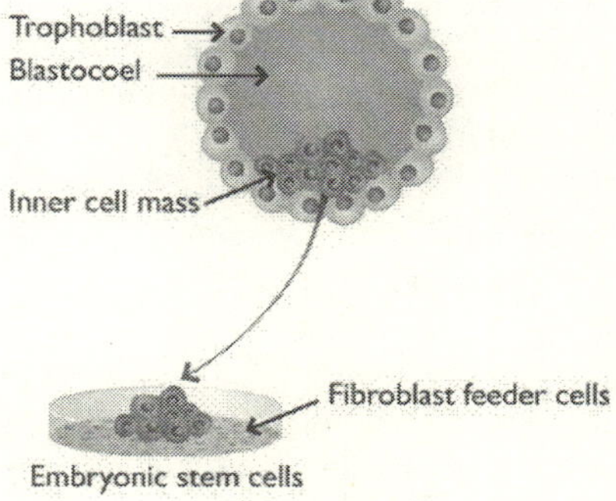

Fig. 1: The blastocyst. Embryonic stem cells are derived from the inner cell mass of the blastocyst. ESC colonies are transferred to fresh fibroblast feeder layer for propagation. Adapted from Chayosumrit [77].

These hESC colonies maintained a normal and stable karyotype after an extended period of culture and subsequently retained their ability to self-renew and differentiate into cells from all three embryonic germ layers. Furthermore, clonal derivation studies have effectively demonstrated the ability of a single hESC to proliferate and maintain pluripotency [30, 31]. ESCs derived from the ICM are considered to be the immortal *ex vivo* counterpart of early embryo stem cells [32], making them an ideal model for developmental biology and an alternative source for generating derivatives suitable for cell replacement therapy.

1.2.2 Maintenance of ESCs on Feeder Fibroblasts

hESCs are typically derived and grown on a mitotically inactivated fetal mouse fibroblast cell feeder layer which allows for the continuous growth of undifferentiated stem cells [33, 34]. hESCs rapidly differentiate in the absence of feeder cells, which further indicate their role in maintaining stemness. The use of mouse fibroblasts however greatly increases the risk for the cross-transfer of xenogeneic pathogens which rules out their use in clinical application. Several reports have described the promotion of human feeder cells to maintain stem cells characteristics. Various human fibroblast sources including fetal and adult fibroblast cells [35, 36] and foreskin fibroblasts [37] have been effectively used to support prolonged undifferentiated hESCs growth. Furthermore, the derivation of autogenic fibroblast-like cells from hESCs has allowed the use of genotypically homogenous feeder cells which prevents the risks of pathogens cross-transfer generated from allogenic feeders [38]. Other groups have also reported the successful maintenance of hESCs on extracellular matrix-coated surfaces in serum-free or feeder-free culture conditions using different exogenous signalling such as high concentrations of bFGF, Noggin or transforming growth factor β1 (TGF-β1) (Fig. **2**) [39-43].

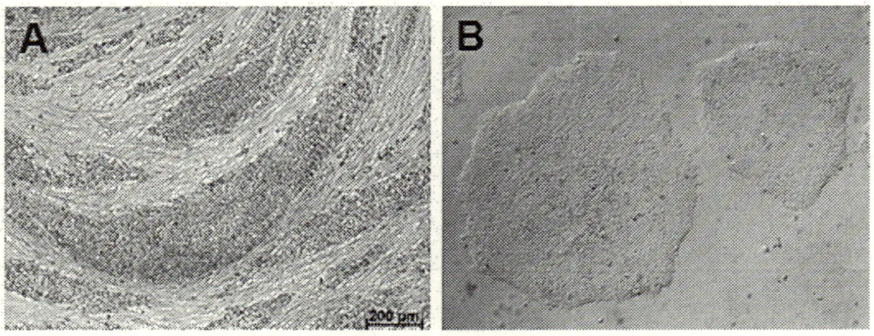

Fig. 2: Human embryonic stem cells cultured on human fibroblast feeder (A) and Matrigel-coated plate (B). Adapted from Chayosumrit [44].

There are two different ways in which fibroblast feeder cells could maintain hESCs pluripotency. Firstly, fibroblast feeder cells express and secrete various exogenous factors involved in the undifferentiated growth and pluripotency of hESCs. Both fibroblast growth factor (FGF-2) and bone morphogenic protein (BMP) inhibitor were highly expressed in the feeder cells and were required to sustain undifferentiated proliferation of human ES cells [45, 46]. Moreover, in the absence of feeder layers, the combinations of activin-A and FGF-2 were found to be essential for maintaining pluripotency [47, 48]. A comparative study between mouse and human feeders reveals significant differences in their capability to produce FGF-2 or activin A and subsequently suggest a greater efficiency of the mouse feeder cells to maintain undifferentiated hESCs [49]. This however is not conclusive as there may be other exogenous factors which remain largely unidentified.

Secondly, fibroblast feeder cells express adhesion molecules and produce extracellular matrices to promote hESC attachment. Several groups have demonstrated the culturing and maintenance of hESC pluripotency in a feeder-free culture with the application of mouse embryonic feeder conditioned medium and matrix protein such as Matrigel or laminin [50, 51]. Even after long-term culturing, hESCs maintained on these feeder-free conditions could still retain its pluripotency. Currently this system is not preferable as these procedures are too costly to maintain and upscale when compared to the conventional feeder system.

Our research group has successfully derived a transgenic human fibroblast cell line that is resistant to puromycin and capable of supporting the undifferentiated growth of hESCs [52]. This transgenic fibroblast cell line provides an effective system for selecting and generating a homogenous population of hESCs after their trangenesis.

1.2.3 Pluripotency of Embryonic Stem Cells

It is necessary to understand the fundamental biology of pluripotent ESCs. Pluripotent ESCs are characterized by a high nuclear to cytoplasmic ratio, the formation of compact multilayered colonies and high alkaline phosphatase activity. The colony morphology of undifferentiated hESCs is compact with a defined edge [53, 54]. In addition, ESCs express high level of telomerase activity which is highly associated with the cells immortality [34]. Pluripotent ESCs also express several specific epitope markers such as stage-specific embryonic antigen-1 (SSEA-1), stage-specific embryonic antigen-3 (SSEA-3), stage specific embryonic antigen-4 (SSEA-4), tissue rejection antigen 1-160 (TRA1-60) and tissue rejection antigen 1-81 (TRA1-81) [55]. SSEA-1 is solely expressed within pluripotent mESCs and differentiated hESCs, whereas SSEA-3, SSEA-4, TRA1-60 and TRA 1-81 are expressed specifically in undifferentiated hESCs. At present the functional significance of all these surface molecules is remains to be determined.

ESC pluripotency can also be demonstrated by the ability to differentiate into tissues or cells of the three germ layers. *In vitro* pluripotency is often demonstrated by the ability of ESCs to form embryo-like aggregates known as embryoid bodies (EBs) which contain derivatives of the three embryonic germ layers [56]. Similarly, *in vivo* pluripotency can be demonstrated by assessing the ability of ESCs to form teratomas following their implantation into a host animal, commonly mice. Such teratomas contain a mixture of partly or fully differentiated cell types from all three embryonic germ layers [26]. *In vivo* pluripotency could also be done through the blastocyst complementation assay which demonstrates mESC contribution to all tissues and organs of the chimeric mice [24, 57]. Chimera formation is not possible with hESCs due to ethical considerations and therefore teratoma formation in mice remains the standard assay for hESCs pluripotency.

Many studies have focused on elucidating the transcription factors and signalling pathways which maintain the pluripotency of ESCs. In mESCs, pluripotency is maintained through the intracellular signalling pathway stimulated by the leukemia inhibitory factor (LIF). LIF operates through the receptor subunit gp130 and induces the activation of the Janus kinase-signal transducer and activator of transcription (Jak-STAT3) pathway [58, 59]. Despite LIF being dispensable in the normal mouse development [60], it is proposed that STAT3 activation is required and sufficient to maintain the undifferentiated state of mESCs. In contrast to mESCs, LIF-induced STAT3 activation is incapable of maintaining stem cell pluripotency in hESCs [33, 34, 61]. This result suggests the existence of other signalling factors which promote the maintenance of hESC characteristics in a STAT3-independent manner.

A transcriptional profiling study has highlighted the role for components of the fibroblast growth factors (FGF), transforming growth factors-beta (TGF-β)/BMP and wingless and integration (Wnt) pathways in the maintenance of hESC pluripotency [62]. The combination of basic FGF (bFGF), feeder firoblasts and serum containing medium has been set as the gold standard for promoting clonal growth of hESCs. Interestingly, higher concentrations of bFGF allow feeder independent growth of hESCs and also suppress BMP signalling known to induce hESCs differentiation [63]. The activation of the TGF-β branch through the signal transducer SMAD2/3 is essential for the maintenance of the undifferentiated state in *ex vivo* mouse blastocycst [64], and in combination with FGF2 could maintain the extended propagation of undifferentiated hESCs in both feeder-containing culture [64] and feeder-free culture system [43, 65]. High concentration of activin-A, *a member* of the TGF-β superfamily, induces the expression of other pluripotency-related pathways (Wnt, FGF) and inhibits BMP signalling which supports hESC growth in the absence of both feeder cells and serum [66]. This study supports the previous report by [64], which suggests a combinatorial role of TGF-β and Wnt signallings to maintain hESC pluripotency. These signalling pathways are thought to work and interact together to form an inter-regulatory network of genes that could control the development of ESC stemness.

Studies in gene expression profiling have found that different hESC lines share significant homology in gene expression and specified highly-expressed key-genes (POU5f1, NANOG, SOX2, GTCM-1, connexin

43, REX1, TDGF1, and LEFTY-A*)* which are considered to be critical for the maintenance of pluripotency [67, 68]. The well-characterized transcription factors POU domain, class 5, transcription factor1 (OCT4), NANOG homeobox (NANOG) and SRY (sex determining region Y)-box 2 (SOX2) are considered to be the master genes that play a critical role in regulating ESCpluripotency. Using chromatin immunoprecipitation coupled with DNA microarrays (ChIP-Chip) analysis [69], promoter regions were identified that were highly co-occupied in hESCs by these three factors. Additionally, these three master genes also bound to their own promoters to form an autoregulatory loop that serve to maintain their expression and other pluripotency-linked genes. Around half of 353 genes that were bound concurrently by OCT4, NANOG and SOX2 were expressed and the rest were repressed. The expressed genes were linked to the components of self-renewal signalling pathways (TGF-β and Wnt signalling), whereas the repressed genes were found to be critical for germ-layer specific differentiation [69]. These three genes work together to positively regulate the network loop by simultaneously promoting the expression of pluripotency-associated genes and inhibiting the expression of differentiation-associated genes. This study was supported by another study that illustrated a transcriptional network involving OCT4, NANOG and another pluripotent factor FOXD3 (Fig. **3**) [70]. The network firstly specifies the function of OCT4 (red lines) to either promote NANOG activity when expressed at a steady level or to repress NANOG and its own promoter when the expression level rises above a steady level. Secondly, FOXD3 (black lines) positively activates NANOG to counter the repressive effect of excess OCT4. Finally, both FOXD3 (black lines) and NANOG (blue lines) also serve as the activators for OCT4 expression. Hence, it is believed that these interactions led to a negative feedback regulation loop between FOXD3, OCT4 and NANOG which serves to retain a steady OCT4 expression and maintain ESC pluripotency.

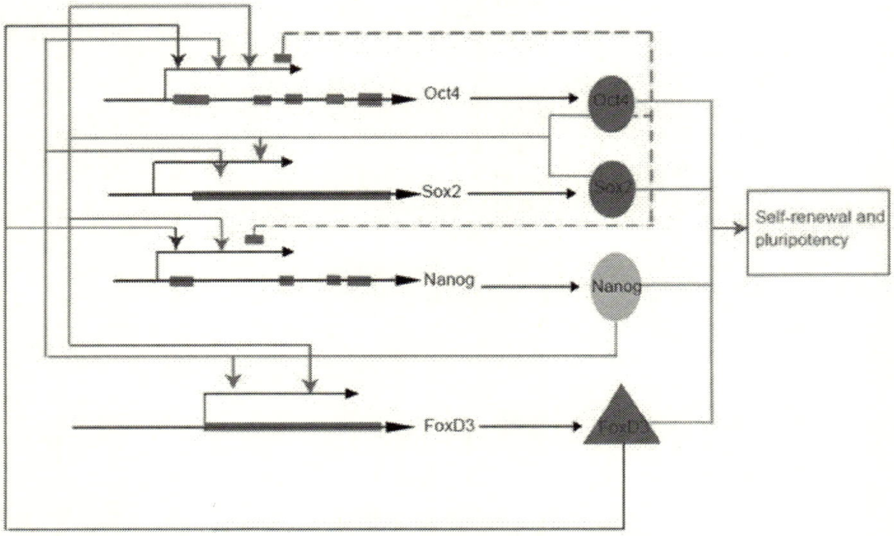

Fig. 3: Interconnected autoregulatory networks between OCT4, SOX2, NANOG and FOXD3 in maintaining ES cell pluripotency and self-renewal. *The model illustrates the bindings of OCT4 (red lines), NANOG (blue lines), SOX2 (red lines) and FOXD3 (black lines) to each other's promoter which eventually lead to the regulation of their expression. The solid lines indicate for a positive regulation of a promoter by the factors. Conversely, the broken lines linking to OCT4 indicate for a negative regulation. Adapted from Pan [70].*

1.2.4 Differentiation potential of hESCs

As mentioned earlier, ESCs have the ability to choose either prolonged self-renewal or differentiation depending on signals and microenvironment. The potential to generate clinically relevant numbers and differentiate to any cell types in the body makes hESCs an alternative source of cells for regenerative medicine for degenerative diseases such as heart failure, type 1 diabetes, stroke, haematological disorders, neurodegenerative disorders, spinal cord injury, osteoarthritis and kidney failure [71, 72] (Fig. **4**). Furthermore, the differentiation to other lineages may provide a tool for understanding molecular mechanisms, early embryonic developmental pathways [73], the pathological basis of genetic disorders, and toxicology or pharmacology testing [74, 75].

Cell fate specification is highly controlled by intrinsic signals and external environment. It is challenging to control the hESC differentiation and to direct the development along specific pathways. There are three basic methods to direct differentiation of hESCs by altering culture conditions: the formation of three dimensional aggregate "embryoid body (EB)", the co-culture of hESCs directly on supportive stromal layers and the culture of hESCs as monolayer on extracellular matrix proteins [72]. Other genetic approaches involve overexpressing tissue specific homeobox transcription factors to induce the differentiation and the use of reporter cell line for lineage tracing study or purification of cells [76, 77].

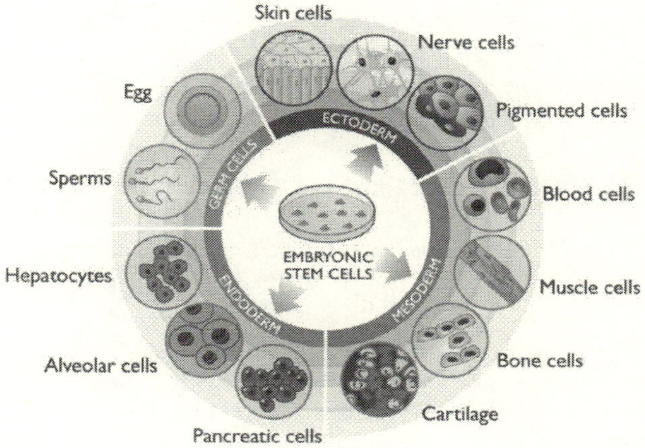

Fig. 4: Differentiation of hESCs to produce desired cell types. hESCs can be propagated to generate clinically relevant numbers and differentiate into any specific lineages to produce terminally differentiated cells for regenerative medicine. Adapted from Chayosumrit [44].

2. INDUCED PLURIPOTENT STEM CELLS

The advent of induced pluripotent stem cell (iPSC) technology has opened up new vistas to generate patient-specific pluripotent stem cells from somatic cells. During the last 5 years, the iPSCs produced from a variety of somatic cell sources are found to be very similar if not identical to embryonic stem cells. Invariably these cells are produced by viral transduction of four transcriptional factors that renders these cells unfit for therapeutic purposes. However, current emphasis is on the development of new methods for generating iPSCs with minimal or no genetic modifications *via* excisable lentiviral and transposon vectors or through repeated application of transient plasmid, episomal, and adenovirus vectors and very recently the use of small molecules, synthetic mRNA and microRNAs (references). The iPSC technology is emerging as an unprecedented opportunity in biomedical research and regenerative medicine. However, to harness the full potential of this technology, a number of issues that need to be resolved pertaining to iPSC safety, stability, culture variability, their comparison with ES cells, the reprogramming mechanisms, and better ways to direct a specific reprogramming process including lineage specifications.

The two most remarkable techniques used in reprogramming are, somatic cell nuclear transfer (SCNT) and iPSC generation. Both of the techniques can reprogram a differentiated cell back to an embryonic state [78, 79]. Nuclear transfer experiments date back to the 1950s where frogs (*Rana pipiens*) were cloned by replacing nuclei of eggs with cells from the late blastula [80] and since then SCNT has been successfully reported in a number of mammalian species and the birth of Dolly the sheep [81]. SCNT although remains the method of choice for generating isogenic pluripotent stem cells, this procedure is technically challenging and inefficient with low success rates and so far not successful in humans because of ethical constrains. The discovery of the iPSC technique using defined transcription factors by Yamanaka invigorated this field and changed the paradigm and shifted the debate away from SCNT program particularly for humans with reduced ethical concerns [2]. Over the past five years, the iPSC technology has significantly advanced. iPSCs have been generated from a broad range of somatic cell types [82-86] and species [87-91], differentiated into various seemingly functional cell types, and used to study basic biology and model human diseases. These studies have opened up new possibilities that iPSCs could be potentially developed as unlimited and patient-matched cell sources for cell-based therapy or drug discovery.

The discovery that somatic cells can be reprogrammed to iPSCs has opened up promising new avenues for regenerative medicine. However, some fundamental questions remain, such as why such cells show variations to pluripotency as determined by tetraploid complementation, which is the defining capability of a bona fide pluripotent stem cell line [92, 93]. This also raises the question how to address the similar issue with human iPSCs that cannot be subjected to a similar stringent test of pluripotency. Are human iPSCs an exact reproduction of human embryonic stem cells (hESCs), or have they inherited the incomplete pluripotency of their mouse iPSC counterparts? Due to the limitation of the use of tetraploid complementation test of pluripotency in human iPSCs, recently a number studies have chosen to compare the gene expression profiles including epigenome, transcriptome, microRNA and genome-wide CpG methylation profiling of such cells with hESCs [94-98]. These studies do indicate that although both iPSCs and hESCs cluster together as far as plurinet genes are concerned but there are also subtle differences between iPSCs and hESCs that have been attributed to a number of factors such as variation in data analyses, laboratory conditions and variations in protocols used for reprogramming. Further studies are required to identify specific genetic and epigenetic signatures that can be used to distinguish and classify iPSC lines as well as analyzing their functionalities, including safety, self-renewal stability, differentiation potentials. This is particularly critical for evaluating human iPSCs for various applications, as there is a lack of authentic and stringent *in vivo* pluripotency assays. Studies have shown that hESCs are significantly different from mESCs in terms of cell behaviours, gene expression, and signalling responses to several key developmental signalling pathways (*e.g.* TGFb, BMP, LIF, FGF, and MAPK) in self renewal and differentiation, but correspond very closely to mouse epiblast stem cells (EpiSCs) that are derived from the postimplantation egg cylinder stage epiblasts of mouse. These observations support the notion that EpiSCs and hESCs (as well as human iPSCs generated with the conventional hESC culture conditions) are intrinsically similar, and suggest that mESCs and EpiSCs/hESCs represent two distinct pluripotent states: the mESC-like state representing the preimplantation ICM and the EpiSC-like state representing late epiblast cells, respectively. Therefore, creating mESC-like human pluripotent cells has attracted increasing interest. It was shown recently that a different type of human iPSCs that functionally behaves similarly to mESCs could be generated and maintained by combining the reprogramming approach and small molecule modulators of TGFb, MAPK and GSK3 [90]. Generation of such mESC-like human pluripotent cells would have significant values in both basic research and practical applications. These cells would be very useful to study the early human development that may not be easily pursued. Given the above considerations, generating reprogrammed cells to be identical to ESCs or with specific properties that distinguish those from ESCs would represent continued areas of interest for various applications.

Both iPSCs and ESCs are pluripotent — they can form any tissue in the body, yet subtle differences exist as revealed recently by Su-Chun Zhang and his colleagues at the University of Wisconsin–Madison who compared the ability of both types of pluripotent cell to form human neurons in a laboratory setting, and found that iPSCs did so with markedly lower efficiency than ESCs [99]. As described above that these differences may arise because of different conditions used for producing iPSCs including unique genetic of the somatic cells used. Therefore, a team led by Konrad Hochedlinger at Massachusetts General Hospital in Boston has now derived iPS and ES cells with identical DNA [92]. The iPSCs were less efficient than the ESCs at incorporating into chimeric mice — a standard test of pluripotency, or 'stemness'. The team added the stem cells into embryos from mice of a different colour; once each mouse matures, the colouring of its coat reveals how much the stem cells contributed to forming its tissue. When the scientists compared genome-wide expression patterns between the two cell types, they discovered that a small stretch of DNA on the long arm of chromosome 12 displayed significantly different gene activity. In this region, two genes and a slew of tiny regulatory sequences called microRNAs were consistently activated in the ES cells and silenced in the iPS cells, regardless of whether the reprogrammed cells came originally from skin, brain, blood or other tissue. Although the function of the key genes is unknown, this region is usually silenced in mouse sperm cells and activated in other types of cell, so reprogramming might somehow mimic the silencing process, the authors speculate. The discovery raises the possibility that human iPSCscarry similar silenced sequences that make them less effective than ESCs.

3. CONCLUSIONS AND PROSPECTS

Recently the prospects of using pluripotent stem cells in human health has increased particularly in drug discovery,toxicological studies and therapeutics. This is largely due to improvement in techniques for their isolation, maintenance, preservation and differentiation. Both the embryonic and adult stem cells are relevant for future studies, however, the former can proliferate indefinitely and produce all cell types in sufficient numbers required for future transpalant studies. With the advent of iPSC, there seems a paradigm shift with majority of the laboratories concentrating in producing these cells. While significant technical progress has been made in generating iPSCs from various somatic cell sources, iPSC research is still in its infancy. There remains a great deal to learn about iPSC safety, the reprogramming mechanisms, and how to change the reprogramming process from its current non-specific process with poor efficiency and slow kinetics to a specific and directed process. Towards this end, a number of recent studies using small molecules, microRNAs, synthetic mRNAs and recombinant proteins are emerging as an alternative to traditional integrating system for reprogramming that may be more acceptable for producing therapeutic cells. With current approaches using integrating system, the iPSC field will have relevance in disease modelling, and identification of new small molecules that modulate reprogramming. It is conceivable that a precise combinatorial action of signalling and epigenetic modifiers may direct a more specific and efficient reprogramming process *in vitro* or *in vivo*. In parallel, improved abilities in differentiating pluripotent stem cells as well as expanding and maintaining their lineage-specific derivatives are also keys to their ultimate applications. There is no doubt that pluripotent stem cell research will continue to offer new opportunities for regenerative medicine for years to come.

REFERENCES

[1] Ratajczak MZ, Machalinski B, Wojakowski W, Ratajczak J, Kucia M. A hypothesis for an embryonic origin of pluripotent Oct-4(+) stem cells in adult bone marrow and other tissues. Leukemia 2007; 21(5): 860-7.

[2] Takahashi K, Yamanaka S. Induction of pluripotent stem cells from mouse embryonic and adult fibroblast cultures by defined factors. Cell 2006 25; 26(4): 663-76.

[3] Hochedlinger K, Jaenisch R. Nuclear reprogramming and pluripotency. Nature 2006; 441(7097): 1061-7.

[4] Limbert C, Path G, Jakob F, Seufert J. Beta-cell replacement and regeneration: Strategies of cell-based therapy for type 1 diabetes mellitus. Diabetes Res Clin Pract 2008; 79(3): 389-99.

[5] Li L, Xie T. Stem cell niche: structure and function. Annu Rev Cell Dev Biol 2005; 21: 605-31.

[6] Gimble JM, Katz AJ, Bunnell BA. Adipose-derived stem cells for regenerative medicine. Circ Res 2007; 100(9): 1249-60.

[7] Kucia M, Reca R, Campbell FR, *et al.* A population of very small embryonic-like (VSEL) CXCR4(+)SSEA-1(+)Oct-4+ stem cells identified in adult bone marrow. Leukemia 2006; 20(5): 857-69.

[8] Zhao Y, Wang H, Mazzone T. Identification of stem cells from human umbilical cord blood with embryonic and hematopoietic characteristics. Exp Cell Res 2006; 312(13): 2454-64.

[9] Kim S, Shin JS, Kim HJ, Fisher RC, Lee MJ, Kim CW. Streptozotocin-induced diabetes can be reversed by hepatic oval cell activation through hepatic transdifferentiation and pancreatic islet regeneration. Lab Invest 2007; 87(7): 702-12.

[10] Sohur US, Emsley JG, Mitchell BD, Macklis JD. Adult neurogenesis and cellular brain repair with neural progenitors, precursors and stem cells. Philos Trans R Soc Lond B Biol Sci 2006;361(1473):1477-97.

[11] Crosnier C, Stamataki D, Lewis J. Organizing cell renewal in the intestine: stem cells, signals and combinatorial control. Nat Rev Genet 2006; 7(5): 349-59.

[12] Hao E, Tyrberg B, Itkin-Ansari P, *et al.* Beta-cell differentiation from nonendocrine epithelial cells of the adult human pancreas. Nat Med 2006; 12(3): 310-6.

[13] National Institutes of Health, US Department of Health and Human Services. Stem Cell Basics: What are adult stem cells? 2009 [30 September, 2009]; Available from: http://stemcells.nih.gov/info/basics/basics4.asp.

[14] Beckmann J, Scheitza S, Wernet P, Fischer JC, Giebel B. Asymmetric cell division within the human hematopoietic stem and progenitor cell compartment: identification of asymmetrically segregating proteins. Blood 2007; 109(12): 5494-501.

[15] Terskikh AV, Easterday MC, Li L, *et al.* From hematopoiesis to neuropoiesis: evidence of overlapping genetic programs. Proc Natl Acad Sci USA 200; 98(14): 7934-9.

[16] Jiang Y, Jahagirdar BN, Reinhardt RL, *et al.* Pluripotency of mesenchymal stem cells derived from adult marrow. Nature 2002; 418(6893): 41-9.

[17] Wright-Kanuth MS, Smith LA. Hematopoietic stem cell transplantation. Clin Lab Sci 2001; 14(2): 112-7.

[18] Couri CE, Oliveira MC, Stracieri AB, *et al.* C-peptide levels and insulin independence following autologous nonmyeloablative hematopoietic stem cell transplantation in newly diagnosed type 1 diabetes mellitus. JAMA 2009; 301(15): 1573-9.

[19] Steiner DF. The proinsulin C-peptide--a multirole model. Exp Diabesity Res 2004; 5(1): 7-14.

[20] Kleinsmith LJ, Pierce GB, Jr. Multipotentiality of single embryonal carcinoma cells. Cancer Res 1964; 24: 1544-51.

[21] Martin GR. Teratocarcinomas and mammalian embryogenesis. Science 1980; 209(4458): 768-76.

[22] Evans MJ, Kaufman MH. Establishment in culture of pluripotential cells from mouse embryos. Nature 1981; 292(5819): 154-6.

[23] Martin GR. Isolation of a pluripotent cell line from early mouse embryos cultured in medium conditioned by teratocarcinoma stem cells. Proc Natl Acad Sci USA 1981; 78(12): 7634-8.

[24] Bradley A, Evans M, Kaufman MH, Robertson E. Formation of germ-line chimaeras from embryo-derived teratocarcinoma cell lines. Nature 1984; 309(5965): 255-6.

[25] Bongso A, Fong CY, Ng SC, Ratnam S. Isolation and culture of inner cell mass cells from human blastocysts. Hum Reprod 1994; 9(11): 2110-7.

[26] Conley BJ, Young JC, Trounson AO, Mollard R. Derivation, propagation and differentiation of human embryonic stem cells. Int J Biochem Cell Biol 2004; 36(4): 555-67.

[27] National Institutes of Health, US Department of Health and Human Services. Stem Cell Basics: What are embryonic stem cells? 2009 [30 September, 2009]; Available from: http://stemcells.nih.gov/info/basics/basics3.asp.

[28] Thomson JA, Itskovitz-Eldor J, Shapiro SS, Waknitz MA, Swiergiel JJ, Marshall VS, *et al.* Embryonic stem cell lines derived from human blastocysts. Science 1998; 282(5391): 1145-7.

[29] Reubinoff BE, Pera MF, Fong CY, Trounson A, Bongso A. Embryonic stem cell lines from human blastocysts: somatic differentiation *in vitro*. Nat Biotechnol 2000; 18(4): 399-404.

[30] Amit M, Carpenter MK, Inokuma MS, *et al.* Clonally derived human embryonic stem cell lines maintain pluripotency and proliferative potential for prolonged periods of culture. Dev Biol 2000; 227(2): 271-8.

[31] Sidhu KS, Ryan JP, Tuch BE. Derivation of a new human embryonic stem cell line, endeavour-1, and its clonal propagation. Stem Cells Dev 2008; 17(1): 41-51.

[32] Smith AG. Embryo-derived stem cells: of mice and men. Annu Rev Cell Dev Biol 2001; 17: 435-62.

[33] Reubinoff BE, Pera MF, Fong CY, Trounson A, Bongso A. Embryonic stem cell lines from human blastocysts: somatic differentiation *in vitro*. Nat Biotechnol 2000; 18(4): 399-404.

[34] Thomson JA, Itskovitz-Eldor J, Shapiro SS, *et al.* Embryonic stem cell lines derived from human blastocysts. Science 1998; 282(5391): 1145-7.

[35] Richards M, Fong CY, Chan WK, Wong PC, Bongso A. Human feeders support prolonged undifferentiated growth of human inner cell masses and embryonic stem cells. Nat Biotechnol 2002; 20(9): 933-6.

[36] Sidhu KS, Walker S, Tuch BE. Derivation and propagation of hESC under therapeutic environment. In: Elefanty A, Gregory AR, Patient R, Schlaeger T, Snyder T, Yamashita Y, Eds. Current Protocols in Stem Cell Biology. USA: John Wiley & Sons, Inc. 2008; 6: 1A.4.1-1A.4.31.

[37] Amit M, Margulets V, Segev H, *et al.* Human feeder layers for human embryonic stem cells. Biol Reprod 2003; 68(6): 2150-6.

[38] Stojkovic P, Lako M, Stewart R, *et al.* An autogeneic feeder cell system that efficiently supports growth of undifferentiated human embryonic stem cells. Stem Cells 2005; 23(3): 306-14.

[39] Amit M, Itskovitz-Eldor J. Maintenance of human embryonic stem cells in animal serum- and feeder layer-free culture conditions. Methods Mol Biol (Clifton, NJ) 2006; 331: 105-13.

[40] Li Y, Powell S, Brunette E, Lebkowski J, Mandalam R. Expansion of human embryonic stem cells in defined serum-free medium devoid of animal-derived products. Biotechnol Bioeng 2005; 91(6): 688-98.

[41] Stojkovic P, Lako M, Przyborski S, *et al.* Human-serum matrix supports undifferentiated growth of human embryonic stem cells. Stem Cells 2005; 23(7): 895-902.

[42] Wang G, Zhang H, Zhao Y, *et al.* Noggin and bFGF cooperate to maintain the pluripotency of human embryonic stem cells in the absence of feeder layers. Biochem Biophys Res Commun 2005; 330(3): 934-42.

[43] Amit M, Shariki C, Margulets V, Itskovitz-Eldor J. Feeder layer- and serum-free culture of human embryonic stem cells. Biol Reprod 2004; 70(3): 837-45.

[44] Chayosumrit M. Encapsulation as a 3D model for human embryonic stem cell propagation and differentiation Sydney: PhD dissertation. Australia, New South Wales: University of New South Wales 2010.

[45] Xu RH, Peck RM, Li DS, Feng X, Ludwig T, Thomson JA. Basic FGF and suppression of BMP signalling sustain undifferentiated proliferation of human ES cells. Nat Methods 2005; 2(3): 185-90.

[46] Kueh J, Richards M, Ng SW, Chan WK, Bongso A. The search for factors in human feeders that support the derivation and propagation of human embryonic stem cells: preliminary studies using transcriptome profiling by serial analysis of gene expression. Fertil Steril 2006; 85(6): 1843-6.

[47] Beattie GM, Lopez AD, Bucay N. *et al.* Activin A maintains pluripotency of human embryonic stem cells in the absence of feeder layers. Stem Cells 2005; 23(4): 489-95.

[48] Vallier L, Alexander M, Pedersen RA. Activin/Nodal and FGF pathways cooperate to maintain pluripotency of human embryonic stem cells. J Cell Sci 2005; 118: 4495-509.

[49] Eiselleova L, Peterkova I, Neradil J, Slaninova I, Hampl A, Dvorak P. Comparative study of mouse and human feeder cells for human embryonic stem cells. Int J Dev Biol 2008; 52(4): 353-63.

[50] Xu C, Inokuma MS, Denham J, *et al.* Feeder-free growth of undifferentiated human embryonic stem cells. Nat Biotechnol 2001; 19(10): 971-4.

[51] Rosler ES, Fisk GJ, Ares X, *et al.* Long-term culture of human embryonic stem cells in feeder-free conditions. Dev Dyn 2004; 229(2): 259-74.

[52] Sidhu KS, Lie KH, Tuch BE. Transgenic human fetal fibroblasts as feeder layer for human embryonic stem cell lineage selection. Stem Cells Dev 2006 ; 15(5): 741-7.

[53] Thomson JA, Itskovitz-Eldor J, Shapiro SS, *et al.* Embryonic stem cell lines derived from human blastocysts. Science 1998; 282(5391): 1145-7.

[54] Reubinoff BE, Pera MF, Fong CY, Trounson A, Bongso A. Embryonic stem cell lines from human blastocysts: Somatic differentiation *in vitro*. Nat Biotechnol 2000; 18(4): 399-404.

[55] Koestenbauer S, Zech NH, Juch H, Vanderzwalmen P, Schoonjans L, Dohr G. Embryonic stem cells: similarities and differences between human and murine embryonic stem cells. Am J Reprod Immunol 2006; 55(3): 169-80.

[56] Zeng X, Miura T, Luo Y, *et al.* Properties of pluripotent human embryonic stem cells BG01 and BG02. Stem Cells 2004; 22(3): 292-312.

[57] Wang ZQ, Kiefer F, Urbanek P, Wagner EF. Generation of completely embryonic stem cell-derived mutant mice using tetraploid blastocyst injection. Mech Dev 1997; 62(2): 137-45.

[58] Smith AG, Nichols J, Robertson M, Rathjen PD. Differentiation inhibiting activity (DIA/LIF) and mouse development. Dev Biol 1992; 151(2): 339-51.

[59] Matsuda T, Nakamura T, Nakao K, *et al.* STAT3 activation is sufficient to maintain an undifferentiated state of mouse embryonic stem cells. EMBO J 1999; 18(15): 4261-9.

[60] Stewart CL, Kaspar P, Brunet LJ, *et al.* Blastocyst implantation depends on maternal expression of leukaemia inhibitory factor. Nature 1992; 359(6390): 76-9.

[61] Humphrey RK, Beattie GM, Lopez AD, *et al.* Maintenance of pluripotency in human embryonic stem cells is STAT3 independent. Stem Cells 2004; 22(4): 522-30.

[62] Sato N, Sanjuan IM, Heke M, Uchida M, Naef F, Brivanlou AH. Molecular signature of human embryonic stem cells and its comparison with the mouse. Dev Biol 2003; 260(2): 404-13.

[63] Xu RH, Peck RM, Li DS, Feng X, Ludwig T, Thomson JA. Basic FGF and suppression of BMP signalling sustain undifferentiated proliferation of human ES cells. Nat Methods 2005; 2(3): 185-90.

[64] James D, Levine AJ, Besser D, Hemmati-Brivanlou A. TGFbeta/activin/nodal signalling is necessary for the maintenance of pluripotency in human embryonic stem cells. Development 2005; 132(6): 1273-82.

[65] Vallier L, Alexander M, Pedersen RA. Activin/Nodal and FGF pathways cooperate to maintain pluripotency of human embryonic stem cells. J Cell Sci 2005; 118: 4495-509.

[66] Xiao L, Yuan X, Sharkis SJ. Activin A maintains self-renewal and regulates fibroblast growth factor, Wnt, and bone morphogenic protein pathways in human embryonic stem cells. Stem Cells 2006; 24(6): 1476-86.

[67] Abeyta MJ, Clark AT, Rodriguez RT, Bodnar MS, Pera RA, Firpo MT. Unique gene expression signatures of independently-derived human embryonic stem cell lines. Hum Mol Genet 2004; 13(6): 601-8.

[68] Bhattacharya B, Miura T, Brandenberger R, *et al.* Gene expression in human embryonic stem cell lines: unique molecular signature. Blood 2004; 103(8): 2956-64.

[69] Boyer LA, Lee TI, Cole MF, *et al.* Core transcriptional regulatory circuitry in human embryonic stem cells. Cell 2005; 122(6): 947-56.

[70] Pan G, Li J, Zhou Y, Zheng H, Pei D. A negative feedback loop of transcription factors that controls stem cell pluripotency and self-renewal. FASEB J 2006; 20(10): 1730-2.

[71] Biswas A, Hutchins R. Embryonic stem cells. Stem Cells Develop 2007; 16(2): 213-21.

[72] Murry CE, Keller G. Differentiation of embryonic stem cells to clinically relevant populations: lessons from embryonic development. Cell 2008; 132(4): 661-80.

[73] Pera MF, Trounson AO. Human embryonic stem cells: prospects for development. Development 2004; 131(22): 5515-25.

[74] Davila JC, Cezar GG, Thiede M, Strom S, Miki T, Trosko J. Use and application of stem cells in toxicology. Toxicolog Sci 2004; 79(2): 214-23.

[75] Rolletschek A, Blyszczuk P, Wobus AM. Embryonic stem cell-derived cardiac, neuronal and pancreatic cells as model systems to study toxicological effects. Toxicol Lett 2004; 149(1-3): 361-9.

[76] Bernardo AS, H-H Cho C, Mason S, *et al.* Biphasic induction of Pdx1 in mouse and human embryonic stem cells can mimic development of pancreatic {beta} cells. Stem Cells 2009; 27(2): 341-51.

[77] Lavon N, Yanuka O, Benvenisty N. The Effect of overexpression of Pdx1 and Foxa2 on the differentiation of human embryonic stem cells into pancreatic cells. Stem cells 2006; 24(8): 1923-30.

[78] Han J CH, Sachdev PS, Sidhu KS. Epigenetic modifications in reprogramming and lineage specifications from embryonic stem cells:current trends and relevance in developing regenerative medicine. In: Singh S, Ed. Stem Cells, Regenerative Medicine and Cancer. New York USA: Nova Science Publishers 2010; pp. 1-27.

[79] Han J, Sachdev PS, Sidhu KS. A combined epigenetic and non-genetic approach for reprogramming human somatic cells. PLoS One 2010; 5(8): e12297.

[80] Briggs R, King TJ. Transplantation of living nuclei from blastula cells into enucleated frogs' eggs. Proc Natl Acad Sci USA 1952; 38(5): 455-63.

[81] Campbell KH, McWhir J, Ritchie WA, Wilmut I. Sheep cloned by nuclear transfer from a cultured cell line. Nature 1996; 380(6569): 64-6.

[82] Kim JB, Sebastiano V, Wu G, *et al.* Oct4-induced pluripotency in adult neural stem Cells Cell 2009; 136(3): 411-9.

[83] Aasen T, Raya A, Barrero MJ, *et al.* Efficient and rapid generation of induced pluripotent stem cells from human keratinocytes. Nat Biotechnol 2008; 26: 1276-84.

[84] Loh YH, Agarwal S, Park IH, *et al.* Generation of induced pluripotent stem cells from human blood. Blood 2009; 113(22): 5476-9.

[85] Pereira CF, Terranova R, Ryan NK, *et al.* Heterokaryon-based reprogramming of human B lymphocytes for pluripotency requires Oct4 but not Sox2. PLoS Genet 2008; 4(9): e1000170.

[86] Takahashi K, Tanabe K, Ohnuki M, *et al.* Induction of pluripotent stem cells from adult human fibroblasts by defined factors. Cell 2007; 131(5): 861-72.

[87] Liao J, Cui C, Chen S, *et al.* Generation of induced pluripotent stem cell lines from adult rat cells. Cell Stem Cell 2009; 4(1): 11-5.

[88] Wu Z, Chen J, Ren J, *et al.* Generation of pig induced pluripotent stem cells with a drug-inducible system. J Mol Cell Biol 2009; 1(1): 46-54.

[89] Liu H, Zhu F, Yong J, *et al.* Generation of induced pluripotent stem cells from adult rhesus monkey fibroblasts. Cell Stem Cell 2008; 3(6): 587-90.

[90] Li W, Wei W, Zhu S, *et al.* Generation of rat and human induced pluripotent stem cells by combining genetic reprogramming and chemical inhibitors. Cell Stem Cell 4(1): 16-9.

[91] Ezashi T, Telugu BP, Alexenko AP, Sachdev S, Sinha S, Roberts RM. Derivation of induced pluripotent stem cells from pig somatic cells. Proc Natl Acad Sci USA 2009; 106(27): 10993-8.

[92] Stadtfeld M, Apostolou E, Akutsu H, *et al.* Aberrant silencing of imprinted genes on chromosome 12qF1 in mouse induced pluripotent stem cells. Nature 2010; 465(7295): 175-81.

[93] Zhao XY, Li W, Lv Z, *et al.* iPS cells produce viable mice through tetraploid complementation. Nature 2009; 461(7260): 86-90.

[94] Chin MH, Mason MJ, Xie W, *et al.* Induced pluripotent stem cells and embryonic stem cells are distinguished by gene expression signatures. Cell Stem Cell 2009; 5(1): 111-23.

[95] Guenther MG, Frampton GM, *et al.* Chromatin structure and gene expression programs of human embryonic and induced pluripotent stem cells. Cell Stem Cell 2010; 7(2): 249-57.
[96] Newman AM, Cooper JB. Lab-specific gene expression signatures in pluripotent stem cells. Cell Stem Cell 2010; 7(2): 258-62.
[97] Polo JM, Liu S, Figueroa ME, *et al.* Cell type of origin influences the molecular and functional properties of mouse induced pluripotent stem cells. Nat Biotechnol 2010; 28(8): 848-55.
[98] Zhou H, Ding S. Evolution of induced pluripotent stem cell technology. Curr Opin Hematol 2010; 17(4): 276-80.
[99] Hu BY, Weick JP, Yu J, *et al.* Neural differentiation of human induced pluripotent stem cells follows developmental principles but with variable potency. Proc Natl Acad Sci USA 2010; 107(9): 4335-40.

CHAPTER 2

Therapeutic Cloning: Derivation and Propagation of Embryonic Stem Cells by Somatic Cell Nuclear Transfer

Zichuan Liu, Shuya Zhou, Shun Zhang, Baolong Xia and Qi Zhou*

State Key Lab of Reproductive Biology, Institute of Zoology, Chinese Academy of Sciences, Beijing, P R. China

Abstract: Embryonic stem cells (ESCs) can grow infinitely and give rise to all types of cells in human body, thus of tremendous therapeutic potentials for a variety of diseases, such as Parkinson's disease, spinal cord injury, and diabetes. Moreover, the combination of gene modification and directed differentiation of ESCs provides perfect tool for disease modelling and drug discovery. However, tissue rejection following ESCs derivatives transplantation greatly hinders its application. Under such circumstances, the idea of "therapeutic cloning" was proposed, indicating the generation of ESCs from SCNT embryos for therapeutic purpose.

Mouse nuclear transfer embryonic stem cells (NT-ESCs) were first established in 2000, and then proved to be able to differentiate either *in vivo* or *in vitro*, and give rise to individual tissues through germ line transmission or tetraploid complementation. Fully reprogrammed NT-ESCs are indistinguishable from ESCs derived from fertilized eggs functionally and substantially. What is more, by deriving NT-ESCs from patient cells, the problem of immune rejection may be avoided. However, the derivation of human NT-ESCs goes with the destruction of clone embryos, leading to fierce ethical disputes. There has not been report of successful establishment of human NT-ES cells so far, and the limited resource of human eggs used for nuclear transfer cumbers the future application of NT-ESCs.

In this chapter, we will introduce therapeutic cloning in two aspects as SCNT and NT-ESCs, and the history, nowadays status and prospects of them will be reviewed.

Keywords: Therapeutic cloning, somatic cell nuclear transfer, NT-ES cells, differentiation, ethical concerns.

1. INTRODUCTION

Reprograming the differentiated somatic cell into pluripotent state by nuclear transfer method in rodents is currently a routine approach in many laboratories. The technologies of SCNT and embryonic stem cells derivation make possible the production of genetically matched cells and tissue for the cure of degenerative diseases. In this chapter, a review of somatic cell nuclear transfer technology will be presented. In favor of the growing knowledge on the application of SCNT on therapeutic therapy, more and more valuable domestic or wild animals can be secured and expanded in quantity. It is also obvious that the incorporation of SCNT in conservation projects can help prevent the decline of certain species of large mammals.

The History of Somatic Cell Nuclear Transfer

Nuclear transfer technology was first proposed at the end of 19th century[1, 2]. In the beginning, most nuclear transfer experiments were conducted on fish or amphibian eggs due to their large size and abundant resources. Rauber described an experiment in which the nucleus of frog egg was replaced by the nucleus of toad cell [3]. Hans Spemann published the outstanding achievement of his primitive nuclear transfer experiment involving salamander embryos in his eBook "*Embryonic Development and Induction*" [4]. In this

***Address correspondence to Qi Zhou:** State Key Lab of Reproductive Biology, Institute of Zoology, Chinese Academy of Sciences, Beijing, P R. China; Tel: 86 10 64807299; Fax: 86 10 64807306; Email: zhouqi@ioz.ac.cn eBook, it was the first time that the nuclear transfer technology was defined and described systematically. Spemann claimed his idea of cloning by transferring the nucleus of differentiated cell into an enucleated fertilized egg. This cloning proposal is also the prototype of current therapeutic cloning theory. However, not until 1952 had Briggs and King achieved Spemann's cloning idea in northern leopard frogs. They succeeded in getting cloned frog from a blastula cell using nuclear transfer method [5]. They also found that the cloning success rate was strongly correlated with the differentiation state of donor cell. Only a few morphologically abnormal tadpoles could be cloned from adult differentiated cells. Their experiments sparked great scientific interest in cloning among the researchers. In 1996, John Gurdon used the nucleus of fully developed adult intestinal cells to clone South African frogs, which proved the somatic cells retained all the genetic information necessary for creating a new organism [6]. Chinese biologist Tong Dizhou also cloned adult fertility fishes using cross-strains nuclear transfer.

As their smaller cell size and less abundant oocyte resource, cloning in mammals is even more technically demanding. The first report about nuclear transfer experiment on mammalian animals occurred in 1975. The cloned embryos, which were derived from fusion of rabbit morula cells and enucleated rabbit eggs, could develop to morula stage. [7]. In 1981, Illmensee and Hoppe first performed nuclear transfer experiment in mouse model. They transferred ICM cells from mouse blastocyst into enucleated oocyte by injection and succeeded in cloning the mouse [8]. Despite the above results obtained, it was still technically demanding to generate cloned animal from differentiated donor cells.

In 1983, the American biologists McGrath and Solter generated live mice by exchanging the pronuclei of two zygotes [9]. Their approach adopted the micromanipulation and electrofusion, which is an early form of Roslin Technique. Thereafter, plenty of mammal animals were cloned with early embryonic cells, including mouse [10], bovine [11], rabbit [12], swine [13], sheep [14] and rhesus macaque[15].

However, the first cloned mammal derived from somatic cell was born in 1996 in Rosline institute, and named as Dolly sheep [16]. In fact, Dolly has two cloned cousins: Morag and Megan, which were derived from nine-day-old sheep embryonic cells.

Ian Wilmut, who was in charge of the Dolly project, was assigned with a bioreactor project in 1986. The purpose of that project was to create a sheep that was able to produce a certain chemical in its milk. The cell cycle stage has been intensively investigated. In their success, they pioneered a new technique of serum starving the donor somatic cell before nuclear transfer. Serum starving could make the adult somatic cell into a suspended state, the G0 state, which could simulate the sperm DNA state. Subsequently, electrofusion was adopted to fuse the donor cell and enucleated oocytes. All these improvements were included in Roslin technique. Dolly died in 2003 due to a respiratory disease but not for any cause of her identity as a clone. The cloning work was continued in Roslin, and another cloned sheep named Polly was created from fetal skin cells that had been genetically altered to include a human gene. The birth of Dolly initiated the public interest on cloning study and debates on ethical concerns. Till now, there have been more than 14 mammalian species cloned by somatic cell nuclear transfer method. The species of cloned mammals include cattle [17], mouse [18], goat [19], pig [20], rabbit [21], cat [22], rat [23], horse [24], mule [25], wolf, dog [26], ferret [27], buffalo [28] and the last one is camel [29]. A new cloning technique which was proved to be more efficient than Roslin technique was created by Wakayama. With this technique, he first produced 50 cloned mice successfully in the world. As this technique was started at University of Hawaii, it was named Honolulu technique. Wakayama injected the fresh cumulus cells, which should be in G0 state, into the enucleated oocytes directly and stimulated the reconstructed embryos by a chemical bath. The efficiency of Honolulu technique is amazing because Wakayama could get 3 cloned fetuses from 100 attempts compared with the Wilmut's one from 277. Although the first cloned mammals were generated from somatic cell using SCNT, most cloned embryos underwent development arrest before or soon after implantation and the success rate for producing live offspring by cloning remained below 5% independent of mammalian species [30].

Low Efficiency of Animal Cloning and Abnormal Phenotype of Cloned Fetus

The animal cloning technique using somatic cell nuclear transfer can largely benefit the animal species conservation, reproduction of superior domestic animals, bioreactor and therapeutic cloning. However, the low success rate and the abnormal phenotype of cloned fetus restricted the application of SCNT technique obviously. For example, the average successful rate of mouse cloning was lower than 5%, it reduced to 0.5% when neural cells were used as somatic donor cells. The survival rate of the cloned bovine was 50% after birth; even some of the survived cloned animals showed abnormalities. Considering the frequency of abnormal gene expression, it could be considered surprising that physically, functionally and histologically normal cloned adults had been produced in several species. The major cause might reside in faulty or incomplete epigenetic reprogramming of the donor nucleus. Additionally, many pathologies of kidney [31] or brain [32] developed in the survived cloned animals. Several symptoms were reported, including diabetes [33], respiratory catastrophe [34], large offspring syndrome (LOS) and obesity [34]. Most interestingly, lots of symptoms happened in placenta of cloned animals. The abnormality of cloned mouse placenta derived from somatic cell nuclear transfer was often found in our lab. Some studies reported that the Acrogranin, Cdx2 and ERR2 genes related to the placenta development were changed from blastocyst by examining the expression of protein and gene in cloned bovine embryos [35]. The elevation of leptin, placental lactogen, TGF and TGFRs's expression detected by RT-PCR could be one of the reasons which led the LOS [36, 37].

As the low efficiency is the last obstacle of the application of cloning, many new methods have been tried, such as activation and enucleation timing, inhibition of cytokinesis, cultural conditions and treatment of histone deacetylase inhibitors (HDAC inhibitors). Although several methods had been proved to be able to improve the cloning success rate in a certain extent, most of them had no effect unfortunately.

Reprogramming and Remodelling During Somatic Cell Nuclear Transfer

In order to improve the success rate of cloning, it must be ascertained what happens to the somatic cell nucleus after being transferred to maternal cytoplasm. In cloned embryos, the genome of somatic cell in differentiated state is transferred to undifferentiated state which is called reprogramming. Although it is known that extensive genomic reprogramming must occur during preimplantation development, and that the biology reagents necessary for studying such a system are in limited supply and rely on model organism. The microarray or other "ome" technical advances make it possible to carry out epigenetic reprogramming with valuable cloned embryos. By definition, "An epigenetic trait is a stable heritable phenotype resulting from changes in a chromosome without in the DNA sequence" [38].

All the genes are examined in cloned bovine embryos. Among the 3353 genes detected, there are 291 genes with their expression level in dead cloned fetus twice that in IVF fetus and the survived cloned fetus, including IGF2, HBA1, HBA2 and SPTB genes, which could cause the abnormal development of placentas; and there are 77 genes with their expression level in dead cloned bovine fetus [39] half as that in IVF fetus and the survived cloned fetus. However, the early reports on these genes are not consistent. In cloned bovine blastocysts, global gene expression pattern are also examined. The individual cloned blastocysts produced by nuclear transfer show less variability than those produced by either natural fertilization or IVF [40]. Nevertheless, most studies prove that relatively normal and correct genome reprogramming can happen after nuclear transfer [41].

DNA Methylation

DNA methylation and histone modification (see also chapter 5) are the fundamental mechanism that can regulate genome reprogramming in epigenetic level. DNA methylation plays an important role during mammalian development, such as cell differentiation, genomic imprinting, X chromosome inactivation, chromosome structure, cell senescence and apoptosis [42]. In vertebrates, methylation of DNA occurs by addition of a methyl group to CpG islands. Meanwhile this gene marked by methlylation has to be repressed. The global DNA methylation is stable maintained during cell cycle only excepted in the ageing process or cancer transformation [42]. Mammalian cells possess two classes of DNA methyltransferase enzymes: DNMT3 family members are responsible for de novo methylation and DNMT1 worked as a maintenance methyltransferase [43].

At the same time as normal early embryonic development in mammals, the global pattern of genomic DNA methylation produces marked changes. Two *de novo* methylation processes exist during mammalian embryonic development. The first one occurres during germ cell development; and the second during embryonic preimplantation development. The level of methylation is very high in male and female gametes. The male gamete demethylates immediately after forming paternal pronuclear of zygotes by an unknown mechanism in moue embryos [44]. The maternal pronuclear demethylates passively during subsequent cell cycle as DNMT1 (which is responsible for maintaining the methylation after replication) is removed from nucleus [45]. Progressive *de novo* methylation occurres at blastocyst stage in mice. Unexpectedly, not all the species display the similar patterns in *de novo* methylation. In sheep and rabbit, the paternal pronuclear does not lose its methylation marks in zygotes, and zygotes genomes are not passively demethylated during the early cell cycles [46, 47]. Although in rat, pig and bovine zygotes, paternal pronuclear demethylates actively following passive demethylation and *de novo* methylation of the TE and ICM cell lineages, and the quantitative and temporal differences of methylation existed among species [48, 49].

SCNT embryos experience similar process-the donor somatic cell nucleus with high level of DNA methylation also undergo demethylation and *de novo* methylation. However, DNA methylation levels of SCNT embryos are higher than that of normal embryos which could be explained by an incompletely somatic cell nuclear reprogramming. In cloned bovine embryos, pronuclear could not demethylate actively before the first DNA replication and the passive demethylation is weak. Although the presence of progressive demethylation of euchromatin caused the hypomethylation in morula and blastocyst stage, heterochromatin keep the process of hypermethylation during development [50, 51]. Histone modification was a landmark of epigenetic regulation. Similar pattern of demethylation in cloned bovine embryos also occurred on cloned mouse embryos [41].

Histone Modification

Histone modification is also landmark for the epigenetic regulation (see chapter 5). Each histone has the potential to be differently regulated by several kinds of covalent modification, including acetylation, methylation, phosphorylation, ubiquitylation and ADP- ribosylation. Among these histone modifications, H3K4 me1/me3 and aceylation are associated with transcriptinally permissive chromatin [52]. More over, histone acetyltion could facilitate cell proliferation and DNA replication complexes formation [53]. On the contrary, modifications of H3K9 me2/me3, H3K27 me and H4K20 me3 are supposed to contribute to construct the heterochromatin in which genes are repressed.

Immediately following fertilization, the arginine-rich protamines in sperm are exchanged for maternal histone presented in occytes, which occurs in 3-5 hours after fertilization in mice. The demethylation of paternal genome and hyper acetylated H4 associated only with paternal chromatin just occurred after the exchange of protamines for histones [54]. Interestingly, most histone modification occurres in paternal chromosome in a step-wise manner, including histone methylation: H3K4 me1, H3K4 me3 [55], H3K9 me1, H3K9 me2, H3K27 me1, H3K27 me2 and H3K27 me2 [56]; histone acetylation: H4K5, H4K12, H4K16, H3K9, H3K14 and H3K18 acetylation; histone phosphorylation on H3S10 [57]. The most plausible hypothesis is that the paternal chromatin is sensitive to the maturation factors presented in the oocytes [57]. In contrast to paternal pronuclear, these histone modifications are absent in maternal pronuclear. After fertilization, striking asymmetry of histone modification occurres on two parental genomes. H3K27 me is enriched in maternal chromatin [58]. However this striking asymmetry presents until the 4-cell stage while all blastomeres nuclei have equivalent level of H3K9, H3K27 and H3K4 methylation.

Upon somatic cell nuclear transfer, the similar exchange of deacetylation for reacetylation histone occurres on somatic nucleus. The stable epigenetic marks in histone modification from adult somatic pattern must be erased within a short period of time and reestablished before MET (Maternal and Embryonic Transmission). Therefore aberrant histone modification often could be found in cloned embryos, such as abnormally high level of H3K9 methylation in the trophectoderm cloned bovine blastocyts [59]. As the hyperacetylation status of somatic nucleus before being transferred into enucleated oocytes, H4K8 and H4K12 acetylation still existed in the genome of cloned emboryos with mild deacetylation [60]. There was accumulating evidence that mouse SCNT has a significant actual improvement with optimal conditional treatment of inhibitor of histone deacetylase (HDAC), like trichostatin A (TSA) [61, 62] and CBHA [63]. Moreover, HDAC inhibitors could significantly improve the efficiency of ntES derivation [61, 63]. These results suggested that the correct reprogramming of the acetylation pattern after nuclear transfer would be helpful to transfer a somatic nucleus to a totipotent state.

Nuclear Remodelling

While the eukaryotic nucleus is highly organized, nuclear architecture and chromosome structure playes a functional role to regulate genes expression. In the early cloned embryos, Lamin A/C which should specially be expressed in somatic cell appeared. In another word, the Lamin B which should be expressed in oocyte and early embryos could not replace Lamin A/C [64]. It was another evidence that the somatic nuclear is not completely remolded. In the early hours after nature fertilization, nuclear reorganization in embryos is associated with important modification of paternal and maternal chromatin in molecular and structural level. Interestingly, it has been shown in mouse embryos that centromere and pericentric heterochromatin regions of chromosomes associated with periphery of NPBs [65]. This characteristic zygotic organization of constitutive heterchromatin could be observed in 1-cell mouse embryos but not at later stage of development or in differentiated somatic cells. It is shown in this study that although remodelling into zygotic-like organization also occurs after somatic cell nuclear transfer, 50% abnormal nuclear remodelling is observed in cloned embryos with more than 3 centromeres disassociate NPBs in 1-cell stage [66]. Interestingly, the TSA treatment, which can improve the cloning success and also this defect in nuclear remodelling [66] which could be explained by histone modification, represents a precise mechanism to establish and maintain of special chromatin structures in a long-term period [67].

All these results from DNA methylation, histone modification and nuclear architecture suggest epigenetic reprogramming is of great importance to the development of somatic cell cloned embryos.

NT-ES Cell (Nuclear Transfer-Embryonic Stem Cells)

Stem cells from embryos, namely embryonic stem cells have been recognized for three decades [68-70]. When Thomson discovered that such cells could also be isolated and cultured *in vitro* in human in 1998[71], a door to medical breakthroughs was opened. Therapeutic cloning (versus reproductive cloning), the goal of which was to produce a healthy copy of a sick person's tissue or organ for transplant, was first proposed more than ten years ago. In this sense, embryonic stem cells are established from the nuclear transfer embryos, while patient's somatic cells are used as nuclear donors. Thereafter, patient-specific ES cells are differentiated into cell types of interest for transplant. Therefore, the problem of organ donation and immune-rejection will be solved by using this technique.

The first human hybrid human clone was created in 1998, by American Cell Technologies. It was created with a man's leg cell, and a cow's egg whose DNA was removed, which was destroyed after 12 days. Embryonic stem cell lines from somatic cells by nuclear transfer was successfully achieved for the first time in bovine[72] and followed by production in mouse[73, 74].

Proceedings in Derivation of NT-ES Cell Lines

Two main techniques are included in the therapeutic cloning, *i.e.*, somatic cell nuclear transfer (SCNT) and embryonic stem cell derivation. As the aforementioned description, SCNT is performed by removing the nucleus of an unfertilized egg (enucleation) and replacing the nucleus with the nucleus of a somatic cell from another person. The egg containing the transferred nucleus is then stimulated to divide until blastocyst stage, and at that time the cells of the inner cell mass can be removed and embryonic stem cells can be derived. During the last decade, significant progress has been made in the process of generating NT-ES cells.

Zona-free (ZF) NT, which performed without the assistance of piezo, was found to be a more efficient method to generate genetically matched NT-ES cell lines in mouse model [75] Sequential nuclear transfer was also tried in therapeutic cloning process to improve the deficiency in NT-ES cells derivation; however this technique was proved invalid when the developmental deficiencies that resided in previous donor generations were improved [76]. Although unremitting efforts have been made in this arena, no currently published report has achieved any success in establishing human embryonic stem cell lines from blastocysts generated by somatic cell nuclear transfer.

Although great progress has been achieved in NT-ES cell research with mouse model, fewer studies are performed in human. The reports of human NT-ES cell lines have been discredited. And there are a lot of difficulties in primate cloning. Thus, the generation of monkey NT-ES cells [77] brought hope for the application of therapeutic cloning in primate, and further in human. Because the oocyte played a key role in reprogramming the epigenetic status of donor cell nuclei, researches about establishing oocyte morphology criterion could simply improve the human SCNT efficiency [78], and further facilitate the application of therapeutic cloning in human.

Comparison of the Therapeutic Potential Between NT-ES Cells and F-ES Cells

There were a series of criterions for evaluating the developmental potency of ES cells, in which the tetraploid complementation was the most stringent test. Pioneers' works from efficient establishing of NT-ES cell lines [72], to producing chimera without germ line transmission [73, 74], then to contributing to germ line [34], and eventually producing "all NT-ES cells mice" illustrated NT-ES cells had equivalent developmental potency compared with ES cells from fertilized embryos (F-ES cells) [79, 80]. However, gene expression aberrations observed in tissues of cloned animals had also raised concerns regarding the therapeutic application of NT-ES cells. Whether the NT-ES cells had an identical therapeutic potential to the F-ES cells was studied as an open question.

Several experiments in mice have demonstrated that the developmental potential of NT-ES cells is identical to that of F-ES cells. The ES cells derived from cloned and fertilized embryos have been reported to be indistinguishable based on their transcriptional profile, including expression of pluripotency markers, presence of tissue-dependent differentially DNA methylated regions, DNA microarray profiles, and high-coverage gene expression profiling [79, 80]. In contrast to the development of cloned animals, it was said that the epigenetic memory could be erased in the NT-ES cells; therefore they became functionally equivalent to ES cells from fertilized embryo [80]. What's more, on the post-transcriptional level, including microRNA and protein expression, NT-ES cells also presented a comparable profile to F-ES cells [81]. Therefore, it was widely considered that the NT-ES cells had an equivalent therapeutic potential with F-ES cells.

In addition, pluripotent stem cells generated from the late epiblast of postimplantation embryos (epiblast stem cells, EpiSCs) was also established successfully [82, 83]. EpiSCs could be derived with the same efficiency from NT embryos and from their fertilized counterparts. However, different from the similarity between F-ES and NT-ES cells, gene expression profile showed divergence between F-EpiSCs and NT-EpiSCs. It was indicated that abnormal epigenetic marks were imposed on cells in NT embryos during the transition from early to late epiblast [83].

Differentiation of NT-ES Cells and Tissue Engineering

Specific differentiation *in vitro* paved the way to therapeutic application of stem cells. NT-ES could differentiate into a wide variety of ectoderm, mesoderm, endoderm and specific cell types, such as neural cells, in which not only the neural stem cells but also the dopaminergic, serotonergic, cholinergic and GABAergic neurons, as well as astroglia and oligodendroglia could be generated [84], and also cardiomyocytes [85]. No differences were observed between NT-ES and ES differentiation *in vitro*. So a concept would be acceptable that protocols of ES differentiation would be applied to NT-ES differentiation.

Besides cell therapy, NT-ES cells also can be used as seed cells in the tissue engineering for their proliferation and differentiation. The main objective of tissue engineering is to reconstruct tissue through cell adhesion, proliferation, differentiation and functioning on scaffold material. A study on using NT-ES cells as seed cells in myocardial tissue engineering has been reported [86]. NT-ES cells mixed with chitosan hydrogel were injected into the left ventricular wall of rat infarction models. Results have shown that, the cell retention after 24 hours, the graft size after 4 weeks, the heart function and the arteriole/venule densities within the infract area have been improved significantly in the NT-ES + chitosan group. It is confirmed in this study that NT-ES cells have therapeutic potential in tissue engineering.

Therapeutic Application of NT-ES Cells

NT-ES cells derived from patient somatic cell nuclear transfer share identical genome with patient, therefore, they are an ideal cell source for auto-transplantation to treat degenerative disorders, such as diabetes, osteoarthritis, and Parkinson's disease. However, their application is limited by cell availability and immune compatibility [87, 88]. Another existing problem is that, although NT-ES cells are tailor-made for the patient, in other words, NT-ES cells generated by SCNT are genetically identical to the person donating the nucleus in terms of the chromosomes, the egg donor also contributes DNA since the mitochondrion residing in the cytoplasm also will be a part of the developing embryo. Thus, the embryo is not 100% identical to the person donating the nucleus, unless it is the same person donating both the egg and the nucleus. Therefore, many issues need to be solved before the clinical application of therapeutic cloning. Several trials using NT-ES for therapeutic application have shown the promising prospect.

Jaenisch and his colleagues have tried NT-ES cells as a cell source to correct gene defect [89]. In a immune-deficient Rag2(-/-) mice model, they used the gene deficient mice somatic cell as nuclear donor to establish an NT-ES cell line. After the Rag2 gene was restored to normal structure in the NT-ES cells by homologous recombination, the repaired NT-ES cells are used to produce mice through tetraploid complementation for bone marrow transplantation or to differentiate into hematopoietic precursors *in vitro* for engrafting. As a result, mature myeloid and lymphoid cells as well as immunoglobulins can be detected 3-4 weeks after transplantation. This study, as they concluded, set a paradigm for the treatment of a genetic disorder by combining therapeutic cloning with gene therapy.

T. Barberi, *et al.* have reported specific neural subtype's differentiation of NT-ES cells and alleviation of Parkinson's disease phenotype by NT-ES cell-derived dopaminergic neurons transplantation [84]. They used co-culture-based ES cell differentiation system that provides both fast, efficient neural induction and the ability to generate a wide range of neural subtypes, and transplanted NT-ES cell-derived dopaminergic neurons into Parkinsonian mice, resulting in the long-term survival of the transplanted dopaminergic neurons and the correction of behavioral deficits. This study has facilitated the use of NT-ES cells in therapeutic cloning in neural diseases.

Another attempt at therapeutic cloning has been reported by V. Tabar's team [90]. As in the previous researches, the adopted NT-ES cell lines were derived from genetically related donor rather than from the affected mouse to be treated, which is unpractical for human therapeutic cloning, they treated multiple Parkinsonian mice with dopaminergic neurons derived from individually matched NT-ES cell lines. The experiment showed therapeutic efficacy and there was a lack of immunological response, which was in accord with previous report in therapeutic cloning in neural diseases.

All these data demonstrate the feasibility of treating individual gene deficient or degenerative disease or tissue-loss mice *via* NT-ES cells and suggest considerable therapeutic potential for the future. It's worthy of predicting that NT-ES cells would pave a novel way for regenerative medicine.

Ethical Concerns About NT-ES Cells

As the intense public debate over stem cell research has become apparent, numerous ethical concerns seriously affect the political, economic and legal contexts in which stem cell research is pursued. The problems, such as principle of human oocyte donation and clear boundary between therapeutic cloning and reproductive cloning *etc.*, need to be dealt with soberly.

'Altered nuclear transfer' (ANT) concept is issue based on the premise that the inactivation of a gene crucial for trophectoderm development will eliminate the potential to form the fetal–maternal interface, but will spare the ICM lineage. Therefore, the feasibility of reproductive cloning is ceased technically. With mouse model, researchers produced Cdx2-deficient cloned blastocysts which were morphologically abnormal, and they lacked functional trophoblast and failed to implant into the uterus. However, they efficiently generated pluripotent embryonic stem cells when explanting the cloned embryos in culture [91].

The advancements of stem cells research also can facilitate regenerative medicine. However, equitable balance should be kept between its risk and benefits to fulfill the bright prospect in therapeutic cloning application.

Ever since the cloning of the sheep Dolly in Scotland, the debate on whether to ban all forms of human cloning for ethical and scientific reasons has been raised among clergy, politicians, doctors and scientists. Restrictions on the research of therapeutic cloning have little chance as they inhibit the development of a technique which holds promise for successful application of pluripotent stem cells in clinical treatment of severe diseases.

Maybe we should cite what Robert had said, who first proposed the concept of therapeutic cloning, "Yes, a human 'pre-embryo' should be treated with respect. But does a blastocyst warrant the same rights and reverence as that accorded a living soul—a parent, a child or a partner—who might die because we failed to move the moral line?" [92].

The policies on human therapeutic cloning were diversified among different countries. Over 30 countries, including France, Germany and Russian Federation, have banned human cloning altogether. Fifteen countries, such as Japan, the United Kingdom and Israel, Australia have banned human reproductive cloning, but permit therapeutic cloning. A few countries such as Hungary and Poland do not explicitly prohibit embryonic stem cell research or therapeutic cloning, partially because their legislation was drafted before embryonic stem cells were first produced (1998). Other countries, similar to the United States, have yet to pass any official legislation concerning human cloning allowing all types of stem cell and cloning research to occur.

Other Approaches Besides SCNT

iPS Cells

The fate of a cell can be altered by forced expression of single tissue-specific transcription factors. In 1987, Gehring and his/her colleagues first reported that, in D. melanogaster larvae, ectopic overexpression of Antennapedia protein by a heat-shock expression vector can alter the body plan of fruit fly and form an additional set of legs instead of antennae [93].

Induced pluripotent stem cells (iPS cells) are a type of pluripotent stem cell artificially derived from a non-pluripotent cell, typically an adult somatic cell, by inducing a "forced" expression of specific genes: Oct4, Klf4, Sox2, cMyc [94]. iPS cells are similar to embryonic stem (ES) cells in many respects, such as the expression of certain pluripotent genes, epigenetic patterns, proliferation, EB and teratoma formation, viable chimera formation, and potency and differentiability, but the full extent of their relation to natural pluripotent stem cells is still being assessed. The mammalian iPSCs were first produced in 2006 from mouse fibroblast by Yamanaka and the first human iPS cell line was in 2007 from human cells. They have been cited as important events in stem cell research, as researchers could obtain pluripotent stem cells without the controversial source of embryos. The issue of graft-versus-host disease and immune rejection is also avoided, like ntES cells, as they are derived entirely from the patient. However, as the use of virus, reprogramming of adult cells to iPS cells may possess significant risks that could limit its use in therapy. In February 2008, scientists announced the discovery of a technique that could be utilized to remove oncogenes after the induction of pluripotency, thereby increasing the potential use of iPS cells in human diseases. In April 2009, Zhou and his colleagues demonstrated that generation of iPS cells was possible without any genetic alteration of the adult cell. A repeated treatment of the cells with certain proteins channeled into the cells *via* poly-arginine anchors was sufficient to induce pluripotency [95].

iPS cells have the obvious advantage of non-necessity of using human embryos. Another critical advantage is that, in theory, they allow the creation of cell lines that are genetically customized to a patient. The issue of immune rejection that the body's immune system recognizes implanted cells or tissues as allogeneic cells and attacks them, is a barrier to the therapeutic application of cell-based therapies. If cells for therapy can be created using a patient's own cells, the issue is likely to be overcome. Besides, iPS technique can be performed in any moderately equipped molecular biology lab and does not require specific micromanipulation skill.

It has become a popular means to reprogram somatic genomes into an embryonic-like pluripotent state, and a preferred alternative to somatic-cell nuclear transfer and somatic-cell fusion with ES cells. However, iPS cell reprogramming remains slow and inefficient, about 0.01–0.1%. Additionally, the mechanism of iPS reprogramming is still unknown.

Cell Fusion and ES Extraction Treatment

Cell fusion involves fusing two or more cell types to form a single entity. In 1983, Cecelia Webster induced the stable expression of muscle-specific genes in human nonmuscle cells. Normal diploid human amniocytes were fused with differentiated mouse muscle cells by using polyethylene glycol. The fusion product, a stable heterocaryon in which the parental cell nuclei remained distinct, did not undergo division and retained a full complement of chromosomes. C Webster interpreted that the expression of genes in the nuclei of differentiated cells was remarkably plastic and susceptible to modulation by the cytoplasm. The isolation of the genes encoding the tissue-specific trans-acting regulators responsible for muscle gene activation should be possible [96].

Cell fusion can generate hybrids or heterokaryons. Hybrids proliferate, causing the nuclei of the original cells to fuse, whereas heterokaryons do not proliferate and therefore, they contain multiple distinct nuclei. Nearly four decades ago, fusion experiments showed that when a differentiated cell such as a mouse fibroblast was fused with a hamster melanocyte or a rat hepatocyte, melanin and tyrosine aminotransferase, respectively, ceased to be synthesized [97-99].

Cell fusion is technically simple. When cell fusion is used to form mixed-species heterokaryons without proliferation, pluripotent genes are activated quickly and can be activated immediately. This approach is therefore particularly well suited for elucidating the molecular mechanisms that control the onset of nuclear reprogramming; secondly, reprogramming by cell fusion is highly efficient. When mouse ES cells are fused with human fibroblasts, up to 70% of heterokaryons (enriched by fluorescence-activated cell sorting) activate the expression of pluripotent genes within 1 day. Thirdly, cell division does not occur in heterokaryons. It also does not occur after nuclear transfer during the time when pluripotent genes are induced, allowing active mechanisms that induce pluripotent gene expression to be studied because this induction is independent of cell division and DNA replication; passive mechanisms may accompany cell division (for example, the dilution of DNA methyltransferases).

This technology may serve to overcome the use of human oocytes to reprogram adult skin cells in the future for production of hESC lines. Research on cell fusion and heterokaryon formation has been utilized to study nuclear reprogramming events and gene expression. Heterokaryon formation may also be a naturally occurring event, as fused cell types have been observed, followed by the transplantation of mouse bone marrow-derived cells into adult mouse brains and liver.

Reprogramming by nuclear transplantation or cell hybridization has provided a need for the emergence of cell-free reprogramming strategies. The cell extract is expected to contain regulatory components which is necessary for diverting the fate of one cell type into another [100]. What's more, extract-based reprogramming approaches have shown that differentiated cells may be "transdifferentiated" into other differentiated cell types, or "dedifferentiated" toward a more pluripotent cell type (Taranger CK, *et al* 2005).

Reports support that within a few days of extract treatment, ATSCs (Adipose tissue stem cells) expressed cardiac-specific proteins. In addition, extracts derived from undifferentiated cells can accomplish reprogramming so that it makes this method more accessible. Extracts of Xenopus eggs were shown to induce expression of pluripotency markers in 293T cells. In 2006, Philippe Collas proved that nuclear and cytoplasmic extracts of undifferentiated human EC cells can reprogram 293T cells to take on properties of undifferentiated cells with a potential for pluripotency [101].

CONCLUSIONS AND PROSPECTS

Therapeutic cloning is a promising way to generate histocompatible cells for transplantation therapy or as a source of patient-specific cells for drug testing. Theoretically it can prevent graft rejection. After the directed cell differentiation of patient-specific NT-ES cells, they can be used for auto-transplantation to treat degenerative disorders such as diabetes, osteoarthritis, and Parkinson's disease. However, the application of therapeutic cloning still has a long way to go. Before the application of the ntES cells in clinical trails, an efficient protocol for human somatic cell nuclear transfer needs to be established, in addition, several critical factors in human ntES cells derivation need to be determined and optimized. No publication of successful generation of human NT-ESCs has emerged till now.

Another existing problem is that, although NT-ES cells are tailor-made for the patient, who provides the chromosome DNA, the mitochondrion in egg donor might also contribute DNA to the NT-ES cells. Thus they are not 100% consistent with DNA of the person donating the nucleus, which is a risk for the application of NT-ES cells in cell therapy for patients.

Besides, undifferentiated embryonic stem cells demonstrate a high risk of developing cancer when they are transplanted into immunocompromised patients. Thus, if undifferentiated embryonic stem cells are to be used, it might be necessary to develop recall mechanisms or, alternatively, to ensure that only differentiated cells will be transplanted, so as to avoid the risk of cancer formation in the recipient.

REFERENCES

[1] Driesch H. Entwicklungsmechanische studien. i. der werth der beiden ersten furchungszellen in der echinodermenentwicklung. experimentelle erzeugung von teil und doppelbildungen. Z Wiss Zool 1891; 53: 160-178 and 183-184.

[2] Loeb J. Ueber eine einfache Methode, zwei oder mehr zusammengewachsene Embryonen aus einem Ei hervorzubringen. Pflügers Arch 1894; 55: 525-530.

[3] Rauber A. Personaltheil und Germinaltheil des Individuum. Zool Anz 1886; 9: 166-171.

[4] Spemann H. Embryonic development and induction. New Haven: Yale Univ. Press 1938.

[5] Briggs R, King TJ. Transplantation of living nuclei from blastula cells into enucleated frogs' eggs. Proc Natl Acad Sci USA 1952; 38(5): 455-63.

[6] Gurdon JB, Brown DD. Cytoplasmic regulation of rna synthesis and nucleolus formation in developing embryos of xenopus laevis. J Mol Biol 1965; 12: 27-35.

[7] Bromhall JD. Nuclear transplantation in the rabbit egg. Nature 1975; 258(5537): 719-22.

[8] Illmensee K, Hoppe PC. Nuclear transplantation in Mus musculus: developmental potential of nuclei from preimplantation embryos. Cell 1981; 23(1): 9-18.

[9] McGrath J, Solter D. Nuclear transplantation in the mouse embryo by microsurgery and cell fusion. Science 1983; 220(4603): 1300-2.

[10] Tsunoda Y, Yasui T, Shioda Y, Nakamura K, Uchida T, Sugie T. Full-term development of mouse blastomere nuclei transplanted into enucleated two-cell embryos. J Exp Zool 1987; 242(2): 147-51.

[11] Prather RS, Barnes FL, Sims MM, Robl JM, Eyestone WH, First NL. Nuclear transplantation in the bovine embryo: assessment of donor nuclei and recipient oocyte. Biol Reprod 1987; 37(4): 859-66.

[12] Stice SL, Robl JM. Nuclear reprogramming in nuclear transplant rabbit embryos. Biol Reprod 1988; 39(3): 657-64.

[13] Prather RS, Sims MM, First NL. Nuclear transplantation in early pig embryos. Biol Reprod 1989; 41(3): 414-8.

[14] Campbell KH, McWhir J, Ritchie WA, Wilmut I. Sheep cloned by nuclear transfer from a cultured cell line. Nature 1996; 380(6569): 64-6.

[15] Meng L, Ely JJ, Stouffer RL, Wolf DP. Rhesus monkeys produced by nuclear transfer. Biol Reprod 1997; 57(2): 454-9.

[16] Wilmut I, Schnieke AE, McWhir J, Kind AJ, Campbell KH. Viable offspring derived from fetal and adult mammalian cells. Nature 1997; 385(6619): 810-813.

[17] Kato Y, Tani T, Sotomaru Y, *et al.* Eight calves cloned from somatic cells of single adult. Science 1998; 282: 2095-2098.

[18] Wakayama T, Perry AC, Zuccotti M, Johnson KR, Yanagimachi R. Full-term development of mice from enucleated oocytes injected with cumulus cell nuclei. Nature 1998; 394(6691): 369-374.

[19] Baguisi A, Behboodi E, Melican DT, *et al.* Production of goats by somatic cell nuclear transfer. Nat Biotechnol 1999; 17: 456-461.

[20] Polejaeva IA, Chen SH, Vaught TD, *et al.* Cloned pigs produced by nuclear transfer from adult somatic cells. Nature 2000; 407: 86-90.

[21] Chesné P, Adenot PG, Viglietta C, Baratte M, Boulanger L, Renard JP. Cloned rabbits produced by nuclear transfer from adult somatic cells. Nat Biotechnol 2002; 20: 366-369.

[22] Shin T, Kraemer D, Pryor J *et al.* A cat cloned by nuclear transplantation. Nature 2002; 415: 859.

[23] Zhou Q, Renard JP, Le Friec G *et al.* Generation of fertile cloned rats by regulating oocyte activation. Science 2003; 302(5648): 1179.

[24] Galli C, Lagutina I, Crotti G. *et al.* A cloned horse born to its dam twin. Nature 2003; 424: 635.

[25] Woods GL, White KL, Vanderwall DK. *et al.* A mule cloned from fetal cells by nuclear transfer. Science 2003; 301: 1063.

[26] Lee BC, Kim MK, Jang G, *et al.* Dogs cloned from adult somatic cells. Nature 2005; 436: 604.

[27] Li Z, Sun X, Chen J, *et al.* Cloned ferrets produced by somatic cell nuclear transfer. Dev Biol 2006; 293: 439-448.

[28] Shi D, Lu F, Wei Y, *et al.* Buffalos (*Bubalus bubalis)* cloned by nuclear transfer of somatic cells. Biol Reprod 2007; 77: 285-291.

[29] Wani NA, Wernery U, Hassan FA, Wernery R, Skidmore JA. Production of the first cloned camel by somatic cell nuclear transfer. Biol Reprod 2010; 82(2); 373-9.

[30] Wakayama T. Production of cloned mice and ES cells from adult somatic cells by nuclear transfer: how to improve cloning efficiency? J Reprod Dev 2007; 53(1): 13-26.

[31] Kato Y, Tani T, Tsunoda, Y. Cloning of calves from various somatic cell types of male and female adult, newborn and fetal cows. J Reprod Fertil 2000; 120(2): 231-7.

[32] McCreath KJ, Howcroft J, Campbell KH, Colman A, Schnieke AE, Kind AJ. Production of gene-targeted sheep by nuclear transfer from cultured somatic cells. Nature 2000; 405(6790): 1066-9.

[33] Hill JR, Winger QA, Long CR, Looney CR, Thompson JA, Westhusin ME. Development rates of male bovine nuclear transfer embryos derived from adult and fetal cells. Biol Reprod 2000; 62(5): 1135-40.

[34] Wakayama T, Yanagimachi R. Mouse cloning with nucleus donor cells of different age and type. Mol Reprod Dev 2001; 58(4): 376-83.

[35] Hall VJ, Ruddock NT, French AJ. Expression profiling of genes crucial for placental and preimplantation development in bovine *in vivo, in vitro*, and nuclear transfer blastocysts. Mol Reprod Dev 2005; 72(1): 16-24.

[36] Ravelich SR, Shelling AN, Wells DN, *et al.* Expression of TGF-beta1, TGF-beta2, TGF-beta3 and the receptors TGF-betaRI and TGF-betaRII in placentomes of artificially inseminated and nuclear transfer derived bovine pregnancies. Placenta 2006; 27(2-3): 307-16.

[37] Ravelich SR, Shelling AN, Ramachandran A, *et al.* Altered placental lactogen and leptin expression in placentomes from bovine nuclear transfer pregnancies. Biol Reprod 2004; 71(6): 1862-9.

[38] Berger SL, Kouzarides T, Shiekhattar R, Shilatifard A. An operational definition of epigenetics. Genes Dev 2009; 23(7): 781-3.

[39] Oishi M, Gohma H, Hashizume K, *et al.* Early embryonic death-associated changes in genome-wide gene expression profiles in the fetal placenta of the cow carrying somatic nuclear-derived cloned embryo. Mol Reprod Dev 2006; 73(4): 404-9.

[40] Smith SL, Everts RE, Tian XC, *et al.* Global gene expression profiles reveal significant nuclear reprogramming by the blastocyst stage after cloning. Proc Natl Acad Sci USA 2005; 102(49): 17582-7.

[41] Yang X, Smith SL, Tian XC, Lewin HA, Renard JP, Wakayama T. Nuclear reprogramming of cloned embryos and its implications for therapeutic cloning. Nat Genet 2007; 39(3): 295-302.

[42] Bird A. DNA methylation patterns and epigenetic memory. Genes Dev 2002; 16(1): 6-21.

[43] Goll MG, Bestor TH. Eukaryotic cytosine methyltransferases. Annu Rev Biochem 2005; 74: 481-514.

[44] Mayer W, Niveleau A, Walter J, Fundele R, Haaf T. Demethylation of the zygotic paternal genome. Nature 2000; 403(6769): 501-2.

[45] Cardoso MC, Sporbert A, Leonhardt H. Structure and function in the nucleus: subnuclear trafficking of DNA replication factors. J Cell Biochem 1999; (Suppl 32-33): 15-23.

[46] Beaujean N, Taylor J, Gardner J, Wilmut I, Meehan R, Young L. Effect of limited DNA methylation reprogramming in the normal sheep embryo on somatic cell nuclear transfer. Biol Reprod 2004; 71(1): 185-93.

[47] Shi W, Dirim F, Wolf E, Zakhartchenko V, Haaf T. Methylation reprogramming and chromosomal aneuploidy in *in vivo* fertilized and cloned rabbit preimplantation embryos. Biol Reprod 2004; 71(1): 340-7.

[48] Haaf T. Methylation dynamics in the early mammalian embryo: implications of genome reprogramming defects for development. Curr Top Microbiol Immunol 2006; 310: 13-22.

[49] Mattick JS, Makunin IV. Epigenetic reprogramming in mammals. Hum Mol Genet 2005. 14 Spec No 1: R47-58.

[50] Bourc'his D, Le Bourhis D, Patin D, *et al.* Delayed and incomplete reprogramming of chromosome methylation patterns in bovine cloned embryos. Curr Biol 2001; 11(19): 1542-6.

[51] Dean W, Santos F, Stojkovic M, *et al.* Conservation of methylation reprogramming in mammalian development: aberrant reprogramming in cloned embryos. Proc Natl Acad Sci USA 2001; 98(24): 13734-8.

[52] Eberharter A, Becker PB. Histone acetylation: a switch between repressive and permissive chromatin. Second in review series on chromatin dynamics. EMBO Rep 2002; 3(3): 224-9.

[53] Luger K, Mäder AW, Richmond RK, Sargent DF, Richmond TJ. Crystal structure of the nucleosome core particle at 2.8 A resolution. Nature 1997; 389(6648): 251-60.

[54] Adenot PG, Mercier Y, Renard JP, Thompson EM. Differential H4 acetylation of paternal and maternal chromatin precedes DNA replication and differential transcriptional activity in pronuclei of 1-cell mouse embryos. Development 1997; 124(22): 4615-25.

[55] Lepikhov K and Walter J. Differential dynamics of histone H3 methylation at positions K4 and K9 in the mouse zygote. BMC Dev Biol 2004; 4: 12.

[56] Santos F, Peters AH, Otte AP, Reik W, Dean W. Dynamic chromatin modifications characterise the first cell cycle in mouse embryos. Dev Biol 2005; 280(1): 225-36.

[57] van der Heijden GW, Derijck AA, Ramos L, Giele M, van der Vlag J, de Boer P. Transmission of modified nucleosomes from the mouse male germline to the zygote and subsequent remodelling of paternal chromatin. Dev Biol 2006; 298(2): 458-69.

[58] Erhardt S, Su IH, Schneider R, *et al.* Consequences of the depletion of zygotic and embryonic enhancer of zeste 2 during preimplantation mouse development. Development 2003; 130(18): 4235-48.

[59] Santos F, Zakhartchenko V, Stojkovic M, *et al.* Epigenetic marking correlates with developmental potential in cloned bovine preimplantation embryos. Curr Biol 2003; 13(13): 1116-21.

[60] Wang F, Kou Z, Zhang Y, Gao S. Dynamic reprogramming of histone acetylation and methylation in the first cell cycle of cloned mouse embryos. Biol Reprod 2007; 77(6): 1007-16.

[61] Kishigami S, Bui HT, Wakayama S, *et al.* Successful mouse cloning of an outbred strain by trichostatin A treatment after somatic nuclear transfer. J Reprod Dev 2007; 53(1): 165-70.

[62] Rybouchkin A, Kato Y, Tsunoda Y. Role of histone acetylation in reprogramming of somatic nuclei following nuclear transfer. Biol Reprod 2006; 74(6): 1083-9.

[63] Dai X, Hao J, Hou XJ, *et al.* Somatic nucleus reprogramming is significantly improved by m-carboxycinnamic acid bishydroxamide, a histone deacetylase inhibitor. J Biol Chem 2010; 285(40): 31002-10.

[64] Hall VJ, Cooney MA, Shanahan P, Tecirlioglu RT, Ruddock NT, French AJ. Nuclear lamin antigen and messenger RNA expression in bovine *in vitro* produced and nuclear transfer embryos. Mol Reprod Dev 2005; 72(4): 471-82.

[65] Martin C, Beaujean N, Brochard V, Audouard C, Zink D, Debey P. Genome restructuring in mouse embryos during reprogramming and early development. Dev Biol 2006; 292(2): 317-32.

[66] Maalouf WE, Liu Z, Brochard V, *et al.* Trichostatin A treatment of cloned mouse embryos improves constitutive heterochromatin remodelling as well as developmental potential to term. BMC Dev Biol 2009; 9: 11.

[67] Lachner M, Jenuwein T. The many faces of histone lysine methylation. Curr Opin Cell Biol 2002; 14(3): 286-98.

[68] Martin GR. Isolation of a pluripotent cell line from early mouse embryos cultured in medium conditioned by teratocarcinoma stem cells. Proc Natl Acad Sci USA 1981; 78(12): 7634-8.

[69] Evans MJ, Kaufman MH. Establishment in culture of pluripotential cells from mouse embryos. Nature 1981; 292(5819): 154-6.

[70] Thomson JA, Kalishman J, Golos TG, *et al.* Isolation of a primate embryonic stem cell line. Proc Natl Acad Sci USA 1995; 92(17): 7844-8.

[71] Thomson JA, Itskovitz-Eldor J, Shapiro SS, *et al.* Embryonic stem cell lines derived from human blastocysts. Science 1998; 282(5391): 1145-7.

[72] Cibelli JB, Stice SL, Golueke PJ, *et al.* Transgenic bovine chimeric offspring produced from somatic cell-derived stem-like cells. Nat Biotechnol 1998; 16(7): 642-6.

[73] Munsie MJ, Michalska AE, O'Brien CM, Trounson AO, Pera MF, Mountford PS. Isolation of pluripotent embryonic stem cells from reprogrammed adult mouse somatic cell nuclei. Curr Biol 2000; 10(16): 989-92.

[74] Kawase E, Yamazaki Y, Yagi T, Yanagimachi R, Pedersen RA. Mouse embryonic stem (ES) cell lines established from neuronal cell-derived cloned blastocysts. Genesis 2000; 28(3-4): 156-63.

[75] Kobolak J, Bodo S, Rungsiwiwut R, *et al.* Generation of mouse embryonic stem cell lines from zona-free nuclear transfer embryos. Cell Reprogram 2010; 12(1): 105-13.

[76] Zhao C, Yao R, Hao J, *et al.* Establishment of customized mouse stem cell lines by sequential nuclear transfer. Cell Res 2007; 17(1): 80-7.

[77] Byrne JA, Pedersen DA, Clepper LL, *et al.* Producing primate embryonic stem cells by somatic cell nuclear transfer. Nature 2007; 450(7169): 497-502.

[78] Yu Y, Mai Q, Chen X, *et al.* Assessment of the developmental competence of human somatic cell nuclear transfer embryos by oocyte morphology classification. Hum Reprod 2009; 24(3): 649-57.

[79] Wakayama S, Jakt ML, Suzuki M. *et al.* Equivalency of nuclear transfer-derived embryonic stem cells to those derived from fertilized mouse blastocysts. Stem Cells 2006; 24(9): 2023-33.

[80] Brambrink T, Hochedlinger K, Bell G, Jaenisch R. ES cells derived from cloned and fertilized blastocysts are transcriptionally and functionally indistinguishable. Proc Natl Acad Sci USA 2006; 103(4): 933-8.

[81] Ding J, Guo Y, Liu S, *et al.* Embryonic stem cells derived from somatic cloned and fertilized blastocysts are post-transcriptionally indistinguishable: a MicroRNA and protein profile comparison. Proteomics 2009; 9(10): 2711-21.

[82] Brons IG, Smithers LE, Trotter MW, *et al.* Derivation of pluripotent epiblast stem cells from mammalian embryos. Nature 2007; 448(7150): 191-5.

[83] Maruotti J, Dai XP, Brochard V, *et al.* Nuclear transfer-derived epiblast stem cells are transcriptionally and epigenetically distinguishable from their fertilized-derived counterparts. Stem Cells 2010; 28(4): 743-52.

[84] Barberi T, Klivenyi P, Calingasan NY, *et al.* Neural subtype specification of fertilization and nuclear transfer embryonic stem cells and application in parkinsonian mice. Nat Biotechnol 2003; 21(10): 1200-7.

[85] Lü S, Liu S, He W, *et al.* Bioreactor cultivation enhances NTEB formation and differentiation of NTES cells into cardiomyocytes. Cloning Stem Cells 2008; 10(3): 363-70.

[86] Lü S, Wang H, Lu W, *et al.* Both the transplantation of somatic cell nuclear transfer- and fertilization-derived mouse embryonic stem cells with temperature-responsive chitosan hydrogel improve myocardial performance in infarcted rat hearts. Tissue Eng Part A 2010; 16(4): 1303-15.

[87] Mombaerts P. Therapeutic cloning in the mouse. Proc Natl Acad Sci USA 2003; 100 (Suppl 1): 11924-5.

[88] Wakayama T. On the road to therapeutic cloning. Nat Biotechnol 2004; 22(4): 399-400.

[89] Rideout WM 3rd, Hochedlinger K, Kyba M, Daley GQ, Jaenisch R. Correction of a genetic defect by nuclear transplantation and combined cell and gene therapy. Cell 2002; 109(1): 17-27.

[90] Tabar V, Tomishima M, Panagiotakos G, *et al.* Therapeutic cloning in individual parkinsonian mice. Nat Med 2008; 14(4): 379-81.

[91] Meissner, A. and Jaenisch R. Generation of nuclear transfer-derived pluripotent ES cells from cloned Cdx2-deficient blastocysts. Nature 2006; 439(7073): 212-5.

[92] Robert V, Planchon O, Lapetite JM, Esteves M. Rainfall is not a direct mortality factor for anopheline larvae. Parasite 1999; 6(2): 195-6.

[93] Schneuwly S, Klemenz R.and Gehring WJ. Redesigning the body plan of Drosophila by ectopic expression of the homoeotic gene Antennapedia. Nature 1987; 325(6107): 816-8.

[94] Takahashi K, Tanabe K, Ohnuki M, *et al.* Induction of pluripotent stem cells from adult human fibroblasts by defined factors. Cell 2007; 131(5): 861-72.

[95] Zhou H, Wu S, Joo JY, *et al.* Generation of induced pluripotent stem cells using recombinant proteins. Cell Stem Cell 2009; 4(5): 381-4.

[96] Blau HM, Chiu CP and Webster C. Cytoplasmic activation of human nuclear genes in stable heterocaryons. Cell 1983; 32(4): 1171-80.

[97] Weiss MC and Chaplain M. Expression of differentiated functions in hepatoma cell hybrids: reappearance of tyrosine aminotransferase inducibility after the loss of chromosomes. Proc Natl Acad Sci USA 1971; 68(12): 3026-30.

[98] Davidson RL, Ephrussi B, Yamamoto K. Regulation of pigment synthesis in mammalian cells, as studied by somatic hybridization. Proc Natl Acad Sci USA 1966; 56(5): 1437-40.

[99] Carlsson SA, Savage RE, Ringertz NR. Behavior of differentiated hen nuclei in the cytoplasm of rat myoblasts and myotubules. Nature 1970; 228(5274): 869-71.

[100] Nair S, P Chacko V, Arnold C, Diehl AM. Hepatic ATP reserve and efficiency of replenishing: comparison between obese and nonobese normal individuals. Am J Gastroenterol 2003; 98(2): 466-70.

[101] Collas P, Taranger CK. Epigenetic reprogramming of nuclei using cell extracts. Stem Cell Rev 2006; 2(4): 309-17.

CHAPTER 3

Characterization of Human Embryonic Stem Cells

Romulo M. Brena*

Epigenome Center, Norris Comprehensive Cancer Center, University of Southern California, Los Angeles, USA

Abstract: Human embryonic stem cells (hESCs) hold great therapeutic promise since they are capable of generating specialized cells that may be utilized to replace damaged tissues in patients suffering from various types of diseases. However, this application is dependent on a comprehensive understanding of the signalling mechanisms involved in ESC lineage restriction, an area in which further scientific proficiency is needed. Moreover, the development of clinical applications using hESCs rests on the formulation of appropriate culture conditions that would allow for the isolation and maintenance of homogeneous, genetically and epigenetically stable cell populations. The stability of these cells is critical to diminish potentially adverse effects following transplantation, such as host rejection, low survival of the grafted cells and maybe even tumor formation. In this chapter I will describe the current techniques for isolating, characterizing and maintaining hESCs in culture. I will address some of the challenges underlying these methods and, when possible, offer alternatives that may help overcome these challenges. Finally, I will discuss the current limitations that have made hESC-based therapy a discipline still in its infancy and we will provide my view of where the field of personalized stem cell-based medicine is likely to go in the future.

Keywords: hESC derivation, hESC characterization, stem cell niche, pluripotency.

INTRODUCTION

Embryonic stem cells (ESCs) are pluripotent cells capable of giving rise to all somatic cell types in the embryo. A key feature of ESCs is their capacity for indefinite self-renewal which, coupled with their plasticity, provide a valuable tool for dissecting the mechanisms involved in cellular differentiation, organogenesis, and, ultimately, embryonic development. These two features of stem cells have paved a new avenue for regenerative medicine aimed at replacing damaged tissues in patients suffering from a host of diseases. A major hurdle towards making stem-cell based regenerative medicine a reality is the fact that the signalling mechanisms involved in lineage restriction are only partially understood. Furthermore, a significant number of stem cell culture protocols call for the use of animal-derived cells, such as mouse embryonic fibroblasts (MEFs), or condition medium, which renders these cells unsuitable for human therapy. The current protocols for *in vitro* propagation of ESCs underscore the need for developing xeno-free culturing systems capable of reproducibly and efficiently maintaining ESCs in an undifferentiated state for prolonged time periods.

Another important element to consider is the role of the stem cell niche, which is increasingly being described as a key component in the regulation and maintenance of stem cell identity. Understanding the physical interactions of stem cells with their surroundings, as well as the biochemical milieu that comprises their microenvironment, will undoubtedly shed light on the mechanisms that underlie the establishment and conservation of their unique phenotype.

Finally, the discovery of new cellular markers capable of identifying and capturing homogeneous stem cell populations with minimal cross-reactivity is central to the successful establishment of well-defined ESC cultures. Unfortunately, several markers currently used for these purposes are not sufficiently specific to achieve these goals and thus bypass some of the potential difficulties outlined above.

***Address correspondence to Romulo M. Brena:** Epigenome Center, Norris Comprehensive Cancer Center, University of Southern California, 1450 Biggy Street, NRT G511, Los Angeles, CA 90033-9601; E-mail: Brena@usc.edu

Kuldip S. Sidhu (Ed)

Despite these challenges, the stem cell field is moving at an amazing speed, with new discoveries reported almost daily. There is no doubt in my mind that the field holds extreme promise for the future of medicine, which will become increasingly evident in the years to come.

The Stem Cell Niche

In biology, a niche is defined as "a habitat that supplies the factors necessary for the existence of an organism or species" [1]. Schofield first proposed the existence of such niche for stem cells in 1978 [2], postulating that maturation of human stem cells (HSCs) was modulated by a "stem cell niche", which allowed for the persistence of stem cells in a quiescent state, while also enabling the production of more mature progeny. The local microenvironment of the stem cell niche is thought to be essential to cell survival and maintenance of cell identity by providing chemical and physical signals that directly influence both proliferation and differentiation.

One of the most extensively studied and best described stem cell microenvironments is the bone marrow (BM). The BM contains at least two distinct stem cell niches, the vascular niche and the osteoblastic niche [3]. While both niches provide many of the cytokines critical for HSC regulation, it is thought that the two microenvironments allow for the specific manifestation of a range of non-overlapping functions. The osteoblastic niche, for example, is involved in maintaining stem cells in a quiescent state, while the vascular niche is thought to encourage proliferation and differentiation [3, 4]. Upon parathyroid hormone stimulation, osteoblasts increase in number with a concomitant Notch-1 mediated expansion of the HSC pool (Burness and Sipkins). Similarly, the HSC pool is also expanded through manipulation of the bone morphogenic signalling pathway [5-7]. Nilsson and colleagues have suggested that factors involved in inducing quiescence are found in the osteoblastic niche. Osteopontin, produced by osteoblasts and secreted into the local microenvironment, plays a central role in the migration of HSCs into this niche and in suppressing their proliferation [8, 9]. Likewise, the tyrosine kinase Tie2 has been proposed to maintain quiescence in this same microenvironment [10].

The vascular niche is critical to the transit of HSCs to the BM. Early experiments demonstrated that $CD34^+$ pluripotent progenitor cells adhere to BM endothelial cells, giving rise to the hypothesis that the vascular niche is involved in cellular transit to and from the circulation (Rafii *et al.*, 1994). More recent studies have shown that the vascular expression of chemotactic and cell adhesion molecules, such as VCAM-1, SDF-1 and E- & P-selectins is important for HSCs homing to the BM [11-13].

It should be noted that a specialized subset of cells present in both niches plays a critical role in defining and maintaining the BM microenvironment. Mesenchymal stem cells (MSC) are pluripotent and capable of differentiating into osteoblasts, adipocytes and fibroblasts, among other cell types. A large body of evidence suggests that MSCs interact with HSCs *in vivo* and *in vitro* [14, 15]. Furthermore, MSCs have been shown to cooperate in the expansion of $CD4^+$ hematopoietic progenitor cells *ex vivo* [16, 17]. Interestingly, MSCs transplanted to subcutaneous sites are capable of recreating a functional hematopoietic microenvironment [14], and this formation progresses *via* a multi-step process that is independent of MSC-organization of sinusoidal vasculature [14].

Altogether, the data discussed thus far make it evident that the stem cell niche in the BM is a complex structure, tightly regulated by the interaction of a number of different cell types that together modulate the quiescent and proliferative states of HSCs *via* the secretion of soluble factors and physical cell-cell interactions.

Wnt Signalling, Epigenetic Regulation and Stemness

It has recently been proposed that ESCs are "trapped" in a specific chromatin state and that this state depends on a wide array of covalent modifications controlled by the activity of chromatin remodelling factors [18]. Recent studies have also highlighted that canonical Wnt signals control not only single targets through local Wnt-responsive elements (WRE), but also global chromatin structure and function. This is evidenced by the fact that several chromatin modifiers are recruited to the C-terminus of β-catenin, a central member of the Wnt signalling pathway [19, 20]. Furthermore, essential Wnt transcription factors strongly induce DNA bending, which may allow for the assembly of large enhanceosome complexes (Mosimann *et al.*, 2009). Put together, these mechanisms have far-reaching effects and may induce, at least in part, the changes required to maintain the stem cell phenotype.

The significance of Wnt activity in ESC biology has been well established in the hematopoietic system *via* reconstitution assays [21]. *In vivo* studies demonstrated that high levels of Wnt3a or activated β-catenin increase cell renewal. If this signals persist, however, a depletion of the ESC pool ensues [22]. Concomitantly, paracrine Wnt signals in the stem cell niche limit ESC proliferation [23]. These observations highlight the role of Wnt activity in maintaining ESCs in a quiescent state, underscoring that normal regulation of the Wnt signalling pathway is central to the reconstructive function of ESCs in the hematopoietic system.

DNA methylation is the most extensively studied and best understood epigenetic modification to have analogous functions in ESC differentiation and tumorigenesis, as promoter hypermethylation is associated with gene silencing in both ESCs and cancer cells [24, 25]. In ESCs, for example, Oct4 expression is regulated *via* DNA methylation of its CpG-rich promoter [26]. In mammary tumors, downregulation of Wnt inhibitors through promoter DNA methylation results in elevated levels of β-catenin, which is thought to be the driving force behind the transition of early lesions to malignant tumors [27]. SOX17 physically interacts with β-catenin and suppresses its activity. Coincidentally, SOX17 expression which is abundant in the normal gut, is downregulated by promoter hypermethylation in colon carcinoma and in stem cells [28, 29]. This observation, one of the many paralleling promoter hypermethylation in ESCs and overt tumors, may provide insight in the mechanisms modulating stemness and Wnt activity.

Derivation of ESCs

During early embryogenesis, multiple mitotic cell divisions of a diploid zygote give rise to a blastocyst. The blastocyst is comprised of an inner cell layer, the embryoblast, and an outer cell layer, the trophoblast. The outer cell mass, also referred to as the trophectoderm, forms the extra-embryonic tissues, including the placenta, the chorion, and the umbilical chord. The inner cell mass (ICM) develops into the embryo proper [30]. In 1975 Sherman examined the growth and differentiation of mouse trophoblast cells and the proliferation of the ICM in long-term cultures [31]. These experiments resulted in the derivation of four cell lines which were maintained for more than 12 months. Significantly, these cell lines were comprised of a variety of cells types, not just undifferentiated ESCs, they were not capable of differentiating into the three germ layers *in vivo* and, ultimately, developed a number of chromosomal abnormalities [31]. Subsequent experiments with embryonal carcinoma stem cells allowed for the development of suitable culture conditions and for the determination of the optimal stage for the isolation of pluripotent ESCs. These advances culminated in the derivation of the first stable mouse embryonic stem cell lines in 1981 [32-34].

With the knowledge afforded by the aforementioned studies, Thomson and colleagues developed a culture system for nonhuman primate ESC lines [35, 36], paving the way for the successful generation of the first hESCs by Thomson and colleagues in 1998 [37] and by Reubinoff and colleagues in 2000 [38]. These hESCs were harvested from human embryos originally produced by *in vitro* fertilization. The hESC lines derived by Thomson were grown on MEFs, remained pluripotent and karyotypically normal and were capable of generating large germ cell teratomas when grafted to severe combined immunodeficient (SCID) mice [39].

Currently, there are hundreds if not thousands of hESC lines derived from donated embryos in laboratories all around the world. The most common techniques to isolate the ICM from the trophectoderm at the blastocyst stage are mechanical dissection and immunosurgery. The first hESC lines were established through immunosurgery, a procedure that requires the use of animal-derived products, such as guinea pig complement and anti-human serum antibodies [38-41]. The use of animal-derived products made these cells unsuitable for human transplantation due to the potential for transfer of pathogens that could initiate the patient's innate immune system, leading to an increase chance of graft rejection. Thus, the development of enzymatic or mechanical methods for the isolation of the ICM from the trophectoderm without the need for animal-derived products is essential if hESCs are intended for human therapy [42-44]. Recently, lasers have been used to create small openings in the zona pellucida that encapsulates the blastocyst, which, followed by laser-assisted isolation of the ICM, has successfully resulted in the establishment of xeno-free hESC lines [45-47].

Generating hESCs from the ICM at the blastocyst stage inextricably requires the destruction of the embryo, which has raised not only ethical but also political concerns. In an attempt to address this issue, a significant amount of research has been devoted to isolating hESCs from various stages of embryonic development without destructing the embryo. The first attempts at removing one cell from the 8-cell morula have resulted in the establishment of hESC lines with varying degrees of success [48-50]. Differentiation of blastomeres to ICM was highly inefficient, with blastomere-derived aggregates giving rise predominantly to trophectoderm-like vesicles. To address this problem and to increase the efficiency of hESC derivation, laminin supplemented culture medium was utilized [51]. The addition of laminin to the culture medium resulted in a derivation efficiency comparable to that attained from processing whole blastocysts. The positive effect of laminin in this process was hypothesized to stem from simulating, at least in part, the natural ICM niche, preventing the polarization of blastomeres into ICM and trophectoderm. Further optimization of this culture procedure resulted in the successful generation of blastomere-derived hESCs in the absence of feeders, eliminating the need for co-culture systems with animal-derived cells or other previously established hESC lines.

Characterization of hESCs

As opposed to most cell lines, which grow in monolayers, hESCs grow in packed colonies with clearly defined borders. Individual hESCs within colonies typically display prominent nucleoli and a high nucleus to cytoplasm ratio. hESCs are characterized by the expression of a host of cell surface markers, such as SSEA-3, SSEA-4 and TRA, and also by the absence of negative markers, such as SSEA-1 [52-56].

The self-renewal capacity of hESCs maintained *in vitro* is typically assessed by telomerase activity and telomere length after prolonged cultivation [57, 58]. As opposed to human somatic cells, which have a limited life span and can only elude senescence through the acquisition of genetic and epigenetic abnormalities [59, 60], the capacity of hESCs for indefinite self-renewal is independent of the disturbance or inactivation of the well known Rb and p53 tumor suppressor pathways [57, 61].

Over the past few years, it has become evident that hESCs can acquire karyotypic abnormalities *in vitro* [54, 62, 63]. It is hypothesized that these karyotypic abnormalities are the result of *in vitro* adaptation and selection, as abnormal rapidly-dividing cells eventually take over the entire diploid culture [62]. Interestingly, the most recurrent chromosomal changes observed in *in vitro* hESC cultures affect genomic regions that harbor genes associated with cell proliferation, anti-apoptosis and pluripotency [62]. These genomic regions are also frequently altered in human cancers [64]. Provided that karyotypically abnormal hESCs may be prone to tumor formation if grafted into human patients, a protocol for screening the genetic integrity of these cells must be established. The main goal of this protocol should be to screen hESCs intended for use in regenerative medicine and to ascertain a normal genetic signature at relatively high resolution. The protocol should include techniques such as fluorescent *in situ* hybridization, single nucleotide polymorphism (SNP) analysis and short tandem repeat analysis, among other methods [54, 63, 65-69]. It should be emphasized that it is not clear how the data generated by these techniques may be interpreted. Is there a tolerable level of genetic abnormalities that may not interfere with the safety and viability of a cell line specifically developed for human therapy, or do any genetic alterations should be grounds for disqualifying these cells? As I have explained in previous sections, it may not be possible to generate and expand hESC *in vitro* without introducing at least some genetic alterations. It should also be noted that various studies have highlighted that hESCs tend to rapidly develop epigenetic abnormalities *in vitro*, which may also predispose them to tumor formation [70-71].

In one of these studies Chung and colleagues described the widespread but yet specific demethylating effect of ascorbic acid on the epigenome of hESCs. Ascorbic acid is a component of almost all commercial stem cell culture media, which suggests that the chemistry of these formulations should be periodically examined and modified based on emerging scientific evidence. Despite the fact that it is almost undisputed that hESCs acquire, at least to some degree, genetic and epigenetic abnormalities in culture, it is not clear what the tolerance, if any, for these changes should be in order to ensure the safety of hESCs for human therapy. Several research groups around the world are attempting to answer this basic question and their recommendations will, without a doubt, foster vigorous debate, which will hopefully result in the establishment of clear guidelines on how the genetic and epigenetic integrity of hESC should be assessed and what genetic and epigenetic changes, if any, may be tolerated in cells prepared for human therapy.

In Vitro Maintenance of hESCs

A hallmark of successfully culturing hESCs is indefinite self-renewal. The first hESC lines relied on MEF feeder layers for propagation and maintenance in an undifferentiated state [38, 39]. However, the advent of xeno-free culture systems replaced of MEFs for human fibroblast cells derived from fallopian tube epithelium [72], fetal foreskin, muscle [73, 74], bone marrow [75], or amniotic epithelium [76]. hESCs may also be maintained in a feeder-free environment if an extracellular matrix, such as fibronectin or matrigel, is supplied. However, cells cultured in the presence of either of these matrices required media conditioned by feeder fibroblasts and supplemented with basic fibroblast growth factor (bFGF) in order to maintain their undifferentiated phenotype [77, 78].

hESCs cultured under feeder-free conditions often give rise to fibroblasts or stromal-like cells. It is hypothesized that these cells serve a supporting role, helping to maintain the hESCs in an undifferentiated state. Two studies have examined the nature of these hESC-derived feeders, and found evidence that in fact they support the growth of their undifferentiated progenitors [79, 80]. Although directly derived from hESCs, these feeder cells senesce after several passages, which poses a limitation to their continuous use. The main obstacle raised by the limited viability of these cells is that derivation of new feeders is not trivial and may introduce high variability into the culture system.

So far, we have examined alternatives for the elimination of exogenous feeders from hESC cultures. However, as detailed above, hESCs themselves give rise to differentiated progeny to serve the purpose of feeder cells. Thus, other approaches are needed in order to entirely eliminate the need for ancillary cells in hESC cultures, which would allow for the establishment of a defined environment for these hESCs to proliferate in. Since MEFs have the capacity to maintain hESC self-renewal, several studies have focused on identifying the secreted factors that contribute or even make possible the maintenance of the self-renewal phenotype [81-84]. Additionally, various research groups have proposed that high concentrations of bFGF and noggin-mediated repression of bone morphogenic protein (BMP) signalling can sustain undifferentiated proliferation of hESCs in serum-free medium [84, 85-87]. Exogenous treatment of hESC cultures with cocktails of human recombinant proteins and signalling molecules, including transforming growth factor beta-1 (TGF-β1) and activin A, have also been successfully implemented [85, 88-89]. Draper and colleagues have reported that maintaining hESCs in a feeder-free culture system may predispose them to developing genetic abnormalities ([90]. However, it is unclear whether this observation applies to all feeder-free culture conditions.

Assessing Pluripotency

The pluripotent nature of hESCs is functionally assessed by examining their ability to differentiate into all three embryonic germ layers (ectoderm, endoderm and mesoderm) both *in vivo* and *in vitro*. The typical *in vivo* test for this purpose is teratoma formation in SCID mice [39-38]. *In vitro*, hESCs are allowed to differentiate into embryoid bodies (EBs), which are cell aggregates grown in suspension culture. EBs are subsequently analyzed *via* immunohistochemistry or by reverse-transcription PCR to measure the expression of genes associated with the formation of the three germ layers [38-39].

The widespread adoption of various reprogramming technologies for the derivation of pluripotent cell lines has precipitated intense debate over the still unresolved issue of how to best evaluate pluripotency in hESCs [91-92]. Quantitative and robust protocols to measure endodermal, mesodermal and neural lineage differentiation are becoming gradually available. Thus, an increasing number of researchers have suggested that these assays could replace the controversial but current "gold standard" for assessing pluripotency: *in vivo* teratoma formation. Methods for generating teratomas and for evaluating their histopathology have been published [93]. However, the inherent complexity of the assay often leads to poor reproducibility of results and an altogether high degree of variability among and within cell cultures [94]. These reasons, particularly poor reproducibility, have discouraged some investigators from adopting teratoma formation as their in-house pluripotency assay.

Recent reports have suggested that specific sites and injection modalities might help improve tumor yield [95-96]. It remains to be determined, however, whether these reports will foster the widespread adoption of the teratoma formation assay in the stem cell research community. Despite the drawbacks of assessing pluripotency *via* teratoma formation, it should be underscored that generating teratomas *in vivo* may provide information not only about differentiation potential, but also about the predisposition of the injected cells to form teratocarcinomas [97-98]. Karyotypically normal hESCs generally give rise to teratomas, which do not contain undifferentiated cells. Karyotypically abnormal cells or cells with relatively small genetic lesions typically give rise to teratocarcinomas that contain undifferentiated cells. These data suggest that the presence of undifferentiated cells within a teratoma may serve as a surrogate indicator that the injected cells may not be genetically normal.

The importance of selecting and adopting a reliable pluripotency assay accessible to most labs involved in stem cell research is highlighted by the fact that the International Stem Cell Initiative (ISCI) is planning to hold a series of discussion groups to assess the reliability of the quantitative protocols mentioned above and put forth recommendations regarding their use.

Induced Pluripotent Stem (iPS) Cells

In 2006 Takahashi and Yamanaka described the reprogramming of somatic cells to a pluripotent stem-like state by retroviral delivery of four genes: Oct4, Sox2, Klf4 and c-Myc [99]. In their initial study, reprogrammed somatic cells were identified *via* the reactivation of the Fbx-15 locus 72 hours after initial retroviral transduction. Although these reprogrammed cells were capable of forming teratomas in which cells representative of all three germ layers could be identified, they were not germ-line competent. Improvements in the reprogramming protocol called for delaying selection of clones for at least seven days after retroviral transduction and for assessing the reactivation of loci known to be associated with stem cell maintenance [100-101]. By selecting for the reactivation of endogenous genes, such as Oct4, which are necessary for the preservation of the stem cell phenotype, germline-competent iPS clones could be identified [100-101]. Interestingly, other studies subsequently reported that iPS cells capable of contributing to chimeras could be isolated based entirely upon morphology, thus removing the need for drug selection [102-103].

Successfully reprogrammed iPS cells generally exhibit expression of stage-specific embryonic antigens (concomitant with the loss of somatic-cell specific markers), X-chromosome reactivation in female cells, telomerase activity, re-activation of endogenous genes essential for maintenance of pluripotency and self-renewal and, notably, silencing of the exogenous factors delivered to initiate the reprogramming process. As described earlier, the re-expression of endogenous genes required to maintain pluripotency involves extensive epigenetic remodelling, particularly the demethylation of key promoters, such as that of Oct4.

Since the inception of iPS cells, extensive efforts have been devoted to developing tests to assess "full reprogramming". In 2009, two studies reported that mouse iPS cells transplanted into tetraploid embryos were able to generate live mice [104-105]. The question of how to best assess full reprogramming in human iPS cells, however, is still being debated since ethical reasons prevent using the formation of chimeras for this purpose. Furthermore, human and mouse iPS cells are not identical, mainly in terms of their developmental status and *in vitro* culture requirements. Mouse iPS cells closely resemble cells of the preimplantation epiblast [106], require LIF and do not express endogenous FGF4 [107]. Human iPS cells, in contrast, are more closely related to cells in the late epiblast and require FGF2 but not LIF to maintain their self-renewal and pluripotency *in vitro* [108]. Furthermore, mouse iPS cells express SSEA-1, while their human counterparts express SSEA-3 and SSEA-4. Altogether, these data indicate that markers utilized to assess reprogramming differ between human and mouse iPS lines.

It should be noted that gene expression profiling has uncovered subtle differences between hESCs and human iPS cells [66, 109]. Of interest is the fact that the longer human iPS cells remain in culture, the more their gene expression profile resembles that of hESCs. Additional studies will be necessary to determine if the identified difference in gene expression between the two types of pluripotent human cells may interfere with the potential clinical applications of human iPS lines.

CONCLUSIONS AND PROSPECTS

Induction of differentiation and lineage restriction of hESCs involve a complex and reciprocal exchange of signals between these cells and their microenvironment. The stem cell microenvironment, or stem cell niche, is an intricate physical and chemical space essential for stem cell survival and maintenance of stem cell identity. Studies over the past decade have shed significant light on the stem cell niche, though many critical questions about its dynamism, composition and function still remain unanswered. Because of our limited knowledge in this realm, it is demanding to control direct stem cell differentiation in experimental systems.

Recent studies have made remarkable progress identifying and discerning the chemical and physical cues provided by animal and human supporting cells (feeders) in *in vitro* stem cell cultures. This knowledge has culminated in the publication of several xeno-free *in vitro* culture protocols, enabling the use of stem cells for therapeutic applications in humans. It is clear that the stem cell field has advanced at a terrific pace, especially over the last 5 years. However, stem cell therapy in humans is still in its infancy and many challenges remain.

Two major difficulties facing the field are the observed genetic and epigenetic instability of stem cells maintained in culture. Is it possible to isolate and grow stem cells *in vitro* without introducing genetic and epigenetic alterations? The answer to this question will dictate much of the future of regenerative medicine. If it were not possible to isolate, grow and differentiate stem cells without altering their genetic and epigenetic profiles, is there a "level of tolerance" for changes that would still render these cells safe for human transplantation? Another concern is the presence of residual undifferentiated cells in otherwise differentiated cultures. These cells could potentially form teratomas or, if genetically and epigenetically abnormal, teratocarcinomas. The development of efficient methods for the detection and isolation of these cells will also be critical for the progress of regenerative medicine.

REFERENCES

[1] Borovski T, De Sousa E M F, Vermeulen L, Medema J P. Cancer stem cell niche: the place to be. Cancer Res 2011; 71(3): 634-639.

[2] Schofield R. The relationship between the spleen colony-forming cell and the haemopoietic stem cell. Blood Cell 1978; 4(1-2): 7-25.

[3] Yin T, Li L. The stem cell niches in bone. J Clin Invest 2006; 116(5): 1195-1201.

[4] Heissig B, Hattori K, Dias S, *et al.* Recruitment of stem and progenitor cells from the bone marrow niche requires MMP-9 mediated release of kit-ligand. Cell 2002; 109(5): 625-637.

[5] Zhang J, Niu C, Ye L, *et al.* Identification of the haematopoietic stem cell niche and control of the niche size. Nature 2003; 425(6960): 836-841.

[6] Adams GB, Chabner KT. Stem cell engraftment at the endosteal niche is specified by the calcium-sensing receptor. Nature 2006; 439(7076): 599-603.

[7] Calvi LM, Adams GB, Weibrecht KW, *et al.* Osteoblastic cells regulate the haematopoietic stem cell niche. Nature 2003; 425(6960): 841-846.

[8] Nilsson SK, Johnston HM, Whitty GA, *et al.* Osteopontin, a key component of the hematopoietic stem cell niche and regulator of primitive hematopoietic progenitor cells. Blood 2005; 106(4): 1232-39.

[9] Stier S, Ko Y, Forkert R, *et al.* Osteopontin is a hematopoietic stem cell niche component that negatively regulates stem cell pool size. J Exp Med 2005; 201(11): 1781-1791.

[10] Arai F, Hirao A, Ohmura M, *et al.* Tie2/angiopoietin-1 signalling regulates hematopoietic stem cell quiescence in the bone marrow niche. Cell 2004; 118(2): 149-161.

[11] Dimitroff CJ, Lee JY, Rafii S, Fuhlbrigge RC, Sackstein R. CD44 is a major E-selectin ligand on human hematopoietic progenitor cells. J Cell Biol 2001; 153(6): 1277-1286.

[12] Frenette PS, Subbarao S, Mazo IB, Von Andrian UH, Wagner DD. Endothelial selectins and vascular cell adhesion molecule-1 promote hematopoietic progenitor homing to bone marrow. Proc Natl Acad Sci USA 1998; 95(24): 14423-14428.

[13] Kahn J, Byk T, Jansson-Sjostrand L, *et al.* Overexpression of CXCR4 on human CD34+ progenitors increases their proliferation, migration, and NOD/SCID repopulation. Blood 2004; 103(8): 2942-2949.

[14] Sacchetti B, Funari A, Michienzi S, *et al.* Self-renewing osteoprogenitors in bone marrow sinusoids can organize a hematopoietic microenvironment. Cell 2007; 131(2): 324-336.

[15] Gronthos S, Zannettino AC, Hay SJ. *et al* Molecular and cellular characterisation of highly purified stromal stem cells derived from human bone marrow. J Cell Sci 2003; 116(9): 1827-1835.

[16] Li N, Feugier P, Serrurrier B, *et al.* Human mesenchymal stem cells improve *ex vivo* expansion of adult human CD34+ peripheral blood progenitor cells and decrease their allostimulatory capacity. Exp Hematol 2007; 35(3): 507-515.

[17] Kadereit S, Deeds LS, Haynesworth SE, *et al.* Expansion of LTC-ICs and maintenance of p21 and BCL-2 expression in cord blood CD34(+)/CD38(-) early progenitors cultured over human MSCs as a feeder layer'. Stem Cell 2002; 20(6): 573-582.

[18] Bibikova M, Laurent LC, Ren B, Loring JF, Fan JB. Unraveling epigenetic regulation in embryonic stem cells. Cell Stem Cell 2008; 2(2): 123-134.

[19] Barker N, Hurlstone A, Musisi H, Miles A, Bienz M, Clevers H. The chromatin remodelling factor Brg-1 interacts with beta-catenin to promote target gene activation. Embo J 2001; 20(17): 4935-4943.

[20] Mosimann C, Hausmann G, Basler K Beta-catenin hits chromatin: regulation of Wnt target gene activation. Nat Rev Mol Cell Biol 2009; 10(4): 276-286.

[21] Reya T, Clevers H Wnt signalling in stem cells and cancer. Nature 2005; 434(7035): 843-850.

[22] Scheller M, Huelsken J, Rosenbauer F, *et al.* Hematopoietic stem cell and multilineage defects generated by constitutive beta-catenin activation. Nat Immunol 2006; 7(10): 1037-1047.

[23] Fleming HE, Janzen V, Lo Celso C, *et al.* Wnt signalling in the niche enforces hematopoietic stem cell quiescence and is necessary to preserve self-renewal *in vivo*. Cell Stem Cell 2008; 2(3): 274-283.

[24] Costello JF. Stem cells: Tips for priming potency. Nature 2008; 454(7200): 45-46.

[25] Nagai K, Ishii H, Miyoshi N, *et al.* Long-term culture following ES-like gene-induced reprogramming elicits an aggressive phenotype in mutated cholangiocellular carcinoma cells. Biochem Biophys Res Commun 2010; 395(2): 258-263.

[26] Ben-Shushan E, Pikarsky E, Klar, A, Bergman Y. Extinction of Oct-3/4 gene expression in embryonal carcinoma x fibroblast somatic cell hybrids is accompanied by changes in the methylation status, chromatin structure, and transcriptional activity of the Oct-3/4 upstream region. Mol Cell Biol 1993; 13(2): 891-901.

[27] Klarmann, GJ, Decker, A, Farrar, WL. Epigenetic gene silencing in the Wnt pathway in breast cancer. Epigenetics 2008; 3(2): 59-63.

[28] Zorn AM, Barish GD, Williams BO, Lavender P, Klymkowsky MW, Varmus HE. Regulation of Wnt signalling by Sox proteins: XSox17 alpha/beta and XSox3 physically interact with beta-catenin. Mol Cell 1999; 4(4): 487-498.

[29] Sinner, D, Rankin, S, Lee, M, Zorn, A M. Sox17 and beta-catenin cooperate to regulate the transcription of endodermal genes. Development 2004; 131(13): 3069-3080.

[30] Gilbert S. Developmental Biology. 2006; Sinauer Associates, Inc. Publishers.

[31] Sherman MI Long term culture of cells derived from mouse blastocysts. Differentiation 1975; 3(1-3): 51-67.

[32] Evans MJ, Kaufman MH. Establishment in culture of pluripotential cells from mouse embryos. Nature 1981; 292(5819); 154-156.

[33] Martin GR. Teratocarcinomas and mammalian embryogenesis. Science 1980; 209(4458): 768-776.

[34] Martin GR. Isolation of a pluripotent cell line from early mouse embryos cultured in medium conditioned by teratocarcinoma stem cells. Proc Natl Acad Sci U S A 1981; 78(12): 7634-638.

[35] Thomson JA, Kalishman J, Golos TG, *et al.* Isolation of a primate embryonic stem cell line. Proc Natl Acad Sci *USA* 1995; 92(17): 7844-848.

[36] Thomson JA, Kalishman J, Golos TG, Durning M, Harris CP, Hearn JP. Pluripotent cell lines derived from common marmoset (Callithrix jacchus) blastocysts. Biol Reprod 1996; 55(2): 254-259.

[37] Thomson JA, Itskovitz-Eldor J, Shapiro SS, *et al.* Embryonic stem cell lines derived from human blastocysts. Science 1998; 282(5391): 1145-1147.

[38] Reubinoff BE, Pera MF, Fong CY, Trounson A, Bongso A. Embryonic stem cell lines from human blastocysts: somatic differentiation *in vitro*. Nat Biotechnol 2000; 18(4): 399-404.

[39] Bosma, GC, Custer RP, Bosma MJ. A severe combined immunodeficiency mutation in the mouse. Nature 1983; 301(5900): 527-530.

[40] Cowan CA, Klimanskaya I, Mcmahon J, *et al.* Derivation of embryonic stem-cell lines from human blastocysts. N Engl J Med 2004; 350(13): 1353-1356.

[41] Ellerstrom C, Strehl R, Moya K, *et al.* Derivation of a xeno-free human embryonic stem cell line. Stem Cells 2006; 24(10): 2170-2176.

[42] Amit M, Itskovitz-Eldor J. Derivation and spontaneous differentiation of human embryonic stem cells. J Anat 2002; 200(3): 225-232.

[43] Genbacev O, Krtolica A, Zdravkovic T. *et al.* Serum-free derivation of human embryonic stem cell lines on human placental fibroblast feeders. Fertil Steril 2005; 83(5): 1517-1529.

[44] Strom S, Inzunza J, Grinnemo KH, *et al.* Mechanical isolation of the inner cell mass is effective in derivation of new human embryonic stem cell lines. Hum Reprod 2007; 22(12): 3051-3058.

[45] Turetsky T, Aizenman E, Gil Y, *et al.* Laser-assisted derivation of human embryonic stem cell lines from IVF embryos after preimplantation genetic diagnosis. Hum Reprod 2008; 23(1): 46-53.

[46] Sidhu KS, Ryan JP, Tuch BE. Derivation of a new hESC line, endeavour-1 and its clonal propagation. Stem Cells and Development 2008; 7(1): 45-54.

[47] Sidhu KS, Ryan JP, Lees JG. and Tuch BE. Derivation of a new human embryonic stem cell line, endeavour-2 and its characterization. *In vitro* Cells and Dev Biol 2010; 46(3-4): 269-275.

[48] Klimanskaya I, Chung Y, Becker S, Lu S J, Lanza R. 'Human embryonic stem cell lines derived from single blastomeres. Nature 2006; 444 (7118): 481- 485.

[49] Klimanskaya I, Chung Y, Becker S, Lu S J, Lanza R. Derivation of human embryonic stem cells from single blastomeres. Nat Protoc 2007; 2(8): 1963-1972.

[50] Strelchenko N, Verlinsky O, Kukharenko V, Verlinsky Y. Morula-derived human embryonic stem cells. Reprod Biomed Online 2004; 9(6): 623 -629.

[51] Chung Y, Klimanskaya I, Becker S, *et al.* Human embryonic stem cell lines generated without embryo destruction. Cell Stem Cell 2008; 2(2): 113-117.

[52] Carpenter MK, Rosler E, Rao MS. Characterization and differentiation of human embryonic stem cells. Cloning Stem Cells 2003;.5(1): 79-88.

[53] Chambers I, Colby D, Robertson M, *et al.* Functional expression cloning of Nanog, a pluripotency sustaining factor in embryonic stem cells. Cell 2003; 113(5): 643-655.

[54] Draper JS, Moore HD, Ruban LN, Gokhale PJ, Andrews PW. Culture and characterization of human embryonic stem cells. Stem Cells Dev 2004; 13(4): 325 -336.

[55] Heins N, Englund MC, Sjoblom C, *et al.* Derivation, characterization, and differentiation of human embryonic stem cells. Stem Cells 2004; 22(3): 367-376.

[56] Nichols J, Zevnik B, Anastassiadis K, *et al.* Formation of pluripotent stem cells in the mammalian embryo depends on the POU transcription factor Oct4. Cell 1998; 95(3): 379-391.

[57] Brandenberger R, Khrebtukova I, Thies RS, *et al.* MPSS profiling of human embryonic stem cells. BMC Dev Biol 2004; 4: 10.

[58] Carpenter MK, Rosler ES, Fisk GJ, *et al.* Properties of four human embryonic stem cell lines maintained in a feeder-free culture system. Dev Dyn 2004; 229(2): 243-274.

[59] Hayflick L, Moorhead PS. The serial cultivation of human diploid cell strains. Exp Cell Res 1961; 25: 585-621.

[60] Brena RM, Costello JF. Genome-epigenome interactions in cancer. Hum Mol Genet 2007; 16(1): R96.

[61] Rosler ES, Fisk GJ, Ares X, *et al.* Long-term culture of human embryonic stem cells in feeder-free conditions. Dev Dyn 2004; 229(2): 259-274.

[62] Baker DE, Harrison NJ, Maltby E, *et al.* Adaptation to culture of human embryonic stem cells and oncogenesis *in vivo*. Nat Biotechnol 2007; 25(2): 207-215.

[63] Mitalipova MM, Rao RR, Hoyer DM, *et al.* Preserving the genetic integrity of human embryonic stem cells. Nat Biotechnol 2005; 23(1): 19-20.

[64] Beerenwinkel N, Antal T, Dingli D, *et al.* Genetic progression and the waiting time to cancer. PLoS Comput Biol 2007; 3(11): e225.

[65] Brimble S N, Zeng X, Weiler DA, *et al.* Karyotypic stability, genotyping, differentiation, feeder-free maintenance, and gene expression sampling in three human embryonic stem cell lines derived prior to August 9, 2001. Stem Cells Dev 2004; 13(6): 585-597.

[66] Sidhu KS. New approaches for the generation of induced pluripotent stem cells. Expert Opin Biol Ther 2011; 11(5): 569-79.

[67] Josephson R, Sykes G, Liu Y, *et al.* A molecular scheme for improved characterization of human embryonic stem cell lines. BMC Biol, 2006; 4: 28.

[68] Maitra A, Arking DE, Shivapurkar N, *et al.* Genomic alterations in cultured human embryonic stem cells. Nat Genet 2005; 37(10):1099-1103.

[69] Mosher JT, Pemberton TJ, Harter K, *et al.* Lack of population diversity in commonly used human embryonic stem-cell lines. N Engl J Med. 2010; 62(2): 183-205.

[70] Chung TL, Brena RM, Kolle G, *et al.* Vitamin C promotes widespread yet specific DNA demethylation of the epigenome in human embryonic stem cells. Stem Cells 2010;.28(10): 1848-1855.

[71] Walter J. An epigenetic Tet a Tet with pluripotency. Cell Stem Cell, 2011; 8(2): 121-122.

[72] Bongso A, Fong CY, Ng SC, Ratnam S. 'Isolation and culture of inner cell mass cells from human blastocysts. Hum Reprod 1994; 9(11): 2110-2117.

[73] Amit M, Margulets V, Segev H, *et al, et al.* Human feeder layers for human embryonic stem cells. Biol Reprod 2003; 68(6): 2150-2156.

[74] Richards M, Fong CY, Chan WK, Wong PC, Bongso A. Human feeders support prolonged undifferentiated growth of human inner cell masses and embryonic stem cells. Nat Biotechnol 2002; 20(9): 933-936.

[75] Cheng L, Hammond H, Ye Z, Zhan X, Dravid G. Human adult marrow cells support prolonged expansion of human embryonic stem cells in culture. Stem Cells 2003; 21(2): 131-142.

[76] Miyamoto K, Hayashi K, Suzuki T, *et al.* Human placenta feeder layers support undifferentiated growth of primate embryonic stem cells. Stem Cells 2004;.22(4): 433-440.

[77] Van Hoof D, Braam SR, Dormeyer W, *et al.* Feeder-free monolayer cultures of human embryonic stem cells express an epithelial plasma membrane protein profile. Stem Cells 2008; 26(11): 2777-2781.

[78] Xu C, Inokuma MS, Denham J, *et al.* Feeder-free growth of undifferentiated human embryonic stem cells. Nat Biotechnol 2001; 19(10): 971-974.

[79] Stojkovic P, Lako M, Stewart R, *et al.* An autogeneic feeder cell system that efficiently supports growth of undifferentiated human embryonic stem cells. Stem Cells 2005; 23(3): 306-314.

[80] Wang Q, Fang ZF, Jin F, Lu Y, Gai H, Sheng HZ. Derivation and growing human embryonic stem cells on feeders derived from themselves. Stem Cells, 2005b; 23(9): 1221-1227.

[81] Cai J, Chen J, Liu Y, *et al.* Assessing self-renewal and differentiation in human embryonic stem cell lines. Stem Cells 2006; 24(3): 516-530.

[82] Chin, AC, Fong WJ, Goh LT, Philp R, Oh SK, Choo AB. Identification of proteins from feeder conditioned medium that support human embryonic stem cells. J Biotechnol 2007; 130(3): 320-328.

[83] Lim JW, Bodnar A. Proteome analysis of conditioned medium from mouse embryonic fibroblast feeder layers which support the growth of human embryonic stem cells. Proteomics 2002; 2(9): 1187-1203.

[84] Xu RH, Peck RM, Li DS, Feng X, Ludwig T, Thomson JA. Basic FGF and suppression of BMP signalling sustain undifferentiated proliferation of human ES cells. Nat Methods 2005; 2(3): 185-190.

[85] Beattie GM, Lopez AD, Bucay N, *et al.* Activin A maintains pluripotency of human embryonic stem cells in the absence of feeder layers. Stem Cells 2005; 23(4): 489-495.

[86] Pera MF, Andrade J, Houssami S, *et al.* Regulation of human embryonic stem cell differentiation by BMP-2 and its antagonist noggin. J Cell Sci 2004; 117(7): 1269-1280.

[87] Wang G, Zhang H, Zhao Y, *et al.* Noggin and bFGF cooperate to maintain the pluripotency of human embryonic stem cells in the absence of feeder layers. Biochem Biophys Res Commun 2005a; 33093): 934-942.

[88] James D, Levine AJ, Besser D, Hemmati-Brivanlou A. TGFbeta/activin/nodal signalling is necessary for the maintenance of pluripotency in human embryonic stem cells. Development 2005; 132(6): 1273-1282.

[89] Lu J, Hou R, Booth CJ, Yang SH, Snyder M. Defined culture conditions of human embryonic stem cells. Proc Natl Acad Sci USA 2006; 103(15): 5688-5693.

[90] Draper JS, Pigott C, Thomson JA, Andrews PW. Surface antigens of human embryonic stem cells: changes upon differentiation in culture. J Anat 2002; 200(3): 249-258.

[91] Ellis J, Bruneau BG, Keller G, *et al.* Alternative induced pluripotent stem cell characterization criteria for *in vitro* applications. Cell Stem Cell 2009; 4(3): 198-199.

[92] Maherali N, Hochedlinger K. Guidelines and techniques for the generation of induced pluripotent stem cells. Cell Stem Cell 2008; 3(6): 595-605.

[93] Gertow K, Przyborski S, Loring JF, *et al.* Isolation of human embryonic stem cell-derived teratomas for the assessment of pluripotency. Curr Protoc Stem Cell Biol 2007; Chapter 1: Unit 1B 4.

[94] Muller FJ, Goldmann J, Loser P, Loring JF. A call to standardize teratoma assays used to define human pluripotent cell lines. Cell Stem Cell 2010; 6(5): 412-414.

[95] Cooke MJ, Stojkovic M, Przyborski SA. Growth of teratomas derived from human pluripotent stem cells is influenced by the graft site. Stem Cells Dev 2006; 15(2): 254-259.

[96] Prokhorova TA, Harkness LM, Frandsen U, *et al.* Teratoma formation by human embryonic stem cells is site dependent and enhanced by the presence of Matrigel. Stem Cells Dev 2009; 18(1): 47-54.

[97] Blum B, Benvenisty N. The tumorigenicity of human embryonic stem cells. Adv Cancer Res 2008; 100: 133-158.

[98] Przyborski SA. Differentiation of human embryonic stem cells after transplantation in immune-deficient mice. *Stem Cells* 2005; 23(9): 1242-1250.

[99] Takahashi K, Yamanaka S. Induction of pluripotent stem cells from mouse embryonic and adult fibroblast cultures by defined factors. Cell 2006; 126(4): 663-676.

[100] Wernig M, Meissner A, Foreman R, *et al. In vitro* reprogramming of fibroblasts into a pluripotent ES-cell-like state. *Nature* 2007; 448(7151): 318-324.

[101] Maherali N, Sridharan R, Xie W. *et al.* Directly reprogrammed fibroblasts show global epigenetic remodelling and widespread tissue contribution. Cell Stem Cell 2007;1(1): 55-70.

[102] Meissner A, Wernig M, Jaenisch R. Direct reprogramming of genetically unmodified fibroblasts into pluripotent stem cells. Nat Biotechnol 2007; 25(10): 1177-1181.

[103] Takahashi K, Okita K, Nakagawa M, Yamanaka S. Induction of pluripotent stem cells from fibroblast cultures. *Nat Protoc* 2007; 2(12): 3081-3089.

[104] Kang L, Wang J, Zhang Y, Kou Z, Gao S. iPS cells can support full-term development of tetraploid blastocyst-complemented embryos. *Cell Stem Cell* 2009; 5(2): 135-138.

[105] Boland M J, Hazen J L, Nazor KL, *et al.* Adult mice generated from induced pluripotent stem cells. *Nature* 2009; 461(7260): 91-94.

[106] Nichols J, Silva J, Roode M, Smith ASuppression of Erk signalling promotes ground state pluripotency in the mouse embryo. *Development* 2009; 136(19): 3215-3222.

[107] Wilder PJ, Kelly D, Brigman K, *et al.* Inactivation of the FGF-4 gene in embryonic stem cells alters the growth and/or the survival of their early differentiated progeny. Dev Biol 1997; 192(2): 614-629.

[108] Yu J, Thomson JA. Pluripotent stem cell lines. Genes Dev 2008; 22(15): 1987-1997.

[109] Chin MH, Mason MJ, Xie W, *et al.* Induced pluripotent stem cells and embryonic stem cells are distinguished by gene expression signatures. Cell Stem Cell 2009; 591: 111-123.

CHAPTER 4

Generation of Pluripotent Stem Cells and their Developmental Potential

Brock J. Conley, Mark Denham, Martin F. Pera and Mirella Dottori*

Centre for Neuroscience, Department of Pharmacology, The University of Melbourne, VIC, Australia

Abstract: The defining characteristic of a pluripotent stem cell is that is has the potential to self renew and the capacity for differentiation into derivatives of the three embryonic germ layers; endoderm, mesoderm and ectoderm, from which all somatic tissues in an organism can be traced. This full potentiality makes them particularly suitable for developing cell replacement therapies or establishing cellular model systems. Although ES cells may be ideal in terms of their pluripotency, from a therapeutic point of view, their great disadvantage is that they are not patient derived. More recently, advances in medical biology have shown that the genomic state of a somatic cell can be altered or 'reprogrammed' to become pluripotent. This is a significant leap forward for obtaining patient-specific pluripotent stem cells that can then be differentiated to the cell type of interest. Generation of reprogrammed somatic cells, known as induced pluripotent stem (iPS) cells, was initially performed by ectopic expression of four transcription factors, however, now this process is becoming more streamlined by improved methods. Through these advances, critical factors and mechanisms involved in driving pluripotency are being determined. This chapter provides a review on how pluripotency is defined and examined *in vitro* and *in vivo*. The various ways in which pluripotent stem cells are generated, particularly in relation to induced pluripotency, are also discussed. Finally an outline of disease-specific pluripotent stem cells is reviewed within the context of their capacity to generate the cell type of interest.

Keywords: Embryonic germ layers, pluripotent stem cells, reporgamming, induced pluripotency.

CHARACTERIZATION OF PLURIPOTENCY

One of the defining characteristics which marks a stem cell as being pluripotent is the expression of pluripotent transcription factors, in particular Oct4 and Nanog, which become down regulated upon differentiation [1-4]. The presence of specific surface markers, such as, stage specific embryonic antigens (SSEA) are also characteristic of pluripotent stem cells, however there are variations of expression across different species. For example, mouse embryonic stem cells (mESC) express SSEA-1, whereas human embryonic stem cells (hESC) do not, but instead express SSEA-3 and SSEA-4 [1, 5]. Other cell surface markers which are specific to hESC are the keratin sulphate-associated antigens TRA-1-60, TRA-1-81 and GCTM-2 [1, 2, 6]. Regular analysis of these markers *via* immunohistochemistry or fluorescent activated cell sorting utilizing specific antibodies is a relative easy assay to undertake regularly to ensure gross maintenance of cells (see also chapter 3).

Pluripotency of stem cells is also assessed *via in vitro* differentiation and subsequent analysis of specific markers representative of each of the three germ layers (endoderm, mesoderm and ectoderm). Methods to directly differentiate pluripotent stem cells to neuroectoderm, endoderm or mesoderm lineages have been established [7-12]. A commonly used method for demonstrating differentiation to all three germ lineages is *via* the formation of embryoid bodies (EBs). EBs are three dimensional structures which recapitulate certain aspects of early embryogenesis and are characterized by the appearance of lineage-specific regions of differentiation representative of the three germ layers. They are formed by placing pluripotent stem cells in suspension cultures in the absence of feeder cells and anti-differentiating agents [13-15].

***Address correspondence to Mirella Dotorri:** Centre for Neuroscience, Department of Pharmacology, The University of Melbourne, Melbourne, Australia; E-mail: mdottori@unimelb.edu.au

Pluripotency of stem cell lines has only been truly demonstrated for mESC, which have been demonstrated to integrate into all embryonic germ layers, including the germ line, following injection into blastocysts. Indeed generation of viable animals derived entirely from the contribution of ES cells, attest to their true pluripotency [16, 17]. This procedure in untenable in humans for obvious ethical reasons and thus alternative strategies for the determination of pluripotency are routinely utilized. These include injection of cells under the kidney or testis capsule of immunocompromised mice to induce teratoma formation, which contain morphological representatives of all three germ layers [1, 2].

DERIVING PLURIPOTENT STEM CELLS

From the Embryo: Embryonic Stem Cells

A number of different stem cell lines have been derived from the mammalian embryo. During embryonic development individual cells progressively lose their pluripotentiality becoming increasingly differentiated and lineage committed. Cells of the early embryo, including zygotes and blastomeres, have the capacity to generate all embryonic and extraembryonic tissues and are termed totipotent. Following compaction and the formation of the blastocyst, the inner cell mass (ICM) maintains the potential to generate all embryonic tissues, however, not all extraembryonic tissues and are termed pluripotent. It is at this stage that ES cells can be isolated and propagated. The ICM subsequently segregates to form the epiblast and hypoblast, which give rise to the embryo proper and yolk sac, respectively. In the mouse, following implantation, the egg cylinder forms and epiblast cells can be isolated and propagated as epiblast stem cells (EpiSCs) [18, 19]. Around mid-gestation pluripotent cell lines have additionally been derived from primordial germ cells and termed embryonic germ cells (EGCs) [20, 21].

The derivation of ES cell lines from mouse blastocysts in 1981 [22, 23] generated much scientific discussion and progress, eventually resulting in non-human primate ES cell line derivation [24, 25] and subsequently hESC derivation [1, 2]. hESC lines derived from the inner cell mass of the human blastocyst exhibit indefinite *in vitro* culture as well as *in vitro* and *in vivo* pluripotency(see also chapter 1). hESC provide a unique opportunity to study early cell fate decisions in human embryogenesis, a process difficult to study *via* other mechanisms, and also introduces the possibility of cellular based therapies, with the provision of readily accessible cells which can theoretically form any human cell type for transplantation and drug discovery applications.

The original methodologies for the derivation and propagation of hESC reflected those established for the mouse, where the outer trophectodermal layer of the blastocyst is removed by immunosurgery, and the ICM subsequently plated onto mitotically inactivated mouse embryonic fibroblasts (MEFs) in the high glucose media containing 20% fetal bovine serum [1, 2]. Developing colonies are subsequently mechanically dissected and morphologically undifferentiated clumps of cell replated on fresh MEF feeder layers and propagated as hESC. Removal of the feeder layer results in spontaneous heterogeneous differentiation. This procedure, or variations of it, is still routinely practiced throughout the world to propagate hESC in an undifferentiated, karyotypically normal state.

Generally, the ethical objection to hESC research has centered on their derivation and the resultant destruction of human embryos with the prevention of potential human life. Original reports describe hESC line derivation from excess frozen embryos, obtained with informed consent, following *in vitro* fertilization (IVF) for clinical purposes [1, 2]. These, and subsequent reports detailing the derivation of hESC lines, required the destruction of human embryos. Alternatively, mESC have been derived utilizing a refined method whereby a single cell biopsy of eight cell stage embryos, similar to that used for pre-implantation genetic diagnosis (PGD) for genetic defects, was used to generate ES or trophoblast cell lines, while the embryos transferred to the oviduct of surrogate mothers generated live young at comparable efficiency to normal embryos [26]. Subsequently, the same laboratory was able to reproduce these results utilizing human embryos at the eight cell stage which are routinely biopsied for PGD [27]. The biopsy of a single blastomere from these embryos and their culture for one day yields enough cellular material for both PGD and ES cell derivation and does not interfere with clinical outcome.

Alternatively, hESCs have been derived albeit at low frequency from so-called 'dead' embryos [28]. In IVF procedures a portion of embryos display arrested *in vitro* development and fail to progress to the stage required for transplantation and thus do not have the potential to develop into a human being. These two studies suggest ways in which hESC lines may be derived in accordance with ethical considerations without the destruction of viable embryos.

Germ Cell Tumours

The first pluripotent stem cell type isolated was from germ cell tumors termed teratocarcinomas, which occur in both mice and humans, and contain derivatives of all three embryonic germ layers [29, 30]. Within teratocarcinomas reside undifferentiated stem cell regions that can be propagated as embryonal carcinoma (EC) cells. Although generally pluripotent, EC cells differ in their developmental potential with ES, EG and EpiS cells, most likely due to severe karyotypic abnormalities and their malignant origin [30]. Indeed most mouse EC cell lines contribute poorly to adult somatic tissues in chimeras and rarely incorporate into the germ line [16, 31, 32].

Somatic Cell Nuclear Transfer

The concept of reprogramming cells to a pluripotent state began with the advent of somatic cell nuclear transfer (SCNT). SCNT, a process whereby nuclei isolated from non-pluripotent cells are transplanted into enucleated oocytes and can give rise to cloned animals, gave credence to the idea that lineage commitment and differentiation during development represent reversible epigenetic changes to the genome. Initial experiments were performed in the frog [33-36] and led to the cloning of Dolly the sheep [37] and subsequently other mammalian species including mouse [38], cow [39], goats [40], pigs [41, 42], guar [43], mouflon [44], cat [45], rabbit [46], horse [47], mule [48], rat [49], African wild cat [50], dog [51], ferret [52], wolf [53], deer [54], sand cat [55], ibex [56], and camel [57]. Human blastocyst embryos produced by SCNT have been reported using human oocytes injected with hESC [58] and adult human fibroblasts [59] with confirmation through DNA and mitochondrial DNA fingerprinting analyses. However, this technology is technically challenging and limited by the availability of suitable human oocytes, the inefficiency of the cloning process, and strong ethical concerns which has led to banning of 'therapeutic' cloning of human cells in many countries (see also chapter 2).

Cell Fusion

Fusion of somatic cell types with pluripotent stem cells such as EC [60], EG or ES cells [61] results in hybrid cells obtaining pluripotent characteristics suggesting that pluripotency is dominant over somatic phenotype in these experiments. Indeed both mESC [62] and hESC [63, 64], when fused with somatic cells, displayed the capacity to activate pluripotent genes and/or reactivate the inactive somatic X chromosome indicative of reprogramming. The resultant tetraploidy of fused cells limits their therapeutic potential, although elimination of some ES cell chromosomes has been demonstrated [65]. Alternatively, exposure of permeabilised somatic cells to ES or EC cell extracts has also been reported to result in activation of pluripotent associated genes and epigenetic reprogramming [65].

Reprogramming: Induced Pluripotent Stem Cells

The generation of the first iPS cells in 2006 was the pinnacle of cumulative pioneering research into cellular reprogramming encompassing ideas and concepts derived from nuclear transfer and cell fusion experiments. In 2006, a screen of pluripotency-associated genes yielded a successful combination of transcription factors, including Oct4, Sox2, Klf4 and c-Myc, which, when ectopically expressed, converted mouse somatic fibroblasts into pluripotent cells [66]. These iPS cells contributed to the germline in chimeric mice [67, 68] and were later shown to generate complete mice following tetraploid embryo complementation [69-71]. The combination of Oct4, Sox2, Klf4 and c-Myc, are commonly referred to as the 'Yamanaka factors'.

The first human iPS cell lines were reported in 2007 [72, 73] and soon thereafter iPS cells were derived from other species including rats [74], Tibetan miniature pigs [75], Rhesus monkey [76] and Marmoset

[77]. Following transduction with reprogramming factors, changes in epigenetic state and reactivation of endogenous pluripotency genes eventually leads to establishment of cell lines reminiscent of hESC. Indeed human iPS cells have been shown similar to hESC in regard to morphology, pluripotent gene expression, epigenetic status and *in vitro* and *in vivo* differentiation potential [72, 73].

Since the first generation iPS cells were derived, significant improvements in efficiency, quality and characterisation have been described. These improvements range from more efficient reprogramming with other starting cell types, to non-integrating methods of transgene transduction, number and combination of reprogramming factors, epigenetic modifying agents and small molecules for selection and enhancement of iPS cells. Each of these factors contribute to the kinetics and efficiency of iPS induction, and thus also their pluripotency. Technical improvements in generating iPS cells has forced us to investigate what are the critical factors that drive and induce pluripotency. For the purposes of this review we will focus primarily on human iPS cell generation and advancements in this field.

The Role of the Yamanaka Factors in Reprogramming

Recently, researchers have been investigating the series of events that are required to take place for efficient reprogramming to occur and how the Yamanaka factors cooperate to achieve this. Determining whether stage specific expression is required or whether transient roles for each of the factors is sufficient for the reprogramming, and if the down-regulation of specific somatic cell genes or the induction of pluripotent transcripts is required in a particular temporal order, will all help elucidate the underlying mechanisms involved in generating iPS cells.

Reprogramming with viral vectors can involve the introduction of up to 20 copies of each gene to obtain suitable expression levels and overcome positional effects (Takahashi and Yamanaka, 2006). However, obtaining the precise levels of Oct4 expression may affect the efficiency in establishing iPS cells. Oct4 expression levels have been shown to be critical to ES cell maintenance with abnormally increased or decreased levels of Oct4 resulting in ES differentiation (Niwa *et al.*, 2000). Similarly, elevated levels of the Sox2 protein induces neuroectoderm differentiation in mESC [78], demonstrating that precise control of levels of Sox2 is also required for maintaining pluripotency. Oct4 and Sox2 function as core transcription factors in the establishment of pluripotency, and Oct4 is the only factor which cannot be dispensable in certain cell types or replaced with the addition of chemical compounds. Together these genes can regulate other pluripotent genes as well as their own self-regulation.

c-Myc is important in the initial events of reprogramming with the down-regulation of somatic cell genes and the upregulation of an ES-like expression profile, brought about by its global epigenetic remodelling effects [79]. Furthermore, c-Myc is involved in cell cycle progression by positively regulating cyclin-Cdk activity [80]. In conjunction with Cnot3 and Trim28, c-Myc also regulates a network of gene targets critical for self-renewal, including the regulatory subunit of telomerase (TERT) [81]. Several of these target genes are separate from those regulated by Nanog, Oct4, and Sox2 (Hu *et al.*, 2009). In mESC, c-Myc plays a key role in the establishment and maintenance of pluripotency with it being a target of the LIF/STAT3 pathway (Cartwright *et al.*, 2005). GSK3β is involved in the degradation of phosphorylated c-Myc, which in turn disrupts the pluripotent state (Cartwright *et al.*, 2005).

Klf4, although dispensable in maintaining pluripotency due to functional redundancy, was identified as a key factor in the reprogramming of somatic cells (Takahashi and Yamanaka, 2006). Klf4 has been shown to regulate Nanog expression and share common targets to Nanog (Jiang *et al.*, 2008). In addition, Klf4 can suppress p53 and prevent the apoptotic events that have been shown to occur through forced expression of c-Myc (Lin *et al.*, 2005; Rowland *et al.*, 2005). Kenpaullone (KP), a GSK3β and CDK1/ cyclin B inhibitor can replace Klf4 during reprogramming (Hanna *et al.*, 2009a; Lyssiotis *et al.*, 2009). Alternatively, CHIR99021 a selective Gsk3 inhibitor can also replace Klf4 (Ying *et al.*, 2008; Hanna *et al.*, 2009a).

Lin28 has been used in combination with the four Yamanaka factors to increase the reprogramming efficiency. Lin28 regulates reprogramming through the increase in cell proliferation, resulting in a greater reprogramming efficiency [82]. Furthermore, Lin28 can block the action of the let-7 family of miRNA, which are upregulated in differentiated cells (Melton *et al.*, 2010).

THE EPIGENETICS OF REPROGRAMMING

Remodelling of somatic cells into a pluripotent stem cell state involves global changes to a cell's chromatin profile, which include demethylation of promoter regions of pluripotent genes and a resultant histone methylation profile indistinguishable from ES cells. These epigenetic changes allow for the reinstatement of an embryonic transcriptional profile. Interestingly, the acquisition of pluripotency after the introduction of the Yamanaka factors has been shown to be a gradual process governed by stochastic epigenetic events resulting from global chromatin changes (Meissner *et al.*, 2007; Wernig *et al.*, 2007; Brambrink *et al.*, 2008). For example, in mouse iPS cells, the expression levels of Nanog have been shown to take several weeks before they reach an ES-like expression state and the continued expression of the transgenes enhances this reprogramming (Wernig *et al.*, 2007; Brambrink *et al.*, 2008).

Of the four factors introduced by Yamanaka *et al.*, c-Myc has been shown to be a major contributing factor in controlling the epigenetic state of a cell (Knoepfler *et al.*, 2006). c-Myc has DNA binding sites spread throughout the genome and which cover approximately 15% of all genes [83], however its ability to regulate widespread gene regulation is not achieved solely by itself. c-Myc can recruit and form chromatin-modifying complexes some of which involve its co-activator TRRAP which interact with the histone acetyltransferases (HATs), GCN5 and Tip60 which, in turn, inducing widespread histone acetylation. In addition, c-Myc interacts with CBP and the chromatin-remodelling complex containing INI1 (Cheng *et al.*, 1999; Vervoorts *et al.*, 2003). It is through these interactions with the HATs that allow the other Yamanaka factors to bind to their specific target sites and drive the reinstatement of pluripotent genes, such as Nanog (Knoepfler *et al.*, 2006). In the context of reprogramming however, c-Myc has been shown to be dispensable, although not without a significant reduction in efficiency and a delay in reprogramming (Nakagawa *et al.*, 2008; Wernig *et al.*, 2008a). This demonstrates that the epigenetic reprogramming can be facilitated by exogenous expression of c-Myc but this exogenous expression is not required and can be achieved by an event initially independent of c-Myc. c-Myc and Klf4 can also be replaced with Nanog and Lin28 (Yu *et al.*, 2007). Altogether, these studies reveal that a variety of events or molecular pathways can result in stochastic epigenetic events that lead to reprogramming, although always requiring the key transcription factor Oct4.

The use of chromatin modifying compounds such as the histone deacetylase (HDAC) inhibitor valproic acid (VPA), suberoylanilide hydroxamic acid (SAHA) and trichostatin A (TSA), have been shown to greatly enhance the yield of reprogrammed cells [84]. VPA can replace the function of c-Myc [84] and BIX-01294 (BIX), a small molecule inhibitor of the G9a histone methyltransferase (Kubicek *et al.*, 2007), can improve the reprogramming efficiency in neural progenitors transduced with only Oct4 and Klf4 [85]. As more epigenetic modifying agents are being identified and characterized, better substitutes will be found for most of the genetic factors as well as mediating greater reprogramming efficiency.

LIMITATIONS TO IPS CELL TECHNOLOGIES WHICH MAY COMPRISE THEIR THERAPEUTIC POTENTIAL

Despite reprogrammed cells holding great promise for regenerative medicine, including patient specific iPS cells and differentiated progeny for disease modelling, drug testing and cell based therapies, numerous concerns remain precluding clinical application. Some of the greater concerns centre around the use of oncogenes (including c-Myc) for reprogramming, genomic disruption by insertion of transgenes and incomplete suppression and potential reactivation of exogenous genes. Indeed retroviral and lentival based reprogramming are still currently the most efficient methods for factor-based delivery to cells (albeit still low, 0.001-1.5% of the starting cell population). Viral vectors randomly integrate into the genome which may affect transcriptional activity of endogenous genes. Furthermore, even though integrated viral vectors generally undergoing gene silencing over time, they have the potential for incomplete suppression thereby resulting in reduced pluripotency, and/or reactivation of exogenous factors. This is particularly illustrated by initial reprogramming experiments in the mouse where by reactivation of transgenes including c-Myc resulted in numerous chimeric mice developing tumours [67, 68]. Therefore it is important to develop strategies using approaches that either 1) fail to integrate into the genome, 2) can be readily and completely removed from the genome or 3) do not utilize vectors at all.

Considering some of the issues mentioned above, minicircle vectors were used as an alternative to exogenously introduce pluripotent genes in somatic cells [86]. Minicircle vectors lack a bacterial origin of replication, benefit from high transfection efficiency and fail to integrate into the genome, remaining episomal, and become diluted through cell division. In a study by Jia and collegues, the reprogramming factors Oct-4, Sox-2, Lin28, Nanog and a green fluorescent protein reporter gene were all introduced into a single cassette separated by self-cleaving peptides into a minicircle vector. This vector was used to reprogram human adipose stem cells at an efficiency of ~0.005% as well as neonatal fibroblasts at lower efficiency [86]. Minicircle-derived iPS cells displayed no genomic integration of transgenes, reactivated endogenous pluripotent genes, displayed a similar global gene expression profile to hESCs and differentiated into derivatives of all three embryonic germ layers *in vitro* and *in vivo* in teratomas [86].

Another study took an alternative approach by using a replication-deficient Sendai virus (SeV) containing Oct-4, Sox2, Klf4 and c-Myc, to reprogram mouse fibroblasts which enables strong gene expression, whilst not integrating into the genome [87]. Following reprogramming, treatment with specific siRNA targeting viral genes was reported to eliminate viral DNA, resulting in completely integration free iPS cells [87]. Similarly, human fibroblasts have been reprogrammed using Sendai virus and displayed reactivation of pluripotency genes, as well as differentiation *in vitro* and *in vivo* [88]. The viral vector decreased over time being virtually undetectable after 15-20 passages, whilst using an antibody detecting virally infected cells enabled their removal by negative selection, resulting in viral free iPS cells [88]. A similar method using a temperature sensitive mutant Sendai virus has also been utilized to reprogram circulating T cells from peripheral human blood at an efficiency of 0.1% with exogenous transgene expression lost after several passages [89].

Alternatively, integrating vectors containing LoxP sites have been utilized for reprogramming with subsequent excision *via* transient expression of Cre recombinase [90, 91]. However, short vector sequences remain in the host genome following excision. *PiggyBac* transposon vectors have also been used in reprogramming mouse and human fibroblasts [92, 93]. These can be integrated and subsequently removed from the host genome by transient expression of transposase. Indeed excision is highly efficient resulting in host genome repair leaving no genetic footprint [92, 93].

Other methods for delivering pluripotent factors in somatic cells that do not rely on the use of vectors have also been explored. Transient and sufficient expression of factors can be achieved using synthetic mRNAs which have modified ribonucleosides supplemented together with interferon inhibitor to enable evasion of cellular viral response [94]. Using synthetic mRNA encoding Oct-4, Sox-2, Klf4, c-Myc and Lin28 (ratio of 3:1:1:1:1) and daily transfection of fetal, post-natal and primary skin fibroblasts for 17 days iPS cells were generated at the highest efficiency of 4.4%. Reprogrammed cells expressed pluripotent genes at comparable levels to hESCs and maintained the potential to differentiate *in vitro* and *in vivo* [94]. Recombinant Oct-4, Sox-2, Klf4 and c-Myc proteins containing transduction domains have also been utilised to reprogram mouse embryonic fibroblasts in the presence of the HDAC inhibitor valproic acid (VPA) following four cycles of treatment [95]. These studies showed reactivation of endogenous genes, methylation analyses and contribution to mouse chimeras and germ tissues to attest their pluripotency [95].

OTHER REGULATORS OF PLURIPOTENCY

MicroRNAs

MicroRNAs (miRNA) are short non-coding RNA that interfere with specific mRNA transcripts. Based on their seed sequence together with the complementary sequence found in the target mRNA, they can suppress pathways by destabilization and translational inhibition of their targets. The correct processing of miRNA is achieved through the Drosha, Dicer and RISC pathway. Drosha in conjunction with Dgcr8 process primary transcripts of miRNA into pre-miRNA transcripts, with Dicer then cleaving the pre-miRNA into short double-stranded RNA fragments. Cell type specific miRNA, including pluripotent specific and progenitor cell specific miRNA transcripts, exist. The precise expression profile of miRNA transcripts has been used to assess iPS cells and their level of reprogramming (Chin *et al.*, 2009; Wilson *et al.*, 2009). Indeed, altered expression of specific miRNA transcripts in ES cells has been shown to cause defects in ES maintenance. Ectopic expression of miR-133 or miR-145 alters the expression of pluripotent genes thereby disrupting the ES cell state and initiating differentiation (Ivey *et al.*, 2008; Xu *et al.*, 2009b). miR-145 directly targets OCT4, SOX2 and KLF4 in hESCs. Furthermore the promoter regions of miR-145 is repressed by OCT4, resulting in a double-negative feedback loop [96]. Other microRNA clusters have also been found to be expressed specifically in ES cells. The miR-371/372/373 and the miR-302 clusters were shown to regulate cell cycles by interfering with cyclin D1 transcripts in hESC, with these miR transcripts being regulated at the promoters by OCT4 and SOX2 (Suh *et al.*, 2004; Laurent *et al.*, 2008; Morin *et al.*, 2008). Recently miRNA profiling of hESC and iPS cells have identified specific miRNA transcripts, miR-92 and miR-141, which regulate pluripotency *via* targeting the p53 pathway [97]. Overall, as the function of various miRNA transcripts are being identified, their role in regulating pluripotency will be better understood and therefore also taken into greater consideration when exploring novel methods to reprogram cells.

Small Molecule Inhibitors

Increased reprogramming efficiency in both mouse and human cells have been achieved by using small molecules inhibitors of the TGFβ receptor (SB431542), MEK (PD0325901), Alk5 (A-83-01) and GSK3 (CHIR99021) pathways [74, 98-100]. Whether these pathways facilitate direct genetic/epigenetic reprogramming or enable support and survival to emerging iPS cells is not necessarily clear, although combined MEK and GSK3 inhibition has been effectively used to propagate mESC in the absence of exogenous growth factors or cytokines highlighting their importance in maintenance of pluripotency [101]. Screens of small molecule compounds have also been investigated to discover substitutes for exogenous genetic factors. In one such screen, benzazepinone kenpaullone was found to replace Klf4 in reprogramming mouse fibroblasts when in combination with Oct-4, Sox2 and c-Myc, although at a lower efficiency [102]. Additionally, a combination of the GSK inhibitor CHIR99021 and the lysine-specific demethylase 1 inhibitor, Parnate have been used in combination with only Oct4 and Klf4 to reprogram human keratinocytes [98]. Together these results suggest that compound use may be highly beneficial in increasing the efficiency of reprogramming and introduces the possibility of deriving iPS cells without direct factor treatment. However, to date, the decreased efficiency of substituting compounds suggests that no single chemical or inhibitor can completely replace the broad action and synergy of a transcription factor.

NOT ALL REPROGRAMMED CELLS ARE PLURIPOTENT

Although reprogramming cells into an iPS cell with defined factors is described as directed, not all reprogrammed cells achieve the same level of reprogramming, even from daughter cells derived from the same infected cell (Mikkelsen *et al.*, 2008). This anomaly points again to the stochastic nature of the epigenetic reprogramming, and as such, has been shown to give rise to partially reprogrammed cells (Takahashi and Yamanaka, 2006). Incomplete reprogramming can result in the aberrant expression of genes that are not expressed in either the starting cell type or a pluripotent cell, trapping it in an intermediate state. Certain genes have been associated with this, such as *Gata6, Sox9* and *Pax7* with the knockdown of these genes facilitating the conversion from an intermediate to a fully reprogrammed state (Mikkelsen *et al.*, 2008).

The intermediate stages of cell reprogramming start with the fibroblast cells down-regulating somatic markers, such as Thy1 and collagens, a mechanism attributed to c-Myc or VPA [79]. This event occurs in the majority of infected cells. It is then followed by the reactivation of SSEA1 in mouse cells, which occurs in only a low percentage of cells and represents an important stage in the reprogramming (Brambrink *et al.*, 2008). In human cells, SSEA4 expression can be seen after 2 days and TRA-1-60 expression is observed in a subset of cells at day 6 (Chan *et al.*, 2009). For mechanisms not completely understood, cells undergoing reprogramming can arrest at various stages resulting in a partially reprogrammed state with various combinations of pluripotent genes being expressed (Chan *et al.*, 2009). Incomplete reprogramming may also be due to drug selection methods for isolating iPS cells. Drug resistance under the Fbx15 locus resulted in partially reprogrammed iPS cells, as they could not contribute to the germline of chimeric mice after blastocyst injection (Takahashi and Yamanaka, 2006). However using the same factors, when drug selection was placed under the Nanog or Oct4 promoters, germline transmission was observed (Maherali *et al.*, 2007; Okita *et al.*, 2007; Wernig *et al.*, 2007).

Partially reprogrammed cells can display an ES-like morphology with the loss of somatic cell gene expression and the expression of some of the pluripotent stem cell markers. Furthermore, retroviral gene expression is not silenced and X chromosome reactivation is not achieved. In mouse cells, partially reprogrammed cells are not LIF responsive and do not form chimeras and therefore are not germline competent (Takahashi and Yamanaka, 2006; Silva *et al.*, 2008). This incomplete reprogramming represents an intermediate state prior to the expression of the endogenous Nanog (Brambrink *et al.*, 2008; Stadtfeld *et al.*, 2008). The halt in reprogramming and lack of Nanog expression cannot be overcome with Oct4, Sox2 or Klf4 expression. The change from Fbx15 or Oct4 reporters for identification of reprogrammed cells to Nanog expression results in the isolation of germline competent pluripotent cells (Takahashi and Yamanaka, 2006; Okita *et al.*, 2007). For mouse cells, the roadblocks to reprogramming can be overcome by using two small molecule inhibitors of the Mek/Erk pathway and GSK3 (Silva *et al.*, 2008), which are used to maintain mouse cells pluripotent (Ying *et al.*, 2008).

One of the final stages of reprogramming in iPS cells is the reactivation of the X chromosome, however to date this has only been reported for mouse iPS cells. In contrast, human iPS cells have been found to maintain X chromosome inactivation [103]. This is similar to hESC, which display X chromosome inactivation variability (Dhara and Benvenisty, 2004; Shen *et al.*, 2008). These species differences may be related to developmental stages of pluripotency with hESC being more developmentally similar to epiblast stem cells (Ying *et al.*, 2008; Nichols and Smith, 2009). However, recently it has been shown that when hESC are established and maintained in physiological low oxygen conditions, the status of the X chromosome remains active (Lengner *et al.*, 2010). The oxidative stress of atmospheric conditions was shown to induce irreversible X chromosome inactivation, while low oxygen was proposed to maintain hESC an earlier state of pluripotency, representing a similar ground state of pluripotency observed in mESC.

As well as epigenetic remodelling and the down regulation of a somatic cell gene profile, the final stages of reprogramming human cells involves the silencing of viral genes with the cells being able to enter a self-sustaining state. Silencing of pro-viral genes and the expression of TRA-1-60, DNMT3B and REX1 has been used as a defining characteristic of fully reprogrammed cells (Chan *et al.*, 2009). However careful characterization of the iPS cells is required for human cells as the stringent criteria of generating chimera and germline competence cannot be tested for.

CAN ALL CELL TYPES BE REPROGRAMMED TO PLURIPOTENCY?

A number of different cell types have been utilized for human reprogramming including fetal, post natal [73] and adult fibroblasts [72], peripheral blood [104], cord blood stem cells [105, 106], cord blood endothelial cells [107], adipose stem cells [108, 109], hepatocytes [110], keratinocytes [111], neural stem cells [112], parthenogenetic neural stem cells [113], dermal papilla [114], umbilical cord matrix and placental amniotic membrane [115]. It is becoming apparent that certain cell types may be more amenable to reprogramming than others due to certain factors including, endogenous expression of some of the reprogramming factors and similarity to ES cell transcriptional profile, global methylation, and inherent age of the cell type. For example, neural stem cells already expressing Sox-2 and c-Myc appear more amenable to reprogramming with fewer factors such as Oct-4 alone or in combination with Nanog in human [116], and Oct-4 alone or in combination with Klf4 in the mouse [85, 117-119].

Although iPS cells have been derived from a large array of varying types of somatic cell, the choice of cell type to be reprogrammed has been shown to affect their differentiation ability and their propensity to form tumors in mice (Miura *et al.*, 2009; Kim *et al.*, 2010; Polo *et al.*, 2010; Tchieu *et al.*, 2010). Among these differences, human and mouse iPS cells have been demonstrated to have a persistent gene expression pattern reminiscent of the somatic cell of origin, including X chromosome inactivation (Ghosh *et al.*, 2010; Kim *et al.*, 2010; Tchieu *et al.*, 2010). However, it was demonstrated in mouse cells that this genetic memory is lost upon continuous passaging (Polo *et al.*, 2010). These findings emphasise that although key hallmarks to defining pluripotency may be met, the complete reprogramming of a cell is a gradual process, and one that requires stringent assays to identify truly reprogrammed iPS cells.

DISEASE-SPECIFIC IPS CELLS

Specific disease states have been targeted with iPS technology. Much of this depends upon the capacity of iPS cells to differentiate into functional lineage-specific cell types *in vivo*. This has been explored for conditions of the nervous system since there are well establish methods for generating neurons from pluripotent stem cells. Indeed, reprogrammed mouse fibroblasts have been shown to differentiate into neural progenitors and functional neural cell types *in vitro* which upon transplantation displayed the capacity for migration, differentiation into glia and neurons and synaptic integration [120]. Furthermore, directed differentiation of iPS cells to mid-brain neurons and their transplantation into a rat model of Parkinson's disease allowed engraftment and functional recovery [120]. Similarly, mouse iPS cells differentiated into electrophysiologically functional neurons, astrocytes, and oligodendrocytes *in vitro*, were transplantation into mice with contusive injury, and shown to contribute to remyelination and induction of axonal regrowth of host 5HT(+) serotonergic fibers, resulting in locomotor function recovery [121].

IPS cells have also been directly examined for their capacity to alleviate cardiac damage. Intramyocardial transplantation of iPS cells into immunocompetent mice suffering infarct resulted in engraftment, integration, expression of cardiac, smooth muscle and endothelial markers, translating to an increase in contractile performance, ventricular wall thickness and electrical stability [122]. Whilst iPS cells injected into immunocompromised mouse hearts consistently formed tumours, injection into immunocompetent hosts failed to develop tumours at the site of injection, or in other major organs analysed, the cells having apparently integrated and differentiated within the heart tissue [122]. Another study reported using mouse iPS cells differentiated into endothelial and endothelial progenitor cells which, when injected into hemophilia A mice, significantly increased their survival [123].

One of the tremendous potentials of iPS technologies is to replace diseased cells that harbor genetic defects with corrected, patient-derived cells. Proof of principle of this potential was elegantly shown using a humanized sickle cell anemia mouse model. Diseased fibroblasts were isolated and reprogrammed into iPS cells where the genetic defect was then corrected by homologous recombination. Corrected iPS cells were subsequently differentiated *in vitro* into hematopoietic progenitors and transplanted into affected mice following irradiation to allow reconstitution of the hematopoietic system [124]. Reconstitution with corrected cells was observed together with a marked improvement of pathological features [124]. These strategies are applicable if the genetic defects do not hamper the ability to generate iPS cells. Such limitations were observed in a study that attempted to generate iPS cells from patients with Fanconi anaemia, an inherited disorder characterized by bone marrow defects. Patient-derived fibroblasts were not conducive to reprogramming as they displayed chromosomal instability, however, upon their genetic correction using lentiviral vectors, iPS cells could then be generated that were disease-free [125].

IPS technologies are not yet at the stage to be directly used in the clinic. Despite their current limitations, iPS cells nevertheless provide a useful cell source for studying molecular and cellular mechanisms of disease pathology *in vitro* and also for establishing drug-screening assays. To date, iPS cells have been derived from patients suffering from Parkinson's Disease, Huntington's disease, adenosine deaminase deficiency-related severe combined immunodeficiency (ADA-SCID), Shwachman-Bodian-Diamond syndrome, Gaucher disease type III, Duchenne, Becker muscular dystrophy, juvenile-onset type 1 diabetes mellitus, Down syndrome/trisomy 21, Lesch-Nyhan syndrome, recessive dystrophic epidermolysis bullosa, beta-thalassemia [91, 126-128] [129]. Additionally, reprogrammed cells derived from patients with spinal muscular atrophy [130], familial dysautonomia [131], Friedreich's ataxia [132, 133] and LEOPARD syndrome [134] have been shown to reflect the disease phenotype *in vitro*. As better systems for differentiating pluripotent stem cells to specific lineages are being developed, the application of using iPS cells to create cellular models of human diseases will also be significantly greater in the near future.

CONCLUSIONS AND PROSPECTS

The fact that a single cell may have the capacity to become all cell types of the body has been for many years mainly a wonder without much understanding of the processes involved. With the breakthroughs and advancements in stem cell technologies over the last decade, these mysteries are now being unraveled and, furthermore, demonstrate that almost any cell type can have this potential given the right conditions. As the technologies for generating pluripotent stem cells and their differentiation to specific functional lineages improves over time, their contribution to the clinic will be immense, whether it is because of their capacity to create robust human cellular model systems and/or their direct use for cell replacement therapies. Moreover, and just as important, we will also gain a greater appreciation and understanding of basic mechanisms in cell biology and development.

REFERENCES

[1] Reubinoff BE, Pera MF, Fong CY, Trounson A, Bongso A. Embryonic stem cell lines from human blastocysts: somatic differentiation *in vitro*. Nat Biotechnol 2000; 18(4): 399-404.

[2] Thomson JA, Itskovitz-Eldor J, Shapiro SS, *et al.* Embryonic stem cell lines derived from human blastocysts. Science. 1998; 282(5391): 1145-7.

[3] Chambers I, Colby D, Robertson M, *et al.* Functional expression cloning of Nanog, a pluripotency sustaining factor in embryonic stem cells. Cell. 2003; 113(5): 643-55.

[4] Mitsui K, Tokuzawa Y, Itoh H, *et al.* The homeoprotein Nanog is required for maintenance of pluripotency in mouse epiblast and ES cells. Cell. 2003 30; 113(5): 631-42.

[5] Savatier P, Lapillonne H, van Grunsven LA, Rudkin BB, Samarut J. Withdrawal of differentiation inhibitory activity/leukemia inhibitory factor up-regulates D-type cyclins and cyclin-dependent kinase inhibitors in mouse embryonic stem cells. Oncogene. 1996; 12(2): 309-22.

[6] Heins N, Englund MC, Sjoblom C, *et al.* Derivation, characterization, and differentiation of human embryonic stem cells. Stem Cells 2004; 22(3): 367-76.

[7] Furth ME, Atala A. Stem cell sources to treat diabetes. J Cell Biochem 2009; 106(4): 507-11.

[8] Micallef SJ, Li X, Elefanty AG, Stanley EG. Pancreas differentiation of mouse ES cells. Current protocols in stem cell biology 2007; Chapter 1: Unit 1G 2.

[9] Schmitt RM, Bruyns E, Snodgrass HR. Hematopoietic development of embryonic stem cells *in vitro*: cytokine and receptor gene expression. Genes Dev 1991; 5(5): 728-40.

[10] Mummery CL, Ward D, Passier R. Differentiation of human embryonic stem cells to cardiomyocytes by coculture with endoderm in serum-free medium. Current protocols in stem cell biology 2007; Chapter 1: Unit 1F 2.

[11] Dottori M, Pera MF. Neural differentiation of human embryonic stem cells. Methods Mol Biol 2008; 438: 19-30.

[12] Kim M, Habiba A, Doherty JM, Mills JC, Mercer RW, Huettner JE. Regulation of mouse embryonic stem cell neural differentiation by retinoic acid. Dev Biol 2009; 328(2): 456-71.

[13] Doetschman TC, Eistetter H, Katz M, Schmidt W, Kemler R. The *in vitro* development of blastocyst-derived embryonic stem cell lines: formation of visceral yolk sac, blood islands and myocardium. J Embryol Exp Morphol 1985; 87: 27-45.

[14] Keller GM. *In vitro* differentiation of embryonic stem cells. Curr Opin Cell Biol 1995; 7(6): 862-9.

[15] Itskovitz-Eldor J, Schuldiner M, Karsenti D, *et al.* Differentiation of human embryonic stem cells into embryoid bodies compromising the three embryonic germ layers. Mol Med 2000; 6(2): 88-95.

[16] Bradley A, Evans M, Kaufman MH, Robertson E. Formation of germ-line chimaeras from embryo-derived teratocarcinoma cell lines. Nature 1984; 309(5965): 255-6.

[17] Nagy A, Rossant J, Nagy R, Abramow-Newerly W, Roder JC. Derivation of completely cell culture-derived mice from early-passage embryonic stem cells. Proc Natl Acad Sci USA 1993; 90(18): 8424-8.

[18] Brons IG, Smithers LE, Trotter MW, *et al.* Derivation of pluripotent epiblast stem cells from mammalian embryos. Nature 2007; 448(7150): 191-5.

[19] Tesar PJ, Chenoweth JG, Brook FA, *et al.* New cell lines from mouse epiblast share defining features with human embryonic stem cells. Nature 2007; 448(7150): 196-9.

[20] Matsui Y, Zsebo K, Hogan BL. Derivation of pluripotential embryonic stem cells from murine primordial germ cells in culture. Cell 1992; 70(5): 841-7.

[21] Resnick JL, Bixler LS, Cheng L, Donovan PJ. Long-term proliferation of mouse primordial germ cells in culture. Nature 1992; 359(6395): 550-1.

[22] Evans MJ, Kaufman MH. Establishment in culture of pluripotential cells from mouse embryos. Nature 1981; 292(5819): 154-6.

[23] Martin GR. Isolation of a pluripotent cell line from early mouse embryos cultured in medium conditioned by teratocarcinoma stem cells. Proc Natl Acad Sci USA 1981; 78(12): 7634-8.

[24] Thomson JA, Kalishman J, Golos TG, *et al.* Isolation of a primate embryonic stem cell line. Proc Natl Acad Sci USA 1995; 92(17): 7844-8.

[25] Thomson JA, Kalishman J, Golos TG, Durning M, Harris CP, Hearn JP. Pluripotent cell lines derived from common marmoset (Callithrix jacchus) blastocysts. Biol Reprod 1996; 55(2): 254-9.

[26] Chung Y, Klimanskaya I, Becker S, *et al.* Embryonic and extraembryonic stem cell lines derived from single mouse blastomeres. Nature 2006; 439(7073): 216-9.

[27] Klimanskaya I, Chung Y, Becker S, Lu SJ, Lanza R. Human embryonic stem cell lines derived from single blastomeres. Nature 2006; 444(7118): 481-5.

[28] Zhang X, Stojkovic P, Przyborski S, *et al.* Derivation of human embryonic stem cells from developing and arrested embryos. Stem Cells 2006; 24(12): 2669-76.

[29] Damjanov I SD. Animal model of human disease: teratoma and teratocarcinoma. Am J Pathol 1976; 83(1): 241-4.

[30] Roach S CS, Bennett W, Pera MF. Cultured cell lines from human teratomas: windows into tumour growth and differentiation and early human development. Eur Urol 1993; 23(1): 82-8.

[31] Brinster RL. The effect of cells transferred into the mouse blastocyst on subsequent development. J Exp Med 1974; 140(4): 1049-56.

[32] Stewart TA, Mintz B. Successive generations of mice produced from an established culture line of euploid teratocarcinoma cells. Proc Natl Acad Sci USA 1981; 78(10): 6314-8.

[33] Briggs R, King TJ. Transplantation of Living Nuclei From Blastula Cells into Enucleated Frogs' Eggs. Proc Natl Acad Sci USA 1952; 38(5): 455-63.

[34] Gurdon JB, Brown DD. Cytoplasmic Regulation of Rna Synthesis and Nucleolus Formation in Developing Embryos of Xenopus Laevis. J Mol Biol 1965; 12: 27-35.

[35] Gurdon JB, Laskey RA, Reeves OR. The developmental capacity of nuclei transplanted from keratinized skin cells of adult frogs. J Embryol Exp Morphol 1975; 34(1): 93-112.

[36] King TJ, Briggs R. Changes in the Nuclei of Differentiating Gastrula Cells, as Demonstrated by Nuclear Transplantation. Proc Natl Acad Sci USA 1955; 41(5): 321-5.

[37] Wilmut I, Schnieke AE, McWhir J, Kind AJ, Campbell KH. Viable offspring derived from fetal and adult mammalian cells. Nature 1997; 385(6619): 810-3.

[38] Wakayama T, Perry AC, Zuccotti M, Johnson KR, Yanagimachi R. Full-term development of mice from enucleated oocytes injected with cumulus cell nuclei. Nature 1998; 394(6691): 369-74.

[39] Kato Y, Tani T, Sotomaru Y, *et al.* Eight calves cloned from somatic cells of a single adult. Science 1998; 282(5396): 2095-8.

[40] Baguisi A, Behboodi E, Melican DT, *et al.* Production of goats by somatic cell nuclear transfer. Nat Biotechnol 1999; 17(5): 456-61.

[41] Onishi A, Iwamoto M, Akita T, *et al.* Pig cloning by microinjection of fetal fibroblast nuclei. Science 2000; 289(5482): 1188-90.

[42] Polejaeva IA, Chen SH, Vaught TD, *et al.* Cloned pigs produced by nuclear transfer from adult somatic cells. Nature 2000; 407(6800): 86-90.

[43] Lanza RP, Cibelli JB, Diaz F, *et al.* Cloning of an endangered species (Bos gaurus) using interspecies nuclear transfer. Cloning 2000; 2(2): 79-90.

[44] Loi P, Ptak G, Barboni B, Fulka J, Jr., Cappai P, Clinton M. Genetic rescue of an endangered mammal by cross-species nuclear transfer using post-mortem somatic cells. Nat Biotechnol 2001; 19(10): 962-4.

[45] Shin T, Kraemer D, Pryor J, *et al.* A cat cloned by nuclear transplantation. Nature 2002; 415(6874): 859.

[46] Chesne P, Adenot PG, Viglietta C, Baratte M, Boulanger L, Renard JP. Cloned rabbits produced by nuclear transfer from adult somatic cells. Nat Biotechnol 2002; 20(4): 366-9.

[47] Galli C, Lagutina I, Crotti G, *et al.* Pregnancy: a cloned horse born to its dam twin. Nature 2003; 424(6949): 635.

[48] Woods GL, White KL, Vanderwall DK, Li GP, Aston KI, Bunch TD, *et al.* A mule cloned from fetal cells by nuclear transfer. Science 2003; 301(5636): 1063.

[49] Zhou Q, Renard JP, Le Friec G, *et al.* Generation of fertile cloned rats by regulating oocyte activation. Science 2003; 302(5648): 1179.

[50] Gomez MC, Pope CE, Giraldo A, *et al.* Birth of African Wildcat cloned kittens born from domestic cats. Cloning Stem Cells 2004; 6(3): 247-58.

[51] Lee BC, Kim MK, Jang G, Oh HJ, Yuda F, Kim HJ, *et al.* Dogs cloned from adult somatic cells. Nature 2005; 436(7051): 641.

[52] Li Z, Sun X, Chen J, *et al.* Cloned ferrets produced by somatic cell nuclear transfer. Develop Biol 2006; 293(2): 439-48.

[53] Kim MK, Jang G, Oh HJ, *et al.* Endangered wolves cloned from adult somatic cells. Cloning Stem Cells 2007; 9(1): 130-7.

[54] Berg DK, Li C, Asher G, Wells DN, Oback B. Red deer cloned from antler stem cells and their differentiated progeny. Biol Reprod 2007; 77(3): 384-94.

[55] Gomez MC, Pope CE, Kutner RH, *et al.* Nuclear transfer of sand cat cells into enucleated domestic cat oocytes is affected by cryopreservation of donor cells. Cloning Stem Cells 2008; 10(4): 469-83.

[56] Folch J, Cocero MJ, Chesne P, Alabart JL, Dominguez V, Cognie Y, *et al.* First birth of an animal from an extinct subspecies (Capra pyrenaica pyrenaica) by cloning. Theriogenology 2009; 71(6): 1026-34.

[57] Wani NA, Wernery U, Hassan FA, Wernery R, Skidmore JA. Production of the first cloned camel by somatic cell nuclear transfer. Biol Reprod 2010; 82(2): 373-9.

[58] Stojkovic M, Stojkovic P, Leary C, *et al.* Derivation of a human blastocyst after heterologous nuclear transfer to donated oocytes. Reprod Biomed Online 2005; 11(2): 226-31.

[59] French AJ, Adams CA, Anderson LS, Kitchen JR, Hughes MR, Wood SH. Development of human cloned blastocysts following somatic cell nuclear transfer with adult fibroblasts. Stem Cells 2008; 26(2): 485-93.

[60] Solter D. From teratocarcinomas to embryonic stem cells and beyond: a history of embryonic stem cell research. Nat Rev Genet 2006; 7(4): 319-27.

[61] Zwaka TP, Thomson JA. A germ cell origin of embryonic stem cells? Development 2005; 132(2): 227-33.

[62] Tada M, Takahama Y, Abe K, Nakatsuji N, Tada T. Nuclear reprogramming of somatic cells by *in vitro* hybridization with ES cells. Curr Biol 2001; 11(19): 1553-8.

[63] Cowan CA, Atienza J, Melton DA, Eggan K. Nuclear reprogramming of somatic cells after fusion with human embryonic stem cells. Science 2005; 309(5739): 1369-73.

[64] Yu J, Vodyanik MA, He P, Slukvin, II, Thomson JA. Human embryonic stem cells reprogram myeloid precursors following cell-cell fusion. Stem Cells 2006; 24(1): 168-76.

[65] Matsumura H, Tada M, Otsuji T, *et al.* Targeted chromosome elimination from ES-somatic hybrid cells. Nat Methods 2007; 4(1): 23-5.

[66] Takahashi K, Yamanaka S. Induction of pluripotent stem cells from mouse embryonic and adult fibroblast cultures by defined factors. Cell 2006; 126(4): 663-76.

[67] Okita K, Ichisaka T, Yamanaka S. Generation of germline-competent induced pluripotent stem cells. Nature 2007; 448(7151): 313-7.

[68] Wernig M, Meissner A, Foreman R, *et al. In vitro* reprogramming of fibroblasts into a pluripotent ES-cell-like state. Nature 2007; 448(7151): 318-24.

[69] Boland MJ, Hazen JL, Nazor KL, *et al.* Adult mice generated from induced pluripotent stem cells. Nature 2009; 461(7260): 91-4.

[70] Kang L, Wang J, Zhang Y, Kou Z, Gao S. iPS cells can support full-term development of tetraploid blastocyst-complemented embryos. Cell Stem Cell 2009; 5(2): 135-8.

[71] Zhao XY, Li W, Lv Z, *et al.* iPS cells produce viable mice through tetraploid complementation. Nature 2009; 461(7260): 86-90.

[72] Takahashi K, Tanabe K, Ohnuki M, *et al.* Induction of pluripotent stem cells from adult human fibroblasts by defined factors. Cell 2007; 131(5): 861-72.

[73] Yu J, Vodyanik MA, Smuga-Otto K, *et al.* Induced pluripotent stem cell lines derived from human somatic cells. Science 2007; 318(5858): 1917-20.

[74] Li W, Wei W, Zhu S, *et al.* Generation of rat and human induced pluripotent stem cells by combining genetic reprogramming and chemical inhibitors. Cell Stem Cell 2009; 4(1): 16-9.

[75] Esteban MA, Xu J, Yang J, *et al.* Generation of induced pluripotent stem cell lines from Tibetan miniature pig. J Biol Chem 2009; 284(26): 17634-40.

[76] Liu H, Zhu F, Yong J, *et al.* Generation of induced pluripotent stem cells from adult rhesus monkey fibroblasts. Cell Stem Cell 2008; 3(6): 587-90.

[77] Wu Y, Zhang Y, Mishra A, Tardif SD, Hornsby PJ. Generation of induced pluripotent stem cells from newborn marmoset skin fibroblasts. Stem Cell Res 2010; 4(3): 180-8.

[78] Kopp JL, Ormsbee BD, Desler M, Rizzino A. Small increases in the level of Sox2 trigger the differentiation of mouse embryonic stem cells. Stem Cells 2008; 26(4): 903-11.

[79] Sridharan R, Tchieu J, Mason MJ, *et al.* Role of the murine reprogramming factors in the induction of pluripotency. Cell 2009; 136(2): 364-77.

[80] Amati B, Alevizopoulos K, Vlach J. Myc and the cell cycle. Front Biosci 1998; 3: 250-68.

[81] Kim J, Chu J, Shen X, Wang J, Orkin SH. An extended transcriptional network for pluripotency of embryonic stem cells. Cell 2008; 132(6): 1049-61.

[82] Hanna J, Saha K, Pando B, *et al.* Direct cell reprogramming is a stochastic process amenable to acceleration. Nature 2009; 462(7273): 595-601.

[83] Fernandez PC, Frank SR, Wang L, *et al.* Genomic targets of the human c-Myc protein. Genes Dev 2003; 17(9): 1115-29.

[84] Huangfu D, Maehr R, Guo W, *et al.* Induction of pluripotent stem cells by defined factors is greatly improved by small-molecule compounds. Nat Biotechnol 2008; 26(7): 795-7.

[85] Shi Y, Do JT, Desponts C, Hahm HS, Scholer HR, Ding S. A combined chemical and genetic approach for the generation of induced pluripotent stem cells. Cell Stem Cell 2008; 2(6): 525-8.

[86] Jia F, Wilson KD, Sun N, *et al.* A nonviral minicircle vector for deriving human iPS cells. Nat Methods 2010; 7(3): 197-9.

[87] Nishimura K, Sano M, Ohtaka M, *et al.* Development of defective and persistent sendai virus vector: a unique gene delivery/expression system ideal for cell reprogramming. J Biol Chem 2010; 286(6): 4760-71.

[88] Fusaki N, Ban H, Nishiyama A, Saeki K, Hasegawa M. Efficient induction of transgene-free human pluripotent stem cells using a vector based on Sendai virus, an RNA virus that does not integrate into the host genome. Proc Jpn Acad Ser B Phys Biol Sci 2009; 85(8): 348-62.

[89] Seki T, Yuasa S, Oda M, *et al.* Generation of induced pluripotent stem cells from human terminally differentiated circulating T cells. Cell Stem Cell 2010; 7(1): 11-4.

[90] Kaji K, Norrby K, Paca A, Mileikovsky M, Mohseni P, Woltjen K. Virus-free induction of pluripotency and subsequent excision of reprogramming factors. Nature 2009; 458(7239): 771-5.

[91] Soldner F, Hockemeyer D, Beard C, *et al.* Parkinson's disease patient-derived induced pluripotent stem cells free of viral reprogramming factors. Cell 2009; 136(5): 964-77.

[92] Woltjen K, Michael IP, Mohseni P, *et al.* piggyBac transposition reprograms fibroblasts to induced pluripotent stem cells. Nature 2009; 458(7239): 766-70.

[93] Yusa K, Rad R, Takeda J, Bradley A. Generation of transgene-free induced pluripotent mouse stem cells by the piggyBac transposon. Nat Methods 2009; 6(5): 363-9.

[94] Warren L, Manos PD, Ahfeldt T, *et al.* Highly efficient reprogramming to pluripotency and directed differentiation of human cells with synthetic modified mRNA. Cell Stem Cell 2010; 7(5): 618-30.

[95] Zhou H, Wu S, Joo JY, *et al.* Generation of induced pluripotent stem cells using recombinant proteins. Cell Stem Cell 2009; 4(5): 381-4.

[96] Xu N, Papagiannakopoulos T, Pan G, Thomson JA, Kosik KS. MicroRNA-145 regulates OCT4, SOX2, and KLF4 and represses pluripotency in human embryonic stem cells. Cell 2009; 137(4): 647-58.

[97] Neveu P, Kye MJ, Qi S, *et al.* MicroRNA profiling reveals two distinct p53-related human pluripotent stem cell states. Cell Stem Cell 2010; 7(6): 671-81.

[98] Li W, Zhou H, Abujarour R, *et al.* Generation of human-induced pluripotent stem cells in the absence of exogenous Sox2. Stem Cells 2009; 27(12): 2992-3000.

[99] Lin T, Ambasudhan R, Yuan X, *et al.* A chemical platform for improved induction of human iPSCs. Nat Methods 2009; 6(11): 805-8.

[100] Maherali N, Ahfeldt T, Rigamonti A, Utikal J, Cowan C, Hochedlinger K. A high-efficiency system for the generation and study of human induced pluripotent stem cells. Cell Stem Cell 2008; 3(3): 340-5.

[101] Ying QL, Wray J, Nichols J, *et al.* The ground state of embryonic stem cell self-renewal. Nature 2008; 453(7194): 519-23.

[102] Lyssiotis CA, Foreman RK, Staerk J, *et al.* Reprogramming of murine fibroblasts to induced pluripotent stem cells with chemical complementation of Klf4. Proc Natl Acad Sci USA 2009; 106(22): 8912-7.

[103] Tchieu J, Kuoy E, Chin MH, *et al.* Female human iPSCs retain an inactive X chromosome. Cell Stem Cell 2010; 7(3): 329-42.

[104] Loh YH, Agarwal S, Park IH, *et al.* Generation of induced pluripotent stem cells from human blood. Blood 2009; 113(22): 5476-9.

[105] Eminli S, Foudi A, Stadtfeld M, *et al.* Differentiation stage determines potential of hematopoietic cells for reprogramming into induced pluripotent stem cells. Nat Genet 2009; 41(9): 968-76.

[106] Giorgetti A, Montserrat N, Aasen T, *et al.* Generation of induced pluripotent stem cells from human cord blood using OCT4 and SOX2. Cell Stem Cell 2009; 5(4): 353-7.

[107] Haase A, Olmer R, Schwanke K, *et al.* Generation of induced pluripotent stem cells from human cord blood. Cell Stem Cell 2009; 5(4): 434-41.

[108] Aoki T, Ohnishi H, Oda Y, *et al.* Generation of induced pluripotent stem cells from human adipose-derived stem cells without c-MYC. Tissue Eng Part A 2010; 16(7): 2197-206.

[109] Sugii S, Kida Y, Kawamura T, *et al.* Human and mouse adipose-derived cells support feeder-independent induction of pluripotent stem cells. Proc Natl Acad Sci USA 2010; 107(8): 3558-63.

[110] Liu H, Ye Z, Kim Y, Sharkis S, Jang YY. Generation of endoderm-derived human induced pluripotent stem cells from primary hepatocytes. Hepatology 2010; 51(5): 1810-9.

[111] Aasen T, Raya A, Barrero MJ, *et al.* Efficient and rapid generation of induced pluripotent stem cells from human keratinocytes. Nat Biotechnol 2008; 26(11): 1276-84.

[112] Kim JB, Greber B, Arauzo-Bravo MJ, *et al.* Direct reprogramming of human neural stem cells by OCT4. Nature 2009; 461(7264): 649-3.

[113] Do JT, Joo JY, Han DW, *et al.* Generation of parthenogenetic induced pluripotent stem cells from parthenogenetic neural stem cells. Stem Cells 2009; 27(12): 2962-8.

[114] Tsai SY, Clavel C, Kim S, *et al.* Oct4 and klf4 reprogram dermal papilla cells into induced pluripotent stem cells. Stem Cells 2010; 28(2): 221-8.

[115] Cai J, Li W, Su H, *et al.* Generation of human induced pluripotent stem cells from umbilical cord matrix and amniotic membrane mesenchymal cells. J Biol Chem 2010; 285(15): 11227-34.

[116] Marchetto MC, Yeo GW, Kainohana O, Marsala M, Gage FH, Muotri AR. Transcriptional signature and memory retention of human-induced pluripotent stem cells. PLoS One 2009; 4(9): e7076.

[117] Kim JB, Zaehres H, Wu G, *et al.* Pluripotent stem cells induced from adult neural stem cells by reprogramming with two factors. Nature 2008; 454(7204): 646-50.

[118] Kim JB, Sebastiano V, Wu G, *et al.* Oct4-induced pluripotency in adult neural stem cells. Cell 2009; 136(3): 411-9.

[119] Silva J, Barrandon O, Nichols J, Kawaguchi J, Theunissen TW, Smith A. Promotion of reprogramming to ground state pluripotency by signal inhibition. PLoS Biol 2008; 6(10): e253.

[120] Wernig M, Zhao JP, Pruszak J, *et al.* Neurons derived from reprogrammed fibroblasts functionally integrate into the fetal brain and improve symptoms of rats with Parkinson's disease. Proc Natl Acad Sci USA 2008; 105(15): 5856-61.

[121] Tsuji O, Miura K, Okada Y, *et al.* Therapeutic potential of appropriately evaluated safe-induced pluripotent stem cells for spinal cord injury. Proc Natl Acad Sci USA 2010; 107(28): 12704-9.

[122] Nelson TJ, Martinez-Fernandez A, Yamada S, Perez-Terzic C, Ikeda Y, Terzic A. Repair of acute myocardial infarction by human stemness factors induced pluripotent stem cells. Circulation 2009; 120(5): 408-16.

[123] Xu D, Alipio Z, Fink LM, *et al.* Phenotypic correction of murine hemophilia A using an iPS cell-based therapy. Proc Natl Acad Sci USA 2009; 106(3): 808-13.

[124] Hanna J, Wernig M, Markoulaki S, *et al.* Treatment of sickle cell anemia mouse model with iPS cells generated from autologous skin. Science 2007; 318(5858): 1920-3.

[125] Raya A, Rodriguez-Piza I, Guenechea G, *et al.* Disease-corrected haematopoietic progenitors from Fanconi anaemia induced pluripotent stem cells. Nature 2009; 460(7251): 53-9.

[126] Dimos JT, Rodolfa KT, Niakan KK, *et al.* Induced pluripotent stem cells generated from patients with ALS can be differentiated into motor neurons. Science 2008; 321(5893): 1218-21.

[127] Park IH, Arora N, Huo H, *et al.* Disease-specific induced pluripotent stem cells. Cell 2008; 134(5): 877-86.

[128] Tolar J, Xia L, Riddle MJ, *et al.* Induced Pluripotent Stem Cells from Individuals with Recessive Dystrophic Epidermolysis Bullosa. J Invest Dermatol 2010; 131(4): 848-56.

[129] Wang Y, Jiang Y, Liu S, Sun X, Gao S. Generation of induced pluripotent stem cells from human beta-thalassemia fibroblast cells. Cell Res 2009; 19(9): 1120-3.

[130] Ebert AD, Yu J, Rose FF, *et al.* Induced pluripotent stem cells from a spinal muscular atrophy patient. Nature 2009; 457(7227): 277-80.

[131] Lee G, Papapetrou EP, Kim H, *et al.* Modelling pathogenesis and treatment of familial dysautonomia using patient-specific iPSCs. Nature 2009; 461(7262): 402-6.

[132] Liu J, Verma PJ, Evans-Galea MV, *et al.* Generation of Induced Pluripotent Stem Cell Lines from Friedreich Ataxia Patients. Stem Cell Rev. 2010 Dec 22. [Epub ahead of print].

[133] Ku S, Soragni E, Campau E, *et al.* Friedreich's ataxia induced pluripotent stem cells model intergenerational GAATTC triplet repeat instability. Cell stem cell 2010; 7(5): 631-7.

[134] Carvajal-Vergara X, Sevilla A, D'Souza SL, *et al.* Patient-specific induced pluripotent stem-cell-derived models of LEOPARD syndrome. Nature 2010; 465(7299): 808-12.

CHAPTER 5

Genome, Epigenome and Transcriptome Profiling of Pluripotent Stem Cells

Henry Chung and Kuldip S. Sidhu*

Stem Cell Laboratory, Faculty of Medicine, University of New South Wales, Sydney, Australia

Abstract: Pluripotent stem cells are unique candidates for regenerative medicine owing to their differential and self renewal capabilities. With the rapid development of stem cell differentiation methodologies, a number of functional cell types have already been generated, though replacement of damaged or diseased cells/tissues/organs remains an exciting and challenging task. The most common pluripotent cell type are embryonic stem cells (ESCs), however in recent years, several pluripotent cell types have been emerged – nuclear transfer ESCs and induced pluripotent stem cells (iPSCs). Both of these cell types were generated for therapeutic cloning, hence facilitating the transition from lab bench to patients. However, before clinical implementation of pluripotent stem cell based therapies, it must firstly pass several tests that evaluate the long-term safety, with respect to genetic stability. Both nuclear transfer ESCs and iPSCs share many common features but due to their different origins, exhibit slight differences at the epigenomic and transcriptomic levels. This chapter attempts to elucidate some of the many differences that occur at each of the genomic, epigenomic and transcriptomic levels.

Keywords: Differentiation, induced pluripotent stem cells, epigenome, transcriptome, DNA methylation, histone modification, nuclear transfer.

INTRODUCTION

The word pluripotent is derived from Latin, '*pluri*', meaning multiple/many and stem cells that are pluripotent, possess the unique capacity to develop into any one of the 200+ cell types in our bodies, in addition to indefinite cellular mitotic division. This makes pluripotent stem cells an ideal candidate for studying developmental biology and also cell based therapies in regenerative medicine. One of the greatest examples of pluripotent stem cells is embryonic stem cells. As the name implies, these particular stem cells are isolated from a pre-implantation embryo, days after conception, more specifically at the blastocyst stage, which occurs in humans, 5-7 days after a sperm fertilizes with the egg. Embryonic stem cells have been generated from multiple animals, including humans (Table **1**) and hold great therapeutic potential in regenerative medicine, especially for the treatment of many diseases (*e.g.* hematological, neurodegenerative, cardiovascular).

Table 1. A timeline for the generation of pluripotent stem cell lines.

Animal	ESC	Year	nt-ESC	Year	iPSC	Year
Mouse	[1]	1981	[2]	2001	[3]	2006
Monkey	[4]	1995	[5]	2007	[6]	2008
Human	[7]	1998	-	-	[8]	2007
Rat	[9]	2008	-	-	[10]	2009

Currently there are 2 other distinct populations of pluripotent stem cells; nuclear transfer embryonic stem cells (ntESC) and induced pluripotent stem cells (iPSCs). Until 2006, somatic cell nuclear transfer (SCNT) was considered the only viable way to produce patient specific pluripotent stem cells. By closely mimicking the development of an early embryo, scientists were able to inseminate eggs, replacing the

***Address correspondence to Kuldip S. Sidhu:** Director, Stem Cell Laboratory, Faculty of Medicine, The University of New South Wales, Level 2, Clinical Sciences Building, Prince of Wales Hospital, Randwick, NSW 2031, Australia; Tel: + 61 2 9382 2622; Fax: +61 2 9382 2622; E-mail: k.sidhu@unsw.edu.au sperm with diploid somatic cells to yield ntESCs. These pluripotent cells had all the potential and characteristics that traditional embryo-derived ESCs shared [2, 11]. Towards the fall of 2006, scientists were able to directly reprogram somatic diploid cells back into a pluripotent state by over expression of genes encoding for key pluripotency-associated transcription factors (CMYC, KLF4, LIN28, OCT4, NANOG and/or SOX2) and were termed induced pluripotent stem cells [3]. Again, this particular pluripotent stem cell population share many common features with embryo-derived ESCs [8, 12-13]. Both these pluripotent stem cells populations were generated for the purpose of therapeutic cloning, that is with the purpose of treating cellular degenerative diseases and immune compatibility in regenerative medicine.

The pluripotency of *in vitro* cultures of embryonic stem cells is largely regulated by extrinsic signalling, transcriptional regulation (*i.e.* intrinsic signalling) and epigenetic modifications. The genome is essentially defined as the genetic blueprint of a cell and is highly conserved within a species. The genome is composed of a random, yet ordered sequence of 4 basic genetic elements – nucleotides; Adenine (A), Thymine (T), Cytosine (C), Guanine (G), and is commonly known as DNA. The transcriptome is transcribed from DNA and is defined as all genes expressed in a cell at a given time. Unlike the genome, the transcriptome is highly variable and is dependent on extrinsic (cell signalling) and intrinsic (epigenetic modifications) cues during many cellular events, such as; differentiation/development, apoptosis, senescence *etc.* Similarly, the highly variable epigenome is defined as the total epigenetic modifications imposed on the DNA and histones (otherwise known as chromatin), which in turn regulates the transcriptional network. There is a logical but interwining relation between the genome, epigenome and the transcriptome.

In order for pluripotent stem cells, such as human ESCs or iPSCs to be used in regenerative medical purposes, one must thoroughly characterize the cells and evaluate their safety, cellular memory and/or similarity, which correspond to genomic, epigenomic and transcriptomic profiling, respectively.

Evaluating the Safety of Pluripotent Cells Through Genomic Profiling

The subtle differences in the human genome provide the basis for genetic variability. There are different types of genetic variations including; single nucleotide polymorphisms (SNPs), chromosomal duplications, translocations and insertions/deletions. Many studies analyze the genome of human ESCs through traditional comparative genomic hybridization (CGH) and karyotyping techniques [7, 14]. Both techniques are considered 'low resolution' from which can only detect variations in the 2 – 10 megabase (Mb) range.

As mentioned previously, pluripotent stem cells, in particular, human ESCs and iPSCs offer one of the most promising candidates for regenerative medicine. Therefore, genomic stability must be thoroughly characterized before its therapeutic potential can be realized.

Karyotypic Instability

There are many examples of human ESC line karyotypic variability (otherwise known as aneuploidy) as early as passage #13 whereas some can remain karyotypically normal even after passage #140 [15-16]. Two possible explanations exists, 1) a genetic predisposition of cells lines to chromosomal instability (whether it be from embryos for human ESCs or somatic cells – iPSCs) or 2) aspects of cell culture, which may include passaging methods (*i.e.* chemical or mechanical), oxygen tension, cryopreservation methods and/or feeder-free culture.

Human ESCs are derived from surplus embryos initially used to treat infertility, but the reason why a couple cannot conceive remains undefined. One possible explanation is the genetic status of the embryo, in fact, over 50% of pre-implantation embryos used to treat infertility have chromosomal abnormalities and are incompatible with embryo development [17]. Even high morphological quality embryos may fail to result in live births and some have speculated the involvement of undiagnosed genetic mutations [18-19].

Extrinsic factors that contribute to genomic instability have been widely studied. *In vitro* culture of human ESCs in the presence of low oxygen levels revealed enhanced cell proliferation and reduced spontaneous differentiation [20]. The same was also reported for the generation of human iPSCs [21] and pig nuclear transfer experiments [22]. The type of passaging method also contributes to chromosomal abnormalities. Commonly used enzymes such as, collagenase, trypsin, accutase and dispase are more likely to give rise to chromosomal abnormalities. The most common abnormality is trisomy of chromosomes #12 and #17 upon extended culture [23-24]. In contrast, tedious mechanical passaging maintains the normal karyotype over the long-term. With the development of defined culture systems for pluripotent stem cells (Table **2**, such as feeder-free), was accompanied with karyotypic abnormalities [25].

Table 2. hESC maintenance under different conditions.

Study	Feeder cell/Matrix	Medium	Feeder-free	Serum-free	Xeno -free
[7]	MEF	DMEM + 20% FBS			
[26]	Matrigel	MEF-conditioned	✓		
[27]	Foreskin fibroblasts	KODMEM + 15% SR		✓	✓
[28]	Fibronectin	KODMEM + 15% SR	✓	✓	
[25]	Matrigel	mTeSR1	✓	✓	
[29]	CELLStart / Geltrex	StemPro	✓		
[30]	Foreskin fibroblasts	HEScGRO			✓

Cytogenetics - Molecular Karyotyping

Conventional Giemsa-stained karyotyping (also known as G-banding) remains one of the most widely used techniques to study chromosomal abnormalities in ESCs, nuclear transfer ESCs and recently, iPSCs. Each chromosome has a distinctive banding pattern that can help identify them. Abnormalities can include, deletion/duplication of whole chromosomes, translocations and very large imbalances of greater than 10 Mb. This limits the amount of genomic information required for 'banking' cell lines, including copy number variations or chromosomal imbalances, which can be achieved using a comparative genomic hybridization (CGH) approach. Furthermore, the recent advancement of array technologies has led to the development of a more sophisticated array CGH, which utilizes an oligonucleotide platform with multiple probe sets. This allows us to detect copy number variants (CNVs) in addition to chromosomal imbalances at a higher resolution of ~10-100 kb [14]. Databases are continuously updated with individual's CNV and their correlation to abnormal phenotypes, such as diseases. The DatabasE of Chromosomal Imbalance and Phenotype in Humans using Ensembl Resources (DECIPER) and the International Standard Cytogenomic Array (ISCA) databases are good resources to cross match pluripotent stem cell's CNVs in order to establish clinically user-friendly cell lines. Additionally, in spite of higher resolutions in array CGH, karyotyping through G-banding is disadvantaged by the amount of labor and expertise required to confidently assign anomalies to cell lines.

Single nucleotide polymorphisms (SNPs) are DNA sequence mutations that occur at the base pair level. Owing to the data from the Human Genome Project, SNP identification through arrays are frequently used to locate genes involved with diseases. The DNA sequence of an individual is what defines their phenotype, such as ethnicity and disease. Even under strict maintenance protocols, human ESCs can acquire chromosomal anomalies upon extended culture [31]. The International HapMap Project generated a database to catalogue the genotype of the major ethnic races; African, Asian and European. The International HapMap Consortium [32], clearly documented SNP allele frequency differences between different ethnicities. And interestingly, a recent study commented on the lack of ethnic diversity in our current available human ESC lines [33]. With the exponential increase in human iPSC lines, generated from all over the world and bearing in mind, donor fibroblasts is an easy accessible source, ethnic diversity will be covered for both major and minor races for future iPSC lines. One important reason to identify SNPs in current (and newly generated) pluripotent cell lines is that it can influence the response to drugs and/or vaccinations, similar to a drug therapy in one person may have an opposing effect on another.

Immune Compatibility Issues

One final genomic profiling believed to be important involves human leukocyte antigen (HLA) genotyping. Highly polymorphic HLA genes encode for major (and minor) histocompatibility complexes that are expressed on the surface of cells and mediates graft rejection upon transplantation *via* the host's immune system in recognizing foreign tissue. To identify suitable cell lines for patient therapy and with a low risk of graft rejection, a bank of pluripotent stem cell lines should be established with HLA genotype thoroughly characterized. However, generating a cell line for every patient would be labor intensive and impractical. One way to circumvent around this is to generate cell lines that are homozygous for common HLA types [34].

There had been accounts of unexpected events due to the lack of appropriate and rigorous validation (*e.g.* brain tumor formation after transplantation of fetal neural stem cells [35]). Though these are individual cases and probably seem insignificant, a pluripotent stem cell line may be used to treat multiple patients, hence the impact of infection/tumor growth would increase dramatically.

In a large scale study with attempt to identify chromosomal integrity, a substantial number of cell lines carried full or partial chromosomal aberrations [36], which would hinder its clinical utilization. To prevent this from happening, rigorous 'full' characterization of clinical pluripotent stem cell lines must be carried out to check for chromosomal abnormalities/instabilities. Additionally, genetic matching of pluripotent stem cell lines would reduce risks of graft rejection.

The Epigenome of Pluripotent Cells – The Instructions and Memory

With the recent advancements in high throughput array technology, scientists are now able to obtain a clearer and wider perspective of how epigenetic modifications regulate the maintenance of pluripotency and commitment into specific embryonic germ lineages. '*Epi*' meaning outside of or on top of the genome, essentially describes heritable gene expression changes that are not related to the DNA sequence. Some of the most well studied modifications include: DNA methylation, histone tail modifications (*i.e.* methylation, acetylation, phosphorylation, ubiquitylation), chromatin remodelling and non-coding RNA. Changes in these modifications during *in vitro* culture may have profound effects, which may affect pluripotent cell differentiation [37], thus limiting their application in regenerative medicine.

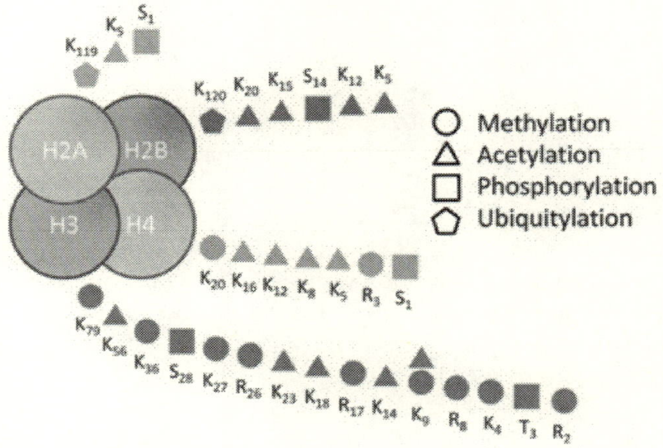

Fig. 1: Different histone tail modifications. A nucleosome unit listing the type of epigenetic modifications imposed on specific amino acid residues on individual histone tails.

Three meters worth of genetic information is cleverly packaged into an average cell nucleic diameter of 6 x 10^{-6} meters. This packaging involves numerous supercoiling into chromatin scaffold fibers. The most basic unit of chromatin is a nucleosome, which consists of an octamer of 4 histone proteins; H2A, H2B, H3, H4 (2 of each) and is wrapped around with negatively charged DNA. The unique ability of pluripotent stem cells to differentiate is also accompanied with dynamic epigenetic modifications that are imposed on nucleosomes – both on DNA and the amino acid tails of each histone (Fig. **1**). Gene expression regulated by epigenetic modifications is thought to establish distinctive domains of active and inactive chromatin structures partly by post-translational modifications of histone proteins. The overall chromatin structure of pluripotent stem cells resides in a decondensed (euchromatic) state and becomes condensed (heterochromatic) during differentiation [38].

DNA Methylation

DNA methylation is a post-replication epigenetic modification that involves the reversible addition of a methyl group ($-CH_3$) to cytosine nucleotides and is largely associated with gene silencing. Bisulfite sequencing remains one of the widely used techniques to study methylation profiles. Sodium bisulfite deals irreversible structural changes to unmethylated cytosine nucleotides, which can be picked up using conventional DNA sequencing methodologies, however one limitation is the number of CpGs that can be analyzed.

Comprehensive high-throughput array-based relative methylation (CHARM) analysis has allowed the simultaneous methylation analysis of over 4.6 million CpGs which includes most CpG islands and nearby sequences, known as shores [39-40]. By using CHARM analysis, Kim and colleagues were able to deduce both statistically and quantitatively the number of differentially methylated regions (DMRs) between the 3 major populations of pluripotent cells – embryo derived ESCs, nuclear transfer ESCs and iPSC. Despite their abilities to differentiate into cells representative of the 3 embryonic lineages, their DMR profile differed dramatically. Nuclear transfer ESCs were most similar to embryo-derived ESCs by having the least number of DMRs (229), in contrast iPSCs with over 5000 DMRs. Interestingly, even between iPSCs generated from different somatic cells yielded ~5000DMRs, suggestive of differential differentiation potential as further projected in the study. Another CHARM analysis also discovered that certain loci in iPSCs remain incompletely reprogrammed and even some abnormally reprogrammed [41].

Furthermore, among the 3 major populations of pluripotent cells, only one of the therapeutically cloned pluripotent cell type closely resembles traditional embryo derived ESCs – nuclear transfer ESCs, in terms of differentiation propensity and methylation profile [40]. The authors mention the existence of an 'epigenetic memory,' also proposed by others [42], in that iPSCs reflect their donor cell type of origin and biases their differentiation towards a certain lineage. However, this memory can be erased with subsequent 're-reprogramming' experiments by using doxycycline induction or initial reprogramming using epigenetic modifying small molecules.

Histone Modifications

One of the earliest techniques used to study epigenetic histone tail modifications, was chromatin immunoprecipitation (ChIP). ChIP involves initial fragmentation of protein cross-linked DNA (*i.e.* chromatin) and the epigenetic modification(s) of interest is enriched by immunoprecipitation with specific antibodies against the modification. These enriched regions can either be detected using Southern blotting or quantitative polymerase chain reaction [43]. Conventional ChIP is limited in analyzing single gene locus, which makes 'genome-wide' epigenomic profiling a tedious task, as with transcriptome profiling using microarrays as opposed to reverse transcription PCR. The development of ChIP-on-chip assays has allowed simultaneous analysis of hundreds of sites harboring epigenetic modification(s), giving the user a more complete perspective between multiple modifications.

Histone methylation has both euchromatic and heterochromatic properties depending on which amino acid is modified. Methylation of histone H3 lysine 4 (H3K4) and lysine 27 (H3K27) are of particular interest and is catalyzed by 2 types of complexes, trithorax and polycomb proteins, respectively [44]. Different states of methylation at H3K4 (mono-, di-, tri-), positively regulates gene transcription by recruiting remodelling enzymes [45], and similarly, methylation of H3K27 generates heterochromatic features, hence negatively regulating gene transcription [46]. The overlapping of these 2 opposing epigenetic markers at a particular locus is termed 'bivalent domains' which establishes the developmental decisions that cells undertake. Initial global analysis of Histone H3 acetylation/methylation were performed in yeast [47]. These bivalent domains have been mapped in mouse ESCs [48], human ESCs [49-50] and iPSCs [51] and consists of large highly conserved noncoding elements, which are clustered within regions coding for important developmental transcription factors. Interestingly, pluripotency related genes demonstrated a conversion of active chromatin markers to a repressive state, while lineage specific genes showed the opposite during ESC differentiation [52-53]. This suggests the changes in global gene expression are directly correlated to the dynamic changes of the epigenome.

The above studies utilized ChIP quantitative PCR and array based technologies, which have limited scope. The depth and coverage of a ChIP analysis can be increased by utilizing short tag-based sequencing technologies. A paired end ditag (PET) approach was developed and was originally used for global transcriptomic analysis [54]. PETs are generated from 18 bp of the 5' and 3' ends of a transcript and can be applied to analyze DNA fragments obtained from ChIP, which is subjected to high throughput sequencing. Using ChIP-PET to profile genome-wide analysis of#H3K4me3 and H3K27me3, H3K4me3 was found to be the most prevalent epigenetic marker across 70% of genes in human ESCs, while smaller regions of H3K27me3 were colocalized in the promoter regions [55]. The different combinations of these 2 epigenetic markers are associated with distinctive gene functional categories. The co-occupying of H3K4me3 and H3K27me3 forms bivalent domains in genes encoding for neuroectoderm, mesoderm and other developmental processes, which are expressed at low levels in pluripotent cells and exists in a 'poised' state ready for activation given the correct signal.

Lastly, by utilizing ultra high throughput and multiplex short tag-based sequencing approaches (i.e. Solexa sequencing), direct sequencing of ChIP enriched DNA fragments can be performed. This technology involves ligating ChIP enriched DNA to adaptors and subjected to cluster amplification by bridging PCR and sequencing by synthesis analysis. By doing so, ChIP sequencing has allowed simultaneous analysis of multiple epigenetic histone modifications at the global level [56-57]. Fortunately, H3K4 and H3K27 trimethylation levels were not distinguishable between human ESCs and iPSCs [58]. Further confirmed by ChIP sequencing, epigenetic profile of H3K27me3 across human ESCs and iPSCs were highly similar [59]. However, another marker, H3K9 was significantly different between the 2 pluripotent cell types, which may contribute to the differences in gene expression between early passages of human ESCs and iPSCs [58].

Both the maintenance of pluripotency and lineage commitment is achieved by dynamic transcription activity, which is instructed with dynamic epigenetic modifications. However with some applications, epigenetic memory can be advantageous, as directed differentiation of pluripotent cells into certain tissues/cells remains a challenge. The precise coordination of gene transcription and epigenetic modification is essential for normal development and/or maintaining pluripotency as demonstrated in gene knockout studies.

Assessing the Similarity/Identity of Different Pluripotent Cells Through Transcriptomic Profiling

Transcriptomics is the study of the total gene set expression and can be utilized in a wide range of analyses. With the development of human ESCs in 1998, more and more hESC lines have been generated and it was made possible to define a 'molecular signature' by comparison with all these lines. This signature is composed of genes that are important for maintaining the undifferentiated state and coincidently are highly expressed (or repressed) in pluripotent stem cells. With the advancement of high throughput technologies, measuring gene transcript levels have enabled extensive transcriptomic profiling of pluripotent cells. There are a number of techniques used; Reverse transcription polymerase chain reaction (RT-PCR), quantitative RT-PCR, DNA microarrays, Expressed sequence tag (EST), Massively parallel signature sequencing (MPSS) and Serial analysis of gene expression (SAGE).

Gene-level Expression Analysis

Reverse transcription polymerase chain reaction remains one of the most commonly used techniques to analyze gene expression. RT-PCR relies on the annealing of short oligonucleotides (*i.e.* primers) to reverse transcribed mRNA transcripts and subsequent amplification to visualize the gene product.

Quantitative RT-PCR has been increasingly popular since it enables real time measurement during a PCR reaction, allowing one to determine quantitative gene transcript levels. By measuring relative transcript levels, a group was able to profile the 17 HUES cell lines according to their increased propensity to differentiate into endoderm, mesoderm and ectoderm lineages [60]. Combined with high throughput technology, the International Stem Cell Initiative (ISCI) was able to profile the molecular signature with quantitative RT-PCR based arrays [61]. Though simple and efficient, these 2 techniques are generally limited to analyzing several genes at once.

Genome-Wide Expression Analysis

Most of the earlier studies were limited in the diversity of cell lines analyzed which led to very little overlap of the molecular signature gene lists. This minimal similarity can be attributed to the different cell culture protocols, sampling and statistical analysis methods. With more commercial competitors/platforms emerging, the costs for conducting genome-wide transcriptome experiments have dramatically decreased, allowing more researchers to access this technology.

International collaborations have begun to emerge in order to characterize multiple pluripotent cell lines across the globe in effort to identify a molecular signature for pluripotency. The International Stem Cell Initiative (ISCI) analyzed over 50 human ESC lines and was able to detect differences in lineage marker expression and several female human ESCs lines expressing XIST (implicating roles in X chromosome inactivation) whereas some did not [61]. Meta analysis studies provide more confident conclusions by analyzing multiple cell lines cultured under a range of conditions, which rules out the possibility of cultural-induced changes or (genetic) variation between cell lines. Different human ESC lines across 38 studies was analyzed to yield a 'consensus human ESC gene list' comprising of 1076 genes over expressed in human ESCs when compared to differentiated cell types [62]. With the development of other pluripotent cell lines, the molecular signature initially being ESC-specific has expanded to include iPSCs as well [63].

Comparative transcriptomics across different species of ESC lines have revealed both universal and species specific mechanisms controlling pluripotency. It was found that a common set of mammalian genes (CCNB1, GDF3, LEFTB, NANOG, OCT4) was highly expressed across human, mouse and monkey ESCs [64]. This is not surprising since important sequence elements in the genome are often evolutionary conserved. One well known species-specific difference is that leukemia inhibitory factor (LIF) is required to maintain the undifferentiated state in murine ESCs, whereas fibroblast growth factor 2 (FGF2) is required for human and monkey ESC maintenance [65].

With respect to meta analysis studies, a recent publication depicted a small cluster of genes, which may become a good indicator of therapeutic grade iPSCs [66]. By analyzing developmental genes containing bivalent domains, which poises them for lineage development, different classes of pluripotent cells have emerged. Mouse iPSCs that were able to give rise to pups through tetraploid complementation (currently the most stringent assay for pluripotency determination) contained a low level expression of this cluster of bivalent domain containing genes, similar to mouse ESCs. In contrast, partially reprogrammed iPSCs and iPSCs that could not contribute to germline, had different methylation profiles and expressed higher levels of 'lineage-decision making' genes, which affected their differentiation potential.

Like the epigenome, the transcriptome of pluripotent cells is highly dynamic in nature, unlike the genome, which may undergo several, if any mutations. This is particularly evident in pluripotent cell differentiation whether directed [67] or spontaneous [68]. Media components and cultural environments largely contribute to different transcriptomic profiles of pluripotent cells. The first human ESC lines were initially cultured in fetal bovine serum (FBS) containing media but were accompanied with high rates of spontaneous differentiation. Later, the development of serum replacement (SR), which widely substituted FBS for better ESC maintenance, was demonstrated both at the cellular and transcript levels [69]. Surprisingly, ~1400 genes were differentially regulated between the 2 culture conditions with metabolism and proliferation related genes upregulated in SR maintained human ESCs. Further studies have now eliminated contaminating feeder cells from cultures, which significantly define another transcriptomic profile of pluripotent cells [25, 70] (See also Table **2**).

However, the focus of this subchapter is to describe the similarities and identities of multiple pluripotent cell types – ESCs, nuclear transfer ESCs and iPSCs. Many people have asked whether the new therapeutically cloned pluripotent cell lines are equivalent to traditional embryo derived ESCs.

Murine SCNT-derived embryos closely resembled IVF-derived embryos at the transcriptomic level [71] and the same for bovine species [72]. Likewise, mouse and human iPSCs have high correlations with their embryo counterparts [3, 8, 73-75]. All these studies suggest, newly derived pluripotent cell types are functionally and transcriptomically similar to embryo-derived stem cells, though not identical. Ongoing studies are now starting to distinguish these pluripotent cell lines (in particular iPSCs) from human ESC lines and the degree of variation is greater than that between human ESC lines. Firstly, late passages of the 4 separate viral vector based iPSCs were more similar to human ESCs than earlier passages – this is believed to occur due to extended culturing and auto-selection of cultural-adapted stem cells [58]. It was also noted that iPSCs with transgenes removed from the genome were more similar to human ESCs than genome-integrated iPSCs [76]. And lastly, iPSCs generated from a single polycistronic vector also were transcriptomically similar to human ESCs [77]. See also Fig. **2**.

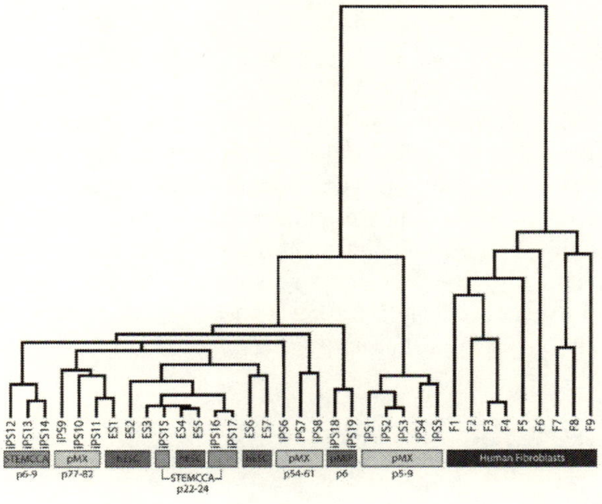

Fig. 2: Hierarchial clustering of multiple human fibroblasts, ESCs and iPSCs. Adapted from [77]. Clustering of global gene expression data of iPSC lines at different passages and derived using different type of vectors. pMX: 4 monocistronic viral vectors, STEMCCA: 1 polycistronic viral vector.

Then the question arises of whether nuclear transfer ESCs or iPSCs are more similar IVF-derived ESCs? Minimal number of studies has addressed such issue, as discussed previously in the epigenome subchapter, Kim and colleagues have demonstrated that nuclear transfer ESCs are more similar to ESCs than transcription factor-based iPSCs with respect to differentiation propensities and methylation profiles [40]. This can be attributable to several factors, 1) the mechanism of reprogramming in nuclear transfer ESCs uses oocytes, which is common with traditional IVF-derived ESC lines, or 2) transcription factor-based reprogramming does not faithfully erase the donor cell (epigenetic) memory.

It is difficult to directly compare the current 3 major pluripotent stem cell populations – ESCs, nuclear transfer ESCs and iPSCs, mainly due to the technical difficulties in generating nuclear transfer ESC lines, especially in humans, where multiple attempts have been made with the furthest culture up to the blastocyst stage [78]. Nevertheless, individual comparisons have been made (*i.e.* ESCs vs. nuclear transfer ESCs or ESCs vs. iPSCs), as discussed above.

By profiling the transcriptomes of multiple human ESCs, one can establish a confident pluripotency 'molecular signature' that can be used to characterize future cell lines. Cross species comparison also yielded interesting information on the core regulatory networks governing pluripotency. And importantly, transcriptomic analysis can help identify the similarity between pluripotent cell types at the genome-wide level. With more array-based platforms emerging, the amount of information that can be obtained in a single run is enormous and one must use proper statistical methods to analyze such as those discussed in [77].

Bringing Pluripotent Stem Cell Therapy to the Bedside

Periodic and extensive testing of human ESC lines have been performed in the past, which looked into short tandem repeats (*i.e.* DNA fingerprinting), HLA typing, SNP genomic analysis, mitochondrial DNA sequencing, differentiation propensity, transfection/growth/cloning efficiencies, chromosomal karyotyping and microarray gene expression analysis [79-80]. The approach used in these studies sets a standard for the rigorous characterization required for translating donor/patient-specific pluripotent cells from bench side to bed side. Different pluripotent cell lines behave in multiple ways, with some having a biased differentiation potential into certain lineages or reacting differently to pharmaceuticals. This suggests that such inherent differences represents a unique 'personality' for the cell line and may be better suited for a particular application over others.

Even though the transcriptome of ESCs and iPSCs are significantly different, upon extended culture, these differences begin to diminish as 'aged' iPSCs were found to be more similar to ESCs. In addition, all pluripotent cell types share common functions including, ability to differentiate into the 3 embryonic germ lineages and have unlimited self renewal. Nevertheless, increasing evidence has suggested the subtle differences between transcriptomes and the vast differences between epigenomes have largely separated the 3 cell types into distinct groups, all of which possess pluripotency potential. It is unfortunate that extensive characterization of the genome, epigenome and transcriptome cannot be carried out between the 3 pluripotent cell types in human cells, since currently, no study has derived human nuclear transfer ESCs. However studies have began to emerge demonstrating (at least in the mouse), that SCNT-derived ESCs epigenetically resembles traditional ESCs, thus indicating improvement is required in the generation of 'true' ESC-like iPSCs.

CONCLUSIONS AND PROSPECTS

Conclusively, many observations regarding differences in pluripotent cell lines adds to the adapted variation we see when these cell lines are subjected to different cultural conditions (*i.e.* type of medium, feeder cells, growth factors, passaging/freezing methods). Only by conducting genome-wide characterization of pluripotent cell's genotype, epigenotype and transcriptome across multiple cell lines cultured in a range of environments can future studies resolve the impact of those effects and transit these cells safely for clinical therapy.

REFERENCES

[1] Martin GR. Isolation of a Pluripotent Cell Line from Early Mouse Embryos Cultured in Medium Conditioned by Teratocarcinoma Stem Cells PNAS 1981; 78(12): 7634-8.

[2] Wakayama T, Tabar V, Rodriguez I, Perry AC, Studer L, Mombaerts P. Differentiation of embryonic stem cell lines generated from adult somatic cells by nuclear transfer. Science 2001; 292(5517): 740-3.

[3] Takahashi K, Yamanaka S. Induction of pluripotent stem cells from mouse embryonic and adult fibroblast cultures by defined factors. Cell 2006; 126(4): 663-76.

[4] Thomson JA, Kalishman J, Golos TG, *et al.* Isolation of a primate embryonic stem cell line. Proc Natl Acad Sci U S A 1995; 92(17): 7844-8.

[5] Byrne JA, Pedersen DA, Clepper LL, *et al.* Producing primate embryonic stem cells by somatic cell nuclear transfer. Nature 2007; 450(7169): 497-502.

[6] Liu H, Zhu F, Yong J, *et al.* Generation of induced pluripotent stem cells from adult rhesus monkey fibroblasts. Cell Stem Cell 2008; 3(6): 587-90.

[7] Thomson JA, Itskovitz-Eldor J, Shapiro SS, *et al.* Embryonic stem cell lines derived from human blastocysts. Science 1998; 282(5391): 1145-7.

[8] Takahashi K, Tanabe K, Ohnuki M, *et al.* Induction of pluripotent stem cells from adult human fibroblasts by defined factors. Cell 2007; 131(5): 861-72.

[9] Buehr M, Meek S, Blair K, *et al.* Capture of authentic embryonic stem cells from rat blastocysts. Cell 2008; 135(7): 1287-98.

[10] Liao J, Cui C, Chen S, *et al.* Generation of induced pluripotent stem cell lines from adult rat cells. Cell Stem Cell 2009; 4(1): 11-5.

[11] Wakayama S, Ohta H, Kishigami S, *et al.* Establishment of male and female nuclear transfer embryonic stem cell lines from different mouse strains and tissues. Biol Reprod 2005; 72(4): 932-6.

[12] Park I-H, Zhao R, West JA, *et al.* Reprogramming of human somatic cells to pluripotency with defined factors. Nature 2008; 451: 141-6.

[13] Aasen T, Raya A, Barrero MJ, *et al.* Efficient and rapid generation of induced pluripotent stem cells from human keratinocytes. Nat Biotech 2008; 26: 1276-84.

[14] Hovatta O, Jaconi M, Tohonen V, *et al.* A teratocarcinoma-like human embryonic stem cell (hESC) line and four hESC lines reveal potentially oncogenic genomic changes. PLoS ONE 2010; 5(4): e10263.

[15] Allegrucci C, L.E.Young. Differences between human embryonic stem cell lines. Human Reprod 2007; 13(2): 103-20.

[16] Catalina P, Montes R, Ligero G, *et al.* Human ESCs predisposition to karyotypic instability: Is a matter of culture adaptation or differential vulnerability among hESC lines due to inherent properties? Mol Cancer 2008; 7: 76.

[17] Harper JC, Coonen E, Handyside AH, Winston RM, Hopman AH, Delhanty JD. Mosaicism of autosomes and sex chromosomes in morphologically normal, monospermic preimplantation human embryos. Prenat Diagn 1995; 15(1): 41-9.

[18] Kahraman S, Bahce M, Samli H, *et al.* Healthy births and ongoing pregnancies obtained by preimplantation genetic diagnosis in patients with advanced maternal age and recurrent implantation failure. Hum Reprod 2000; 15(9): 2003-7.

[19] Gianaroli L, Magli MC, Ferraretti AP, Munne S. Preimplantation diagnosis for aneuploidies in patients undergoing *in vitro* fertilization with a poor prognosis: identification of the categories for which it should be proposed. Fertil Steril 1999; 72(5): 837-44.

[20] Ezashi T, Das P, Roberts RM. Low O2 tensions and the prevention of differentiation of hES cells. Proc Natl Acad Sci U S A 2005; 102(13): 4783-8.

[21] Yoshida Y, Takahashi K, Okita K, Ichisaka T, Yamanaka S. Hypoxia enhances the generation of induced pluripotent stem cells. Cell Stem Cell 2009; 5(3): 237-41.

[22] Pan DK, Zhang YH, Sun XZ, *et al.* Cloned pigs derived from somatic cell nuclear transfer embryos cultured *in vitro* at low oxygen tension. Chinese Sci Bull 2006; 51(7): 839-44.

[23] Hoffman LM, Carpenter MK. Characterization and culture of human embryonic stem cells. Nat Biotechnol 2005; 23(6): 699-708.

[24] Draper JS, Smith K, Gokhale P, *et al.* Recurrent gain of chromosomes 17q and 12 in cultured human embryonic stem cells. Nat Biotechnol 2004; 22(1): 53-4.

[25] Ludwig TE, Levenstein ME, Jones JM, *et al.* Derivation of human embryonic stem cells in defined conditions. Nat Biotech 2006; 24: 185-7.

[26] Xu C, Inokuma MS, Denham J, *et al.* Feeder-free growth of undifferentiated human embryonic stem cells. Nat Biotech 2001; 19: 971-4.

[27] Amit M, Margulets V, Segev H, *et al.* Human Feeder Layers for Human Embryonic Stem Cells. Biol of Reprod 2003; 68: 2 150-6.

[28] Amit M, Shariki C, Margulets V, Itskovitz-Eldor J. Feeder layer- and serum-free culture of human embryonic stem cells. Biol Reprod 2004; 70(3): 837-45.

[29] Swistowski A, Peng J, Han Y, Swistowska AM, Rao MS, Zeng X. Xeno-free defined conditions for culture of human embryonic stem cells, neural stem cells and dopaminergic neurons derived from them. PLoS ONE 2009; 4(7): e6233.

[30] Chin AC, Padmanabhan J, Oh SK, Choo AB. Defined and serum-free media support undifferentiated human embryonic stem cell growth. Stem Cells Dev 2010; 19(6): 753-61.

[31] Narva E, Autio R, Rahkonen N, *et al.* High-resolution DNA analysis of human embryonic stem cell lines reveals culture-induced copy number changes and loss of heterozygosity. Nature Biotechnology 2010; 28(4): 371-U103.

[32] A haplotype map of the human genome. Nature 2005; 437(7063): 1299-320.

[33] Mosher JT, Pemberton TJ, Harter K, *et al.* Lack of population diversity in commonly used human embryonic stem-cell lines. N Engl J Med 2010; 362(2): 183-5.

[34] Taylor CJ, Bolton EM, Pocock S, Sharples LD, Pedersen RA, Bradley JA. Banking on human embryonic stem cells: estimating the number of donor cell lines needed for HLA matching. Lancet 2005; 366(9502): 2019-25.

[35] Amariglio N, Hirshberg A, Scheithauer BW, *et al.* Donor-derived brain tumor following neural stem cell transplantation in an ataxia telangiectasia patient. PLoS Med 2009; 6(2): e1000029.

[36] Mayshar Y, Ben-David U, Lavon N, *et al.* Identification and classification of chromosomal aberrations in human induced pluripotent stem cells. Cell Stem Cell 2010; 7(4): 521-31.

[37] Allegrucci C, Denning C, Priddle H, Young L. Stem-cell consequences of embryo epigenetic defects. Lancet 2004; 364(9429): 206-8.

[38] Meshorer E, Misteli T. Chromatin in pluripotent embryonic stem cells and differentiation. Nat Rev Mol Cell Biol 2006; 7(7): 540-6.

[39] Irizarry RA, Ladd-Acosta C, Carvalho B, *et al.* Comprehensive high-throughput arrays for relative methylation (CHARM). Genome Res 2008; 18(5): 780-90.

[40] Kim K, Doi A, Wen B, *et al.* Epigenetic memory in induced pluripotent stem cells. Nature 2010; 467(7313): 285-90.

[41] Doi A, Park IH, Wen B, *et al.* Differential methylation of tissue- and cancer-specific CpG island shores distinguishes human induced pluripotent stem cells, embryonic stem cells and fibroblasts. Nat Genet 2009; 41(12): 1350-3.

[42] Ghosh Z, Wilson KD, Wu Y, Hu S, Quertermous T, Wu JC. Persistent donor cell gene expression among human induced pluripotent stem cells contributes to differences with human embryonic stem cells. PLoS ONE 2010; 5(2): e8975.

[43] Guccione E, Martinato F, Finocchiaro G, *et al.* Myc-binding-site recognition in the human genome is determined by chromatin context. Nat Cell Biol 2006; 8(7): 764-70.

[44] Schuettengruber B, Chourrout D, Vervoort M, Leblanc B, Cavalli G. Genome Regulation by Polycomb and Trithorax Proteins Cell 2007; 128(4): 735-45.

[45] Santos-Rosa H, Schneider R, Bernstein BE, *et al.* Methylation of histone H3 K4 mediates association of the Isw1p ATPase with chromatin. Mol Cell 2003; 12(5): 1325-32.

[46] Cao R, Wang L, Wang H, *et al.* Role of histone H3 lysine 27 methylation in Polycomb-group silencing. Science 2002; 298(5595): 1039-43.

[47] Bernstein BE, Humphrey EL, Erlich RL, *et al.* Methylation of histone H3 Lys 4 in coding regions of active genes. Proc Natl Acad Sci U S A 2002; 99(13): 8695-700.

[48] Bernstein BE, Mikkelsen TS, Xie X, *et al.* A Bivalent Chromatin Structure Marks Key Developmental Genes in Embryonic Stem Cells Cell 2006; 125(2): 315-26.

[49] Ku M, Koche RP, Rheinbay E, *et al.* Genomewide analysis of PRC1 and PRC2 occupancy identifies two classes of bivalent domains. PLoS Genet 2008; 4(10): e1000242.

[50] Pan G, Tian S, Nie J, *et al.* Whole-genome analysis of histone H3 lysine 4 and lysine 27 methylation in human embryonic stem cells. Cell Stem Cell 2007; 1(3): 299-312.

[51] Buecker C, Chen HH, Polo JM, *et al.* A murine ESC-like state facilitates transgenesis and homologous recombination in human pluripotent stem cells. Cell Stem Cell 2010; 6(6): 535-46.

[52] Kim H, Jang MJ, Kang MJ, Han YM. Epigenetic signatures and temporal expression of lineage-specific genes in hESCs during differentiation to hepatocytes *in vitro*. Hum Mol Genet 2010; 20(3): 401-12.

[53] Golob JL, Paige SL, Muskheli V, Pabon L, Murry CE. Chromatin remodelling during mouse and human embryonic stem cell differentiation. Dev Dyn 2008; 237(5): 1389-98.

[54] Ng P, Wei CL, Sung WK, *et al.* Gene identification signature (GIS) analysis for transcriptome characterization and genome annotation. Nat Methods 2005; 2(2): 105-11.

[55] Zhao XD, Han X, Chew JL, *et al.* Whole-genome mapping of histone H3 Lys4 and 27 trimethylations reveals distinct genomic compartments in human embryonic stem cells. Cell Stem Cell 2007; 1(3): 286-98.

[56] Barski A, Cuddapah S, Cui K, *et al.* High-resolution profiling of histone methylations in the human genome. Cell 2007; 129(4): 823-37.

[57] Mikkelsen TS, Ku M, Jaffe DB, *et al.* Genome-wide maps of chromatin state in pluripotent and lineage-committed cells. Nature 2007; 448: 553-60.

[58] Chin MH, Mason MJ, Xie W, *et al.* Induced pluripotent stem cells and embryonic stem cells are distinguished by gene expression signatures. Cell Stem Cell 2009; 5(1): 111-23.

[59] Hawkins RD, Hon GC, Lee LK, *et al.* Distinct epigenomic landscapes of pluripotent and lineage-committed human cells. Cell Stem Cell 2010; 6(5): 479-91.

[60] Osafune K, Caron L, Borowiak M, *et al.* Marked differences in differentiation propensity among human embryonic stem cell lines. Nat Biot 2008; 26(3): 313-5.

[61] Adewumi O, Aflatoonian B, Ahrlund-Richter L, *et al.* Characterization of human embryonic stem cell lines by the International Stem Cell Initiative. Nat Biot 2007; 25(7): 803-16.

[62] Assou S, Le Carrour T, Tondeur S, *et al.* A meta-analysis of human embryonic stem cells transcriptome integrated into a web-based expression atlas. Stem Cells 2007; 25(4): 961-73.

[63] Müller FJ, Laurent LC, Kostka D, *et al.* Regulatory networks define phenotypic classes of human stem cell lines. Nature 2008; 455(7211): 401-5.

[64] Byrne JA, Mitalipov SM, Clepper L, Wolf DP. Transcriptional profiling of rhesus monkey embryonic stem cells. Biol Reprod 2006; 75(6): 908-15.

[65] Wei CL, Miura T, Robson P, *et al.* Transcriptome profiling of human and murine ESCs identifies divergent paths required to maintain the stem cell state. Stem Cells 2005; 23(2): 166-85.

[66] Boue S, Paramonov I, Barrero MJ, Izpisua Belmonte JC. Analysis of human and mouse eprogramming of somatic cells to induced pluripotent stem cells. What is in the plate? PLoS ONE 2010; 5(9): e12664.

[67] Cao F, Wagner RA, Wilson KD, *et al.* Transcriptional and functional profiling of human embryonic stem cell-derived cardiomyocytes. PLoS ONE 2008; 3(10): e3474.

[68] Bhattacharya B, Cai J, Luo Y, *et al.* Comparison of the gene expression profile of undifferentiated human embryonic stem cell lines and differentiating embryoid bodies. BMC Dev Biol 2005; 5: 22.

[69] Skottman H, Stromberg AM, Matilainen E, Inzunza J, Hovatta O, Lahesmaa R. Unique gene expression signature by human embryonic stem cells cultured under serum-free conditions correlates with their enhanced and prolonged growth in an undifferentiated stage. Stem Cells 2006; 24(1): 151-67.

[70] Rosler ES, Fisk GJ, Ares X, *et al.* Long-term culture of human embryonic stem cells in feeder-free conditions. Dev Dyn 2004; 229(2): 259-74.

[71] Wakayama S, Jakt ML, Suzuki M, *et al.* Equivalency of Nuclear Transfer-Derived Embryonic Stem Cells to Those Derived from Fertilized Mouse Blastocysts Stem Cells 2006; 24: 2023-33.

[72] Smith SL, Everts RE, Tian XC, *et al.* Global gene xpression profiles reveal significant nuclear reprogramming by the blastocyst stage after cloning PNAS 2005; 102(49): 17582-7

[73] Lowry WE, Richter L, Yachechko R, *et al.* Generation of human induced pluripotent stem cells from dermal fibroblasts. Proc Natl Acad Sci USA 2008; 105(8): 2883-8.

[74] Maherali N, Sridharan R, Xie W, *et al.* Directly reprogrammed fibroblasts show global epigenetic remodelling and widespread tissue contribution. Cell Stem Cell 2007; 1(1): 5 5-70.

[75] Yu J, Vodyanik MA, Smuga-Otto K, *et al.* Induced pluripotent stem cell lines derived from human somatic cells. Science 2007; 318(5858): 1917-20.

[76] Soldner F, Hockemeyer D, Beard C, *et al.* Parkinson's disease patient-derived induced pluripotent stem cells free of viral reprogramming factors. Cell 2009; 136(5): 964-77.

[77] Chin MH, Pellegrini M, Plath K, Lowry WE. Molecular analyses of human induced pluripotent stem cells and embryonic stem cells. Cell Stem Cell 2010; 7(2): 263-9.

[78] French AJ, Adams CA, Anderson LS, Kitchen JR, Hughes MR, Wood SH. Development of human cloned blastocysts following somatic cell nuclear transfer with adult fibroblasts. Stem Cells 2008; 26(2): 485-93.

[79] Josephson R, Sykes G, Liu Y, *et al.* A molecular scheme for improved characterization of human embryonic stem cell lines. BMC Biol 2006; 4: 28.

[80] Ware CB, Nelson AM, Blau CA. A comparison of NIH-approved human ESC lines. Stem Cells 2006; 24(12): 2677-84.

CHAPTER 6

Modelling Differentiation of Pluripotent Stem Cells to the Three Germ Layers

Michael B. Morris*

Bosch Institute, School of Medical Sciences, Kolling Institute of Medical Research, Centre for Developmental and Regenerative Medicine, University of Sydney, Australia

Abstract: Embryonic stem cells (ESCs) serve as a model for development and as a potential source of cells for the treatment of a variety of human conditions. ESCs can undergo differentiation to each of the 3 germ layers and, upon successive elaboration, can generate all of the cell types of the developing embryo and adult. Multiple interconnected layers of regulatory networks/circuits control ESC self renewal, pluripotency, and differentiation. This complex biological system can only be partially understood and manipulated to advantage using reductive experimental approaches. Instead, complex systems analysis tools, which inherently take into consideration the nonlinear, non-intuitive, and highly interconnected nature of the system must be applied. In recent years, a combination of the development of high-throughput screening methods for measuring gross system properties, array technologies for measuring global system changes at various molecular levels, and mathematical algorithms and computer software for data handling and modelling has brought complex systems analysis to the fore. This Chapter highlights recent examples of complex systems applications to ESCs undergoing the earliest stages of differentiation to the germ layers *via* intermediate pluripotent populations.

Keywords: Complex systems, systems biology, regulatory networks, molecular circuits, emergent properties, bioinformatics, lineage commitment.

INTRODUCTION

Pluripotent embryonic stem cells (ESCs) are of intense interest for two main reasons: Firstly, they serve as an *in vitro* model of development [1, 2] because of their capacity to differentiate to all cell types from the developing embryo to the adult. Thus, understanding the mechanisms that direct the differentiation of ESCs to specific cell types sheds light on the closely related cascade of events driving normal and abnormal development, and provides hypotheses for testing mechanisms of development *in vivo*. Secondly, they are a potential source of therapies for a range of incurable or poorly treatable human diseases and injuries [3-6]. Successful development of cell therapies will also depend on directing the differentiation of ESCs to specific cell types, which can then be tested in animal models and, ultimately, in clinical trials.

A large body of work has accumulated identifying the roles of, for example, individual genes, proteins such as morphogens, microRNAs and signalling pathways required both for the development of mouse embryos [7-11] and differentiating ESCs [12-17]. This largely reductionist approach, however, does not necessarily establish if the regulatory element under consideration is a key regulator or a much 'lesser' player in these processes, nor how the regulatory element cooperates with other regulators. Reductionist approaches also tend to focus on elements already thought of as being important and therefore can lead to a partial or complete blinding to the presence and role of other master regulatory elements and circuits.

Instead, it is better to think of ESC differentiation and development as representing complex systems which are highly dynamic temporally and spatially [18-21]: Mechanistic regulation is multi-layered and includes (but is not limited to) signalling activity, epigenetic changes to chromatin, RNA activity and turnover, post-

***Address correspondence to Michael B. Morris:** Embryonic Stem Cell Laboratory, Level 11 Kolling Building, Royal North Shore Hospital,St Leonards 2065, Australia; Tel: +61–2–9926 4877; E-mail: michaelmorris@med.usyd.edu.au

Kuldip S. Sidhu (Ed)

translational modification of proteins, and microRNA. Mechanistic regulation also relies on molecular networks (or circuits) that work within and across these layers. Furthermore, the major regulatory elements and their circuits frequently act in highly cooperative, nonlinear and non-intuitive ways and across timescales often as short as a few minutes to as long as a few days. Yet identifying the major regulators and their cooperative circuitry is the key to understanding these complex systems and controlling them to our advantage.

Three factors are changing the landscape of exploring, understanding and modelling complex biological systems such as ESCs at the molecular level. The first is the advent of array data for a range of cellular phenomena such as gene expression, kinome activity, microRNA expression, and epigenetic marks. These array data contain details of the molecular regulatory circuitry of the complex system. The second is the increasingly sophisticated nature of medium- to high-throughput screens suitable for use with ESCs – 'inputs' to the system are changed and 'emergent properties' of the system in response to those changes are quantified [22-26]. The third is developments in mathematics to provide highly effective tools for interrogating all this information and using the results to model the regulatory circuitry that controls the complex system [27-31].

This "systems biology" or "complex systems analysis" approach has been applied extensively to adult stem cells [32], non-mammalian development [20], and other biological systems [33, 34] (and even to detecting emerging research fronts in regenerative medicine [35]). It has resulted in detailed testable network models which illuminate key aspects of complex biological systems. Surprisingly, complex systems analysis has been little used to understand ESC differentiation and mammalian development but interest is growing in using this approach to provide insight into the intricate behaviour of these cells. Here we discuss recent examples of complex systems analysis applied to ESCs undergoing the early stages of differentiation to the germ layers.

'Traditional' Bioinformatics is Not Enough

Bioinformatics-Based Analysis of 'Stemness' in ESCs

In recent years, bioinformatics approaches have been used to interrogate large data arrays such as microarrays from ESCs. For example, several seminal studies were used to the compare the genome-wide gene expression status of ESCs compared to various other stem cell and differentiated cell types in order to identify 'stemness' genes [36-38]. Similar types of array studies have been performed on preimplantation embryos undergoing step-wise development from the unfertilised egg through to the blastocyst [39, 40].

These and related studies have applied relatively simple (heuristic) rules to data analysis and deriving information, resulting principally in generating lists of up- and down-regulated genes. For example, genes expressed in both ESCs and other stem-cell populations but whose expression was downregulated in differentiated cell types were classed as candidate stemness genes [36-38]. In some cases, additional categorisation of the lists was employed, including ontogenetic and chromosome clustering [36-38].

Limitations of Heuristic Rules

These heuristic approaches do have some value but, coupled with the faults in design of these experiments, the potential is limited. There are several reasons for this:

1. Aside from the ESCs themselves, some of the other stem cell populations selected were clearly impure (*e.g.*, adult neural stem cells [41]), and the non-stem cell populations, being terminally differentiated, were therefore ontogenetically greatly removed from pluripotent stem cells. It is perhaps not surprising that the lists of 'stemness' genes from the 3 studies were almost completely mutually exclusive, even though in two of the studies the same microarray and nearly the same cell types were used [42, 43]. This highlights the need for good experimental design when array data are to be collected regardless of whether the data are to be analysed using traditional bioinformatics approaches or by more sophisticated complex systems analyses. A comparison of array data collected from ESCs compared to homogeneous populations of the 3 germ layers may have proved more revealing even if only heuristic rules had been employed for the analysis.

2. Genes whose expression is maintained beyond the pluripotent stage may have been inappropriately excluded from the list of 'stemness' genes. That is, the fact that a gene is expressed in pluripotent ESCs and at similar levels in one or more partially or completely differentiated cell types should not, of itself, exclude it from consideration of a critical role in pluripotency. *Sox2* is a notable example: *Sox2* is expressed in ESCs and expression is maintained in many other cell types derived from them including the pluripotent primitive ectoderm, the germ layer definitive ectoderm, and multipotent neurectoderm [44]. Similarly, in the developing embryo, *Sox2* is widely expressed including pluripotent, neural and non-neural cells [45, 46]. Despite this widespread expression in non-pluripotent cell types *Sox2* does indeed play a crucial role in ESC self-renewal and pluripotency yet it was not found in the list of ESC 'stemness' genes in any of the three studies. Its formal exclusion in at least one of the studies occurred precisely because of this widespread expression [37]. This highlights one of the limitations of over-simplistic rules: Very typically in complex biological systems a single gene can have one or more functions in a particular cell type and the same or different function(s) in a different cell type. It is not expression or lack of expression *per se* that counts but the activity of the component in the context of integrated networks.

3. There is no discrimination in the lists – essentially the lists serve as a catalogue of candidate parts with no information on which parts are irrelevant, which are more important or less important, how they go together, how they work individually and collectively, and what they control. Instead, it would be useful to have hierarchy in which distinction is made between master regulatory genes, functional regulatory genes (which are subservient to the master genes), and 'bystander' genes (which are expressed but play no role in 'stemness'). It would also be advantageous to extract additional information 'hidden' in the array data including: The role(s) of the master and functional genes; the 'connectivities' between the genes (*i.e.*, the circuitry); the types of circuits involved (*e.g.*, feedback, feed-forward, amplifying, suppressing, *etc.*); the mechanisms of nonlinearity, such as cooperativity, within and between circuits; and the hierarchy of the circuits. This highlights the need for moving beyond traditional bioinformatics to computationally driven complex systems analysis.

In short, the array data collected for these studies (even assuming they had been collected using an appropriate experimental design) intrinsically contained information that could be used to understand the genetic networks that underpin pluripotency but the analyses were insufficiently sophisticated to reveal it. Without further (systems-based) analysis, this leads to an immediate and very restrictive experimental bottleneck because the most obvious way forward is the reductive approach of selecting genes on the list one at a time with a goal largely limited to confirming a role in 'stemness'. Furthermore, selecting candidates for further experiments is largely random, although it can be informed by other results in the literature, including those from embryo studies.

Such a piecemeal approach will only very slowly build up a picture of the molecular machinery, and despite the effort the picture will likely remain stubbornly and disappointingly incomplete. Indeed, it is notable that from the 3 studies only one gene, *Esg-1*, was targeted for further experimentation [38]. Its selection was based on known expression patterns in early embryos and ESCs. *Esg-1* was confirmed as being expressed in pluripotent cells (and in other multipotent embryonic cells) but only a small amount of additional information was gained about where it fitted into the regulatory network of pluripotency [38].

Bioinformatics Applied to Differentiating ESCs

Similar issues exist for studies in which ESCs are directed in their differentiation towards specific cell types. In a recent example, mouse ESCs were directed to neurectoderm and microarrays were performed on cells poised to commit to neurectoderm compared to cells which had actually differentiated to neurectoderm [47]. The analysis of the array data resulted in a list of 104 candidate genes for controlling this transition but the list contained only candidates whose expression changed between the two cell types; *i.e.*, it ignored genes which were expressed in both cell types but which might nevertheless have played a critical role in the transition. Ultimately, only one gene, *Zfp521*, was selected for further very detailed analysis. A model was proposed which placed *Zfp521* in a simple regulatory framework where expression may be controlled by *Oct6*. In turn (based in part on ChIP analysis of Flag-tagged Zfp521 protein), *Zfp521* appears to control the expression of a small number of early neurectodermal genes [47].

Zfp521 appears to act as a non-redundant regulator gene (perhaps even a master gene) controlling the transition between pluripotent cells and neurectoderm but what are the details of its relationship with the other candidate genes on the list? What are the other components of the circuit it is a part of? What does the circuit actually do, and where does it lie in the hierarchy of circuits that control commitment to neurectoderm? A systems biology approach applied to the microarray and ChIP data would likely have provided a more detailed model of the regulatory network controlling this step in differentiation, helping to establish hypotheses which better directed further experimentation. In other words, a more complete picture of this network might have emerged with the same amount of effort but using a different approach.

The examples above, and others like them, highlight both the value of collecting array data and the apparent difficulties in moving to a more comprehensive understanding of the mechanisms underlying complex phenomenon such as pluripotency, differentiation and development when the analysis of the array data is restricted to little more than the generation of lists. The situation is made even more complex when pluripotent cells are directed to differentiate to non-pluripotent cells because not only will key regulators generally have a number of roles in addition to pluripotency but their contribution to each of these roles can change in the various cell types that exist along the differentiation pathway. In extreme cases, some functions can be dropped completely and/or completely new functions added (*e.g.*, [17]).

What Can Complex Systems Analysis Do?

In the last few years, researchers in the ESC field have increasingly turned to the use of complex systems analysis tools to interrogate array data and generate detailed and experimentally testable models. As shown by the application of complex systems analysis to a wide variety of other biological systems [20, 27, 32-34, 48], several advantages emerge in addition to those that can be gained through more reductive approaches:

1. Master regulators in the form of specific genes, signalling pathway intermediates, epigenetic regulators, microRNAs and morphogens, *etc.*, can be identified.
2. Functional regulators, which are controlled by master regulators, can be identified.
3. Regulatory networks (or circuits) can be constructed containing these master and functional regulators and the 'connectivities' between them.
4. The dynamic properties of circuits can be measured: Quantitative changes to the activity of a circuit can be measured in response to quantitative changes to the concentration of one or more inputs to the circuit.
5. Circuits can be assigned to one or more identifiable and measurable emergent properties of the system. Emergent properties can include, for example, differentiation down a particular lineage, self-renewal, maintenance or loss of pluripotency, growth, proliferation, apoptosis, cell shape change, motility, *etc.*
6. Circuits can be grouped and arranged hierarchically with individual circuits having an identifiable intrinsic property such as amplification, negative feedback, suppression, cooperativity, *etc.* The grouped circuits might lie within a regulatory layer; *e.g.*, a collection of gene regulatory networks that control the cell cycle and proliferation. Alternatively, the collection of circuits might cross regulatory layers; *e.g.*, signalling circuits that control both genetic and epigenetic networks that regulate a key developmental step in an organism.
7. The models can be tested for various properties such as:

   Robustness – What is the size of change in an input (such as the concentration of a morphogen or set of morphogens) required to significantly affect an emergent property (such as rate of proliferation)?

   Noise – How much random fluctuation (such as stochastic changes in the expression of a master gene) can be tolerated before the emergent property of the system changes? In addition, can noise be used usefully by the system?

   Redundancy – To what extent, if at all, can a regulatory component (such as a transcription factor) be replaced (*e.g.*, by a closely related transcription factor) whilst maintaining the emergent properties of the system?

Multifunctionality – *E.g.*, does a signalling component and/or the signalling pathway in which it resides play a role in, say, lineage commitment and also play in other emergent properties of the system?

8. Model testing can be done *in silico* without, at least initially, recourse to wet-lab experiments: How do the molecular circuits respond when the value of an input or set of inputs is systematically changed, and what effect is this predicted to have on the emergent properties of the system? What happens when a circuit component is partly or completely inactivated or the whole circuit removed? The findings can be used to formulate experimentally testable hypotheses.

   Experimentally, one can ask similar questions: What happens to the activity of the molecular components and the system as a whole when inputs like morphogen concentrations are changed? What happens if a signalling pathway is specifically inhibited, if a component is mutated such that it loses some or all of its functions, or if a transcription factor gene is knocked down or out? The results can serve to confirm aspects of the model or be used to improve it.

9. Complex systems models are amenable to incorporation of experimental data derived from the literature, even if the data have been collected from a system very different to the one under consideration.

In short, complex systems analyses provide the possibility of modelling, understanding and manipulating the gamut of regulatory networks controlling a biological system and, in turn, the properties that emerge from it.

Regulatory Networks Controlling ESCs

Core Pluripotency Network and its Complexity

Recent work strongly highlights the complexity of various regulatory networks that lie at the heart of ESC biology [20, 49-55].

Progress has been made in identifying the core transcriptional network that mediates pluripotency using a combination of microarray data, chromatin immunoprecipitation arrays for protein–DNA interactions and epigenetic status, proteomics/protein interactome data, and knockdown experiments [50, 56-63]. At the core of the network lies (at least) 3 transcription factor genes – *Oct4*, *Nanog*, *Sox2* – which form a triple feedback system involving the genes themselves, their transcription factor proteins, paired interaction of the transcription factors, and transcription factor binding to the genes' promoters [59-61, 64, 65]. The promoters of several hundred other genes have binding sites for one or more of these transcription factors [50, 58, 60-63]. In other words, *Oct4*, *Nanog*, and *Sox2* appear to act as mutually regulating master genes at the core of circuitry controlling a large variety of functional genes which collectively maintain pluripotency. The (downstream) functional genes include other transcription factor genes (such as *Rest*, *Foxd3*, *Klf4*, and *Tbx3* [56, 60]) but also genes whose functions cross over into other layers of regulatory control. For example, the polycomb group (PcG) is controlled by this core circuit and through modification of histone marks it serves to control the state of pluripotency, repress differentiation, and predispose some genes to be expressed once pluripotency is lost [56, 57].

This mutual feedback core circuitry involving just a very small number of genes, as seen here with *Oct4*, *Nanog*, and *Sox2*, is a common feature of a very wide variety of developmental processes [20].

Complex Systems Analysis of a Core Pluripotency Sub-Circuit

Mathematical modelling has shown that pluripotency behaves like a noise-driven excitable system centred on a minimal circuit containing *Oct4* and *Nanog* [66]. In this model, *Oct4* and *Nanog* regulate their own expression, and each regulates the other's expression in a feed-forward manner. In addition, *Oct4* has a negative feedback effect on *Nanog* if *Oct4* expression gets too high. (Each of these conditions is consistent with experimental findings in the literature.) These conditions generate a relatively stable attractor state, in which most pluripotent cells reside at any given time. In this state, levels of Nanog and Oct4 protein are high (but not too high) and fluctuate little. However, the system is noisy and excitable: *I.e.*, the system responds to a small spike in noise above a threshold value in a highly nonlinear way; in this case, the response takes the form of a large decrease in the expression of *Nanog*. Cells that spontaneously downregulate expression of *Nanog* in this way 'wander away' from the attractor state. They tend to have both lower Nanog and Oct4 protein levels and the levels vary much more widely both between cells and within single cells over time. In this state, these cells (5–20% of the pluripotent population depending on the culture conditions) are poised or predisposed to differentiate in response to the correct signals. However, if they do not receive these they tend to return to the attractor state over the course of a few days. This model fits experimental findings well as shown by results obtained with ESCs and pluripotent primitive ectoderm-like P19 cells [66].

Thus, the pluripotency regulatory network is noisy and appears deliberately to be so [66]: *Nanog*- and *Oct4*-dependent expression noise both permits a percentage of pluripotent cells to become predisposed to differentiation should the correct signals arrive and interferes with the differentiation program to prevent that differentiation and allow a return to the more stable pluripotent attractor state should those signals not arrive [66].

Networks of this kind are associated with transcriptional noise in bacteria [67, 68] and may underpin the probability of differentiation of Sca1-expressing haematopoietic progenitors [69].

Differentiation-Mediated Changes in ESC Networks

In a study by Lu *et al.*, one of the master regulators in this core transcriptional network, *Nanog*, was knocked down in ESCs resulting in a generalised loss of pluripotency and differentiation which was followed for 5 days [49]. This study examined changes in 4 regulatory 'layers': gene expression (using microarray), epigenetic modification (reflected by histone H3 acetylation), active transcription (based on chromatin-bound RNA polymerase II), and post-translational modification (nuclear localisation of proteins measured with mass spectrometry). One of the strengths of this work is the use of high-throughput system-wide data gathering for each of these regulatory layers, and the results not only highlight the presence of cellular complexity within and between layers but also the need to analyse the data in ways which comes to grips with this complexity. To facilitate analysis and interpretation, a software package, *GATE*, was developed [49, 70]. *GATE* has a clustering tool that highlights which genes undergo similar time-dependent changes in regulation across the 4 regulatory layers. Scatter plot movies can also be made in which time-dependent changes in a gene can be compared between pairs of regulatory layers. For example, in the first 3 days of the differentiation protocol, the expression of *Oct4* and *Sall1* increases but the nuclear localisation of the proteins does not and instead remains the same or decreases.

Currently, this work lies well short of its potential to generate detailed molecular networks that lie within and between regulatory layers controlling pluripotency and commitment (the loss of pluripotency and the onset of differentiation). Nevertheless, *GATE* provides valuable *in silico* clustering tools which can be used to infer common regulatory circuits, and these can be tested experimentally. The study has also been used to provide testable hypotheses of the hierarchy of regulatory change during ESC differentiation; for example, that changes in the expression of key transcription factors largely precede chromatin remodelling. This work also adds detail, together with data from the literature, to our understanding of the core regulatory pluripotency network and changes to it during differentiation.

In analogous work, the cell signalling networks of ESCs underpinning self-renewal, growth, and differentiation were investigated in response to all 16 possible combinations of 4 extracellular cues (or inputs) over a 5 day period [31, 52, 71]. The inputs comprised two growth factors, LIF and FGF4, and two extracellular matrix proteins, laminin and fibronectin, each at a single fixed concentration. A kinome analysis quantifying the phosphorylation status of ~30 signalling intermediates was carried out on the undifferentiated ESCs and at 3 timepoints over the following 5 days for each of the 16 combinations – a total of ~1500 values. Thus, in this work, the effects of combinations of inputs (growth factors and matrix proteins) to the complex system was quantified in terms of changes in molecular events (signalling activity) and changes in the emergent properties of the system (self-renewal, growth and differentiation).

Unlike the *Nanog* knockdown study (see above) where analysis is currently largely stalled at inferring connections through clustering analysis, this work used complex system analysis tools to generate a detailed model of signalling circuitry. The most successful approach was to use a Bayesian network learning algorithm, which generates a model *via* an iterative process with the highest probability of fitting the available data [31]. (This is essentially a reverse engineering approach.) The model depicts causal relationships (represented by connectivities) between the parameters – in this case between the 4 extracellular cues, the activity of the 31 signalling proteins, and 8 emergent properties centred around self renewal, growth rates and differentiation.

As would be expected for a complex biological system, the signalling model shows both expected and unexpected connectivities (crosstalk) between signalling intermediates in response to the system inputs. Expected and unexpected effects of the inputs and changes in signalling pathway activities on the emergent properties of the system are also observed. The model also captures linear, nonlinear and multistate relationships between signalling intermediates and identifies signalling intermediates most critically associated with each of the emergent properties. Many of these associations are not intuitively obvious and are only uncovered because of the sophistication of the analysis. Thus, collectively, the results highlight both intuitive and non-intuitive relationships between extracellular cues, intracellular signalling, and emergent cellular responses. Furthermore, the model provides quantitative predictions which can be tested experimentally, the results of which can be used to confirm or refine aspects of the model. This Bayesian network approach is also amenable to the inclusion of new data such as the activities of additional signalling intermediates.

Problems Associated with Random Differentiation Experiments

Despite this progress, some important issues arise from these *Nanog* knockdown and cell-signalling studies [31, 49, 52, 71]. Firstly, differentiation of the ESCs was not directed but was allowed to occur in a random fashion. Thus, over the course of the experiment (5 days for both studies), a variety of cell types, both embryonic and extraembryonic, will have formed. This random differentiation is likely to compromise the interpretation of the results and the veracity of the models. Secondly, it was not determined at what timepoint(s) pluripotency was lost and differentiation began and in any case, given the uncontrolled nature of the differentiation, individual cells will likely have lost pluripotency at different rates. In particular, those ESCs committing to embryonic lineages presumably first passed through a second pluripotent population, the primitive ectoderm [44, 72]. Thus, many of the observed changes (or at least some fraction of the observed changes) in regulatory components and circuitry at the earlier timepoints of these experiments may reflect the transition from one pluripotent state to another and not a commitment to lineage differentiation. This situation is complicated even further given that pluripotency may in fact be a continuum containing a number of metastable states, with the population representing each state being less than (sometimes substantially less than) homogeneous [15, 66, 73, 74]. This is particularly problematic for human ESCs [73].

Combining Array Data to Model ESC Pluripotency, Commitment, and Differentiation

Several studies highlight the value of combining array data derived from different layers of regulatory control. Together with the use of rapid screening tests to quantify the emergent properties of the system, this approach appears to provide greater certainty for model building [49, 51, 56, 57, 75]. Some examples of this approach are discussed below.

Genetic and Epigenetic Regulation of Commitment to Differentiation

Novel regulatory networks controlling ESC fate were derived by allowing mouse ESCs to differentiate spontaneously (through removal of LIF) and also to differentiate down the neural pathway (removal of LIF + addition of retinoic acid) [51].

Over the course of 5 days, cells were collected and microarrays were performed to establish temporal changes in expression of genes – with genes downregulated over the timecourse being candidates for involvement in pluripotency and genes upregulated being candidates for involvement in differentiation. These timecourse gene-expression data were integrated with previously published ChIP-chip studies [60, 76]. The ChIP-chip data provided lists of genomic binding locations for the transcription factors expressed from the 3 core pluripotency genes (*Oct4*, *Nanog*, and *Sox2*) and two PcG proteins expressed from *Eed* and *Phc1*. Additional microarrays were performed to obtain genome-wide changes in expression following partial knockdown of *Oct4* and *Sox2*.

Together, these array data resulted in a highly enriched set of 110 genes which were (i) bound by combinations of Oct4, Nanog and Sox2 protein, (ii) regulated up or down in expression over the 5-day timecourse, and (iii) similarly regulated following knockdown of *Oct4* or *Sox2*. Furthermore, 43 regulatory networks, involving the PcG proteins Eed and Phc1 cooperating with Oct4 and Nanog, were predicted to repress key developmental genes responsible for differentiation to all 3 germ layers and extraembryonic trophectoderm.

The predicted role of some of the genes in the enriched set was assessed further using gene expression knockdown combined with simple tests for pluripotency using alkaline phosphatase staining and/or a fluorescence-based high-content imaging assay.

Several advantages accrue from using this combination of techniques:

1. The integration of timecourse gene-expression data (microarray) and transcription regulator binding data (ChIP-chip) provides two independent methods for predicting which functional genes are regulated by master genes. Neither method alone provides this level of certainty.
2. Knockdown of master genes combined with additional microarray analysis can serve to help confirm the relationships predicted in 1.
3. Confirmation that functional genes predicted to play a role in maintaining pluripotency or promoting differentiation can be supported by knockdown of these genes combined with medium to high-throughput pluripotency/differentiation assays.
4. Differentiation induced by LIF removal - retinoic acid can be used as (an admittedly crude) way to help to establish which genes are involved in a generalised loss of pluripotency and which are required for lineage-specific differentiation.

These regulatory models that arise from this work serve to confirm that the molecular circuitry in ESCs robustly retains self-renewal and pluripotency and at the same time keeps the system poised to respond rapidly to appropriate differentiation signals.

Nevertheless, some problems appear to remain: LIF removal results in a slow loss of pluripotency during which many cells presumably transit through pluripotent primitive ectoderm before undirected differentiation to multipotent germ layers and beyond. Indeed, overt differentiation did not appear to occur until after day 3 but the study considered that 'commitment to differentiation' began as soon as *Oct4*, *Sox2* and *Nanog* expression began to drop even marginally (at day 1). However, other studies have shown that primitive ectoderm derived from ESCs maintains high levels of expression of these 3 factors but generally at significantly reduced levels compared to ESCs [44, 77, 78]. Similarly, epiSCs (derived from primitive ectoderm of the embryo) can express *Sox2* and *Nanog* at levels significantly lower than ESCs [78, 79]. Furthermore, even complete loss of *Nanog* expression does not automatically mean loss of pluripotency and differentiation [15], and in cells where *Nanog* expression is reduced *Oct4* expression also tends to be reduced and the expression of both can vary widely over time and from cell to cell [66]. Despite such reduced and variable expression of core pluripotency factors, both ESC-derived primitive ectoderm and epiSCs remain pluripotent, are self-renewing under appropriate culture conditions, and can be reverted to ESCs [73, 77, 78, 80]. This highlights the metastable nature of pluripotent states.

Thus, many of the observed changes in gene expression (both core pluripotency genes and other genes) during early timepoints following LIF withdrawal may largely reflect a readjustment of the pluripotency regulatory network to accommodate the formation of a second metastable pluripotent state, primitive ectoderm. If this is the case, 'commitment' in this and other LIF-withdrawal studies may be better defined as 'maintaining pluripotency but losing ESC self-renewal' and not a commitment to lineage differentiation.

This study also limits the selection of candidate genes involved in pluripotency to those whose expression is downregulated upon commitment, and limits candidate genes involved in differentiation to those whose expression is upregulated upon commitment. Perhaps this restriction usefully limits the number of candidates that need to be considered but again it ignores a salient and very common feature of complex biological systems: That components can play multiple roles in any given situation and can continue to have multiple but changing roles as the system changes.

Shedding light on loss of pluripotency and lineage commitment may ultimately be better served by starting with a homogeneous primitive ectoderm population, which presumably lies much closer to, if not right at, the point of real lineage commitment. Separate experiments can be performed in which differentiation is directed down each of the three germ layers through appropriate modification of the culture conditions. In this way, it may be possible to produce better models of the circuitry involved in the commitment to differentiate, and to subdivide commitment circuitry into that required for 'generalised' commitment (*i.e.*, where any of the three germ layers are formed) and 'specific' commitment (where only one or two of the germ layers are formed).

In the mouse system, several possibilities exist for the primitive ectoderm starting population including epiSCs [78, 79], ESCs first converted to a homogeneous population of primitive ectoderm [72, 77, 81], and embryonal carcinoma cell lines such as P19 [82]. Unfortunately, for human ESCs the situation is not as clear cut both because of the heterogeneity they display using current culturing conditions and because it is not clear where along the continuum of pluripotency they actually lie [73].

Mutually Embracing Circuits are a Feature of Complex Systems

Other modelling data also show that the regulatory networks controlling pluripotency are finely balanced, and in previously unexpected ways. The PcG proteins were originally thought to act in cooperation with the core gene regulatory network of *Oct4*, *Sox2* and *Nanog* to silence developmental genes in pluripotent cells [76]. However, knockdown of the PcG gene *Pcl2*, whose protein product is associated with the Polycomb Repressive Complex 2 (PRC2), makes the pluripotent state even more robust further inhibiting differentiation in response to changes in external cues [56, 57].

Analysis of combinations of array data covering different regulatory layers (in this case gene-expression, promoter occupancy, and histone modifications) together with knockdown studies and high throughput screens measuring the changing properties of the system has provided insight into this apparent contradiction: In pluripotent cells, the Pcl2-PRC2 complex does indeed appear to suppress differentiation networks through deposition of the repressive histone mark 3meH3K27. Simultaneously, it deposits the same repressive mark on genes in the extended pluripotency network, specifically at the promoters of *Klf4*, *Foxd3* and *Tbx3*.

The extended pluripotency network and the core network of *Oct4*, *Sox2* and *Nanog* are in a mutual embrace: The core network promotes the expression of genes in the extended network (despite the presence of the repressive mark) and the products of the extended network, in turn, feed forward to maintain the core network. The core network and the Plc2-PRC2 network are also in a mutual embrace: The core network upregulates the expression of genes in the PcG complex and the complex in turn acts as a rheostat on the extended pluripotency network so that it dampens (but by no means completely represses) the activity of the core network.

Together this combination of circuits in pluripotent cells, involving transcription factor genes, chromatin remodelling genes, and their protein products, acts to suppress differentiation networks and keep the core pluripotency network dominant but nevertheless delicately poised to respond to the correct signals to initiate differentiation.

When *Pcl2* expression is knocked down, ESCs become highly resistant to differentiation. This is because the repressive histone mark is removed only from selected genes, specifically including those of the extended pluripotency network. Because of this, expression of genes in this extended network is increased. This feeds forward to the core network resulting in increased expression of *Oct4*, *Sox2* and *Nanog*. In contrast, the repressive mark remains on developmental genes and the differentiation networks remain repressed.

Thus, knockdown of *Pcl2* resets the system to favour pluripotency even more – a non-intuitive result that would likely have remained unexplained without consideration of pluripotency as a complex biological system controlled by nonlinear interactions between a wide variety of circuits.

Again, it is not clear if, how, or by how much the activities of these circuits change as ESCs first transit to primitive ectoderm and how they might be adjusted to permit metastabilisation and self-renewal of this second pluripotent state. Nor is it clear what further (perhaps more dramatic) changes would then be required to occur to permit overt differentiation. The authors describe differentiation occurring as a result of network "collapse". However, this may only apply to the latter stages where overt differentiation finally does takes places, whereas transit through earlier stages consisting of one or more pluripotent states may depend on equally important but more subtle network changes.

Metabolic Circuits Controlling ESC Fate

Differentiation of stem cells may also depend on metabolic activity and access to nutrients. For example, the haematopoietic stem-cell niche *in vivo* appears to be nutrient poor, helping to sustain self-renewal. Outside the niche, access to nutrients increases, metabolism and cellular activity increases, and this is proposed to stimulate differentiation [83].

Analogously, ESC self-renewal *versus* differentiation may depend on metabolic status, with self-renewal being a state of relative metabolic inactivity compared to ESCs undergoing differentiation. To investigate this, translation state array analysis (TSAA) was used [75]. This combines density gradient centrifugation, to separate polyribosome complexes from ribosome-free transcripts, with microarray analysis. TSAA allows ribosome loading and transcript abundance to be measured on a genome-wide scale and with it the impact of translation on gene expression.

The results showed that ESCs are metabolically surprisingly quiet, despite very rapid cell cycling, compared to ESCs undergoing differentiation as a result of LIF withdrawal. Both cell populations contained nearly the same amount of ribosomal RNA but the differentiated cells had greatly increased mRNA transcript levels and polysome content and, as a result, exhibited a generalised increase in translation efficiency compared to ESCs. By contrast, ESCs, though they contained nearly the same amount of ribosomal RNA machinery, contained fewer transcripts which were also generally more inefficiently loaded onto ribosomes.

Using microarray analysis, the transcripts of several thousand genes were found to be elevated in differentiated cells compared to ESCs and the TSAA results showed that the majority of these were also translated more efficiently. Other subsets were also identified, including a set of 104 genes whose transcript levels remained the same in both populations but whose translation efficiency was significantly increased upon differentiation, indicating altered control of protein expression exclusively at the level of translation.

On the basis of these and other data, a model was proposed to account for this switch to metabolic upregulation upon differentiation. The model contains both mTOR-independent and mTOR-dependent arms. mTOR is proposed globally to upregulate Cap-dependent translation *via* phosphorylation of 4EBP1 as well as upregulate the translation of specific mRNAs *via* selected mediators such as DAZL and GRSF1 [75].

These results are consistent with the emerging paradigm that ESCs are tuned to respond to differentiation signals. In this case, the ESCs appear to be poised at the level of translation and metabolism: They contain ribosomal machinery but tend to use it inefficiently but presumably can rapidly access the excess capacity to meet the metabolic demands of differentiation.

It is not clear from these results whether globally upregulated metabolism/protein translation is solely the province of differentiated ESCs or if it also occurs, at least to some extent, in pluripotent states such as primitive ectoderm along the way to differentiation. The pathways and factors proposed to play important roles in the model are upregulated very early following LIF removal (within 12–48 h) [75], at a time when primitive ectoderm is likely to be abundant. Furthermore, primitive ectoderm produced *in vitro* proliferates much more rapidly than ESCs [77, 81, 84], presumably demanding increased metabolic activity. Similarly, *in vivo*, the hatching blastocyst implants and the cells of the inner cell mass convert to primitive ectoderm accompanied by a massive proliferative burst [85].

Recently, the metabolite L-proline was shown to stimulate the differentiation of mouse ESCs to primitive ectoderm [77, 81]. This mechanism includes upregulation of the mTOR pathway [81] and is dependent, at least in part, on the metabolic/redox status of the cell [77]. A host of other L-proline stimulated signalling pathways are also involved (unpublished data). These data suggest that increased metabolic activity during ESC differentiation begins as early as the transition to pluripotent primitive ectoderm and is maintained and/or increased upon overt differentiation.

CONCLUSIONS AND PROSPECTS

This relatively small body of work emphasises that ESCs and their differentiation, like other biological phenomena, do indeed represent a complex system possessing commonly observed features including molecular hierarchy, circuitry, and nonlinearity, and in which individual components and circuits can take on both expected and unexpected functions underpinning the emergent properties of the system. Furthermore, it emphasises that understanding this complexity is the key to controlling the properties of the system and effectively manipulating it to one's advantage.

In particular, the ESC system is now seen to be both robust and poised, resistant to noise but also dependent on noise, and controlled by a myriad of external cues and conditions that in turn regulate intricate interconnected circuitry within and across various regulatory layers. In this sense it seems remarkable that seemingly very naïve protocols are able to maintain the pluripotent state and, in some cases, direct differentiation with reasonable efficiency to specific cell types. Substantial improvements are, however, needed particularly in terms of directing differentiation to specific cell types that are biotechnologically and clinically relevant, and also in developing stepwise differentiation protocols which generate a temporal series of embryologically relevant populations which can be interrogated to understand the complex system that is development. These improvements will come much more rapidly with a systems biology approach, much more slowly with a purely reductive one.

REFERENCES

[1] Niwa H. Mouse ES cell culture system as a model of development. Dev Growth Differ 2010; 52(3): 275-83.

[2] Murry CE, Keller G. Differentiation of embryonic stem cells to clinically relevant populations: Lessons from embryonic development. Cell 2008; 132(4): 661-80.

[3] Christou YA, Moore HD, Shaw PJ, Monk PN. Embryonic stem cells and prospects for their use in regenerative medicine approaches to motor neurone disease. Neuropathol Appl Neurobiol 2007; 3395): 485-98.

[4] Kim BG, Hwang DH, Lee SI, Kim EJ, Kim SU. Stem cell-based cell therapy for spinal cord injury. Cell Transplant 2007; 16(4): 355-64.

[5] Ronaghi M, Erceg S, Moreno-Manzano V, Stojkovic M. Challenges of Stem Cell Therapy for Spinal Cord Injury: Human Embryonic Stem Cells, Endogenous Neural Stem Cells, or Induced Pluripotent Stem Cells? Stem Cells 2010; 28(1): 93-99.

[6] Kabelitz D, Geissler EK, Soria B, Schroeder IS, Fandrich F, Chatenoud L. Toward cell-based therapy of type I diabetes. Trends Immunol 2008; 29(2): 68-74.

[7] Gangloff YG, Mueller M, Dann SG, *et al.* Disruption of the mouse mTOR gene leads to early postimplantation lethality and prohibits embryonic stem cell development. Mol Cell Biol 2004; 24(21): 9508-16.

[8] Malas S, Postlethwaite M, Ekonomou A, *et al.* SOX1-deficient mice suffer from epilepsy associated with abnormal ventral forebrain development and olfactory cortex hyperexcitability. Neuroscience 2003; 119(2): 421-32.

[9] Nichols J, Zevnik B, Anastassiadis K, *et al.* Formation of pluripotent stem cells in the mammalian embryo depends on the POU transcription factor Oct4. Cell 1998; 95(3): 379-91.

[10] Wassarman KM, Lewandoski M, Campbell K, *et al.* Specification of the anterior hindbrain and establishment of a normal mid/hindbrain organizer is dependent on Gbx2 gene function. Development 1997; 124(15): 2923-34.

[11] Ralston A, Rossant J. Cdx2 acts downstream of cell polarization to cell-autonomously promote trophectoderm fate in the early mouse embryo. Dev Biol 2008; 31392): 614-29.

[12] Morizane A, Doi D, Kikuchi T, Nishimura K, Takahashi J. Small-Molecule Inhibitors of Bone Morphogenic Protein and Activin/Nodal Signals Promote Highly Efficient Neural Induction From Human Pluripotent Stem Cells. J Neurosci Res 2011; 89(2): 117-26.

[13] Niwa H, Miyazaki J, Smith AG. Quantitative expression of Oct-3/4 defines differentiation, dedifferentiation or self-renewal of ES cells. Nat Genet 2000; 24(4): 372-76.

[14] Pevny LH, Sockanathan S, Placzek M, Lovell-Badge R. A role for SOX1 in neural determination. Development 1998; 125(10): 1967-78.

[15] Chambers I, Silva J, Colby D, *et al.* Nanog safeguards pluripotency and mediates germline development. Nature 2007; 450(7173): 1230-34.

[16] Snyder M, Huang XY, Zhang JJ. Stat3 is essential for neuronal differentiation through direct transcriptional regulation of the Sox6 gene. FEBS Lett 2011; 585(1): 148-52.

[17] Zhang KJ, Li LY, Huang CY, *et al.* Distinct functions of BMP4 during different stages of mouse ES cell neural commitment. Development 2010; 137(13): 2095-105.

[18] Kitano H. Looking beyond the details: a rise in system-oriented approaches in genetics and molecular biology. Curr Genet 2002; 4191): 1-10.

[19] Kitano H. Systems biology: A brief overview. Science 2002; 295(5560): 1662-64.

[20] Davidson EH. Emerging properties of animal gene regulatory networks. Nature 2010; 468(7326): 911-20.

[21] Alon U. Network motifs: theory and experimental approaches. Nat Rev Gen 2007; 8(6): 450-61.

[22] Flaim CJ, Teng D, Chien S, Bhatia SN. Combinatorial signalling microenvironments for studying stem cell fate. Stem Cells Dev 2008; 17(1): 29-39.

[23] Yamauchi F, Okada M, Kato K, Jakt LM, Iwata H. Array-based functional screening for genes that regulate vascular endothelial differentiation of Flk1-positive progenitors derived from embryonic stem cells. Biochim Biophys Acta-Gen Subj 2007; 1770(8): 1085-97.

[24] Rosenthal A, Macdonald A, Voldman J. Cell patterning chip for controlling the stem cell microenvironment. Biomaterials 2007; 28(21): 3208-16.

[25] Mei Y, Saha K, Bogatyrev SR, *et al.* Combinatorial development of biomaterials for clonal growth of human pluripotent stem cells. Nat Mat 2010; 9(9): 768-78.

[26] Fang YQ, Wong WQ, Yap YW, Orner BP. Stem cells and combinatorial science. Combinatorial Chem High Throughput Screening 2007; 10(8): 635-51.

[27] Eldar A, Elowitz MB. Functional roles for noise in genetic circuits. Nature 2010; 467(7312): 167-73.

[28] Figueiredo MAT. Adaptive sparseness for supervised learning. IEEE Trans Pattern Anal Machine Intel 2003; 25: 1150-59.

[29] Geva-Zatorsky N, Dekel E, Batchelor E, Lahav G, Alon U. Fourier analysis and systems identification of the p53 feedback loop. Proc Natl Acad Sci USA 2010; 107(30): 13550-55.

[30] Lestas I, Vinnicombe G, Paulsson J. Fundamental limits on the suppression of molecular fluctuations. Nature 2010; 467(7312): 174-78.

[31] Woolf PJ, Prudhomme W, Daheron L, Daley GQ, Lauffenburger DA. Bayesian analysis of signalling networks governing embryonic stem cell fate decisions. Bioinformatics 2005; 21(6): 741-53.

[32] Kirouac DC, Zandstra PW. Understanding cellular networks to improve hematopoietic stem cell expansion cultures. Curr Opin Biotechnol 2006; 17(5): 538-47.

[33] Milo R, Shen-Orr S, Itzkovitz S, Kashtan N, Chklovskii D, Alon U. Network motifs: Simple building blocks of complex networks. Science 2002; 298(5594): 824-27.

[34] Lee TI, Rinaldi NJ, Robert F, *et al.* ranscriptional regulatory networks in Saccharomyces cerevisiae. Science 2002(5594); 298: 799-804.

[35] Shibata N, Kajikawa Y, Takeda Y, Sakata I, Matsushima K. Detecting emerging research fronts in regenerative medicine by the citation network analysis of scientific publications. Technol Forecast Soc Change 2011; 78: 274-82.

[36] Ivanova NB, Dimos JT, Schaniel C, Hackney JA, Moore KA, Lemischka IR. A stem cell molecular signature. Science 2002; 298(5593): 601-04.

[37] Ramalho-Santos M, Yoon S, Matsuzaki Y, Mulligan RC, Melton DA. "Stemness": Transcriptional profiling of embryonic and adult stem cells. Science 2002; 298(5593): 597-600.

[38] Tanaka TS, Kunath T, Kimber WL. *et a.l.* Gene expression profiling of embryo-derived stem cells reveals candidate genes associated with pluripotency and lineage specificity. Genome Res 2002; 12(12): 1921-28.

[39] Hamatani T, Carter MG, Sharov AA, Ko MSH. Dynamics of global gene expression changes during mouse preimplantation development. Dev Cell 2004; 6(1): 117-31.

[40] Wang QT, Piotrowska K, Ciemerych MA, *et al.* A genome-wide study of gene activity reveals developmental signalling pathways in the preimplantation mouse embryo. Dev Cell 2004; 6(1): 133-44.

[41] Ivanova NB, Dimos JT, Schaniel C, *et al.* Response to Comments on " 'Stemness': transcriptional profiling of embryonic and adult stem cells" and "A stem cell molecular signature". Science 2003; 302(5644): 393d.

[42] Evsikov AV, Solter D. Comment on " 'Stemness': Transcriptional profiling of embryonic and adult stem cells" and "A stem cell molecular signature" (II). Science 2003; 302(5644): 393c.

[43] Fortunel NO, Otu HH, Ng HH, *et al.* Comment on " 'Stemness': Transcriptional profiling of embryonic and adult stem cells" and "A stem cell molecular signature" (I). Science 2003; 302(5644): 393b.

[44] Harvey NT, Hughes JN, Lonic A, Yap C, Long C, Rathjen PD, Rathjen J. Response to BMP4 signalling during ES cell differentiation defines intermediates of the ectoderm lineage. J Cell Sci 2010; 123(10): 1796-804.

[45] Pevny LH, LovellBadge R. Sox genes find their feet. Curr Opin Genet Dev 1997; 7(3): 338-44.

[46] Wood HB, Episkopou V. Comparative expression of the mouse Sox1, Sox2 and Sox3 genes from pre-gastrulation to early somite stages. Mech Dev 1999; 86(1-2): 197-201.

[47] Kamiya D, Banno S, Sasai N, *et al.* Intrinsic transition of embryonic stem-cell differentiation into neural progenitors. Nature 2011; 470(7335): 503-U92.

[48] Kirouac DC, Ito C, Csaszar E, *et al.* Dynamic interaction networks in a hierarchically organized tissue. Mol Syst Biol 2010; 6: 417.

[49] Lu R, Markowetz F, Unwin RD, *et al.* Systems-level dynamic analyses of fate change in murine embryonic stem cells. Nature 2009; 462(7271): 358-326.

[50] Zhou Q, Chipperfield H, Melton DA, Wong WH. A gene regulatory network in mouse embryonic stem cells. Proc Natl Acad Sci USA 2007; 104(42): 16438-43.

[51] Walker E, Ohishi M, Davey RE, *et al.* Prediction and testing of novel transcriptional networks regulating embryonic stem cell self-renewal and commitment. Cell Stem Cell 2007; 1(1): 71-86.

[52] Prudhomme W, Daley GQ, Zandstra P, Lauffenburger DA. Multivariate proteomic analysis of murine embryonic stem cell self-renewal versus differentiation signalling. Proc Natl Acad Sci USA 2004; 101(9): 2900-05.

[53] Rada-Iglesias A, Bajpai R, Swigut T, Brugmann SA, Flynn RA, Wysocka J. A unique chromatin signature uncovers early developmental enhancers in humans. Nature 2011; 470(7333): 279-83.

[54] Sinkkonen L, Hugenschmidt T, Berninger P, *et al.* MicroRNAs control de novo DNA methylation through regulation of transcriptional repressors in mouse embryonic stem cells. Nat Struct Mol Biol 2008; 15(3): 259-67.

[55] Mikkelsen TS, Ku MC, Jaffe DB, *et al.* Genome-wide maps of chromatin state in pluripotent and lineage-committed cells. Nature 2007; 448(7153): 553-60.

[56] Walker E, Chang WY, Hunkapiller J, *et al.* Polycomb-like 2 Associates with PRC2 and Regulates Transcriptional Networks during Mouse Embryonic Stem Cell Self-Renewal and Differentiation. Cell Stem Cell 2010; 6(2): 153-66.

[57] Walker E, Manias JL, Chang WY, Stanford WL. PCL2 modulates gene regulatory networks controlling self-renewal and commitment in embryonic stem cells. Cell Cycle 2011; 10(1): 45-51.

[58] Kim JW, Chu JL, Shen XH, Wang JL, Orkin SH. An extended transcriptional network for pluripotency of embryonic stem cells. Cell 2008; 132(6): 1049-61.

[59] Wang JL, Rao S, Chu JL, *et al.* A protein interaction network for pluripotency of embryonic stem cells. Nature 2006; 444(7117): 364-68.

[60] Loh YH, Wu Q, Chew JL, *et al.* The Oct4 and Nanog transcription network regulates pluripotency in mouse embryonic stem cells. Nat Genet 2006; 38(4): 431-40.

[61] Ivanova N, Dobrin R, Lu R, *et al.* Dissecting self-renewal in stem cells with RNA interference. Nature 2006; 442(7102): 533-38.

[62] Chen X, Xu H, Yuan P, *et al.* Integration of external signalling pathways with the core transcriptional network in embryonic stem cells. Cell 2008; 133(6): 1106-17.

[63] Muller FJ, Laurent LC, Kostka D, *et al.* Regulatory networks define phenotypic classes of human stem cell lines. Nature 2008; 455(7211): 401-5.

[64] Rodda DJ, Chew JL, Lim LH, *et al.* Transcriptional regulation of Nanog by Oct4 and Sox2. J Biol Chem 2005; 280(26): 24731-37.

[65] Okumura-Nakanishi S, Saito M, Niwa H, Ishikawa F. Oct-3/4 and Sox2 regulate Oct-3/4 gene in embryonic stem cells. J Biol Chem 2005; 280(7): 5307-17.

[66] Kalmar T, Lim C, Hayward P, *et al.* Regulated fluctuations in nanog expression mediate cell fate decisions in embryonic stem cells. PLoS Biol 2009; 7: e1000149.

[67] Acar M, Mettetal JT, van Oudenaarden A. Stochastic switching as a survival strategy in fluctuating environments. Nat Genet 2008; 40(4): 471-75.

[68] Suel GM, Kulkarni RP, Dworkin J, Garcia-Ojalvo J, Elowitz MB. Tunability and noise dependence in differentiation dynamics. Science 2007; 315(5819): 1716-19.

[69] Chang HH, Hemberg M, Barahona M, Ingber DE, Huang S. Transcriptome-wide noise controls lineage choice in mammalian progenitor cells. Nature 2008; 453(7194): 544-47.

[70] MacArthur BD, Lachmann A, Lemischka IR, Ma'ayan A. GATE: software for the analysis and visualization of high-dimensional time series expression data. Bioinformatics 2010; 26(1): 143-44.

[71] Prudhomme WA, Duggar KH, Lauffenburger DA. Cell population dynamics model for deconvolution of murine embryonic stem cell self-renewal and differentiation responses to cytokines and extracellular matrix. Biotechnol Bioeng 2004; 88(3): 264-72.

[72] Pelton TA, Sharma S, Schulz TC, Rathjen J, Rathjen PD. Transient pluripotent cell populations during primitive ectoderm formation: correlation of *in vivo* and *in vitro* pluripotent cell development. J Cell Sci 2002; 115(2): 329-39.

[73] Pera MF, Tam PPL. Extrinsic regulation of pluripotent stem cells. Nature 2010; 465(7299): 713-20.

[74] Hayashi K, Lopes S, Tang F, Surani MA. Dynamic Equilibrium and Heterogeneity of Mouse Pluripotent Stem Cells with Distinct Functional and Epigenetic States. Cell Stem Cell 2008; 3(4): 391-401.

[75] Sampath P, Pritchard DK, Pabon L, *et al.* A hierarchical network controls protein translation during murine embryonic stem cell self-renewal and differentiation. Cell Stem Cell 2008; 2(5): 448-60.

[76] Boyer LA, Lee TI, Cole MF, *et al.* Core transcriptional regulatory circuitry in human embryonic stem cells. Cell 2005; 122(6): 947-56.

[77] Casalino L, Comes S, Lambazzi G, *et al.* Control of embryonic stem cell metastability by L-proline catabolism. J Mol Cell Biol 2011; 3(2): 108-22.

[78] Guo G, Yang J, Nichols J, *et al.* Klf4 reverts developmentally programmed restriction of ground state pluripotency. Development 2009; 136(7): 1063-69.

[79] Tesar PJ, Chenoweth JG, Brook FA, *et al.* New cell lines from mouse epiblast share defining features with human embryonic stem cells. Nature 2007; 448(7150): 196-98.

[80] Hanna J, Markoulaki S, Mitalipova M, *et al.* Metastable pluripotent states in NOD-mouse-derived ESCs. Cell Stem Cell 2009; 4(6): 513-24.

[81] Washington JM, Rathjen J, Felquer F, *et al.* L-Proline induces differentiation of ES cells: a novel role for an amino acid in the regulation of pluripotent cells in culture. Am J Physiol Cell Physiol 2010; 298(5): C979-981.

[82] Marikawa Y, Tamashiro DAA, Fujita TC, Alarcon VB. Dual roles of Oct4 in the maintenance of mouse P19 embryonal carcinoma cells: As Negative regulator of Wnt/beta-catenin signalling and competence provider for brachyury induction. Stem Cells Dev 2011; 20(4): 621-33.

[83] Cross M, Alt R, Niederwieser D. The case for a metabolic stem cell niche. Cells Tissues Organs 2008; 188(1-2): 150-59.

[84] Rathjen J, Lake JA, Bettess MD, Washington JM, Chapman G, Rathjen PD. Formation of a primitive ectoderm like cell population, EPL cells, from ES cells in response to biologically derived factors. J Cell Sci 1999; 112(5): 601-12.

[85] Snow MHL. Gastrulation in mouse - growth and regionalization of epiblast. J Embryology Exp Morph 1977; 42: 293-303.

CHAPTER 7

Stem Cells and Translational Medicine: Ethics, Law, and Policy

Loane Skene*

Melbourne Law School, University of Melbourne, Parkville, VIC, Australia

Abstract: This paper describes the law and the ethical and administrative guidelines that govern the development and use of human stem cells in Australia. It explains the history of the regulatory framework, particularly the process of periodic review of the relevant legislation in light of changes in the science and community views. It highlights a number of ethical issues that have been contentious and the regulatory responses to date.

Keywords: Stem cells, ethics, law, policy, translation.

INTRODUCTION

In Australia, the law relating to stem cell and translational medicine depends on the source of the cells used in the research, and whether the cells are to be placed in a human patient. If the cells come from human embryos, there is specific legislation (both federal and state) requiring the research to be undertaken only under licence, with ethical oversight and regular reporting to the licensing committee and to the federal parliament (primarily the *Prohibition of Human Cloning for Reproduction Act 2002 (Cth)* and the *Research involving Human Embryos Act 2002 (Cth),* as amended in 2006) [1]. There are many specific statutory offences. Research facilities are open to inspection and there are criminal penalties for non-compliance. This legislation was developed after extensive community consultation. It was required to be reviewed three years after it came into effect (the Lockhart Review); and it was reviewed again in 2011 (the Heerey Review). The second review was scheduled to take place by the end of 2010, or 'as soon as possible' after that date, see text below. In addition, there are ethical guidelines that apply to all types of medical research, in particular the *National Statement on Ethical Conduct for Human Research (National Statement)* [2] published by the National Health and Medical Research Council (NHMRC); and the *Australian Code for the Responsible Conduct of Research*, published jointly by the NHMRC, the Australian Research Council and Universities Australia [3].

The federal and state legislation does not apply if the stem cells are not derived from human embryos. If the cells are 'adult' stem cells, derived from human somatic cells, or they are human induced pluripotent stem (hiPS) cells, they are subject only to the regulatory constraints that apply to all research involving human bodily material – ethical guidelines (particularly the *National Statement*), good laboratory practice and the like. The same principles apply to human stem cell lines and products derived from them.

If stem cells or products derived from them are placed in human patients, either in clinical trials or treatment, ethical review must be undertaken, with ongoing monitoring and researchers must comply with legislation on quality control (the *Therapeutic Goods Act 1989)* [4]. Doctors, other health practitioners and their institutions are also bound by common law principles in the law of negligence and contract to take reasonable care in treating patients, who could sue for compensation if they suffer injury or loss as a result of the treatment. In cases of serious unprofessional conduct or harm to patients, disciplinary action against registered practitioners, and even criminal prosecution, may be initiated, but these are rare and will not be discussed.

***Address correspondence to Loane Skene:** Melbourne Law School, University of Melbourne, Parkville, VIC, Australia 3010; E-mail: l.skene@unimelb.edu.au

Other legislation and case law may also be relevant, such as the law on occupational health and safety, environment protection, the protection of animals used in research and teaching, quarantine, and patents. These are not discussed in this chapter. The law described is principally the law in Australia (though there is some reference to the law in other countries) and much of the description is in general terms as the aim of the chapter is to explain the regulatory structure in relation to the underlying ethics and policy.

THE POTENTIAL OF STEM CELL RESEARCH AND THERAPIES

Many people strongly support stem cell research and therapies, even if they involve the creation and destruction of early human embryos. They believe that human stem cell technology holds great promise for human health care. It can provide information about the operation and function of cells in the body, the development of early human embryos and possible abnormalities that may cause or contribute to birth defects. In future, the technology may be able to do much more. It may be used to test new drugs by creating a 'disease in a dish', extracting and multiplying cells from patients with particular diseases. One day, it may be possible to treat patients with conditions like heart disease, diabetes, spinal and brain injuries, and Alzheimer's disease by transplanting stem cells derived from their own bodies to stimulate the repair of diseased tissue; and tissue-matched body parts. Many animal experiments have shown the 'proof of concept' for stem cell therapies, including embryonic cells. In fact, stem cell transplants are not so different from human bone marrow transplants that have been used for more than 30 years, but, unlike those transplants of cells donated by other people, the cells obtained from the patient's own body would not be rejected by the person's immune system and require a life-time of immuno-suppressive drugs.

Research Involving Human Embryos

Research involving human embryos has been politically sensitive in Australia, as in many other countries, because of the concerns than many people have about the destruction of human embryos for the purpose of research [5]. These concerns are expressed in various ways. People may say that human embryos should not be deliberately destroyed for research, and even more, they should not be created with the intention that they will be destroyed, because embryos are human persons, or potential persons, and they are entitled to the same respect and legal protection as other people. Alternatively, people may accept that human embryos are not human persons but still believe that they have special moral significance and that they should not be treated 'instrumentally'; they should be respected and not used for another purpose. These arguments are made whether the embryos are created specifically for research, or whether they are 'excess' embryos that were formed initially for fertility treatment, but are no longer needed for that purpose. Excess embryos from fertility programs cannot legally be stored indefinitely. If they are not donated to another person for fertility treatment or for research, they must be removed from storage and allowed to 'succumb'. However, many opponents of embryo research believe that, even if these embryos must by law be allowed to succumb, that does not mean that they can then be ethically used in research. That would not accord proper respect to the embryos; or it would use them instrumentally; or possibly it might 'validate' their wrongful destruction, or create the impetus for the creation and destruction of more human embryos in the future.

Other opponents of human embryo research express concern about its potential implications for women. If embryo research is permitted, women undergoing fertility treatment may be under pressure to donate their excess embryos, and even more problematically, to donate their eggs, for use in research. Even if there are strict procedures to ensure that all donations are voluntary and fully informed, this may place an unfair burden on women at a sensitive time when they are undergoing fertility treatment. It may also impact unfairly on women who are especially vulnerable because of economic and other circumstances.

Concerns are expressed about the 'slippery slope' – if human embryo research is permitted, where will it lead? Could scientists create people who are genetically identical (reproductive cloning); or human-animal hybrids; or people whose genes are manipulated to achieve a particular outcome? Will eggs and sperm be sold? If stem cell treatments are developed, how do we know whether they will they be safe and not cause tumours and other adverse outcomes? What treatments have come from human embryo research? Has it aroused false hopes? Would the money that is being spent on embryo research be better spent on other treatments that have been shown to have more potential in human health care, such as 'adult' stem cell treatments, many of which are already being used?

In view of the divergence of views relating to human embryo research, many of which are deeply held, and the often heated debates, Australia, like the United Kingdom, passed detailed legislation. In 2002 after extensive community consultation, the federal parliament passed two Acts - the *Prohibition of Human Cloning Act 2002 (Cth)* (the *Prohibition Act*) and the *Research Involving Human Embryos Act 2002 (Cth)* (the *Research Act*) [6], both of which received Royal Assent on 19 December 2002. The NHMRC is responsible for administering the two Acts, through the Embryo Research Licensing Committee, one of its principal committees. When the legislation came into effect, the NHMRC instituted a regulatory framework under the legislation that includes licensing, and also appropriate monitoring and compliance arrangements. Some of these are described below.

The two Acts that were passed in 2002 were designed to enable scientists to do research on early human embryos and the stem cells derived from them, while trying to take account of the concerns of people who opposed the use of human embryos in research. Scientists would know that they could lawfully undertake particular activities as long as they obeyed 'the rules'.

The *Research Act* allowed certain research on human embryos and human genetic material to be conducted, provided that the researcher obtained a license from the NHMRC Embryo Research Licensing Committee and complied with strict reporting requirements, subject to conditions set out in the Act. The research also had to be approved and monitored by an institutional Human Research Ethics Committee (HREC). To obtain a licence, scientists had to justify the use of human embryos and use as few embryos as possible to achieve the aims of their research (while at the same time, using sufficient embryos for their results to be credible). Under the 2002 *Research Act* when it was first passed, the only human embryos that could be used in research were embryos that had been formed in fertility treatment programs but were no longer needed by the couples whose gametes (sperm and eggs) were used to create them and who wanted to donate them for research. Scientists were not permitted to create human embryos specifically for research.

The *Prohibition Act* prohibited human cloning and a range of other research activities involving human bodily material and provided substantial penalties for breach of its provisions. It totally banned activities such as human cloning to breed identical people; combining human and animal gametes (sperm and eggs) to breed half-human and half-animal hybrids; allowing a human research embryo to develop longer than 14 days; and buying and selling human eggs. All of these provisions are still in the current legislations (see the *Research Act* and the *Prohibition Act*, both amended by the *Prohibition of Human Cloning for Reproduction and the Regulation of Human Embryo Research Amendment Act 2002 (Cth) 2006*) [7].

'Mirror' legislation was passed in the states and territories to ensure that there was effective legislation throughout the country. There would then be no doubt about potential activities falling outside the jurisdiction of the federal parliament. This followed an agreement in 2002 between the members of the Council of Australian Governments (COAG, comprising the federal government and the governments of the states and the territories) to introduce nationally consistent legislation [8]. Under the agreement, the Commonwealth would pass the 2002 *Research Act* and the 2002 *Prohibition Act* and the states would pass legislation in similar terms. That occurred in all the states and in the ACT [9]. The Northern Territory has not yet passed any Act but it is apparently drafting legislation. Since then, however, the initial consistency in the Australian legislation has been undermined because the 2006 amendments to the federal legislation that implemented the Lockhart Committee's recommendations (see below) have not been adopted in all the states and territories. COAG representatives agreed in 2007 to endeavour to introduce corresponding legislation in all jurisdictions but the proposed legislation was rejected by the Western Australian parliament [10].

Thus, since the 2002 legislation came into effect, Australian scientists have been legally permitted to undertake research involving donated human embryos, provided they obtain a licence from the federal licensing committee, have their research approved and monitored by an HREC and report regularly to the licensing committee and to Parliament. The NHMRC has published guidelines to assist in this process [11], together with an Information Kit for applicants, licence holders and Human Research Ethics Committees [12]. It should be noted that, although these are guidelines and not laws passed by Parliament, they have statutory authority because the NHMRC is a statutory body established under its own Act, the *National Health and Medical Research Council Act 1992 (Cth)* [13]. Also, clinics undertaking fertility services must be registered by the Reproductive Technology Accreditation Committee (RTAC), which was established by the Fertility Society of Australia. This is a national body representing scientists, doctors, researchers, nurses, consumer groups, patients and counsellors in reproductive medicine in Australia and New Zealand [14]. Scientists undertaking research on human embryos must be accredited by the Fertility Society of Australia: *Research Act* ss 8 and 11 (offence to use a human embryo *in any way* without accreditation). Compliance with the NHMRC guidelines and RTAC's *Code of Practice* and accreditation requirements is a condition of a licence to undertake fertility services [15], and it is linked to federal funding [16]. Since the NHMRC and RTAC are federal authorities, their guidelines apply across Australia though they may be complemented by legislation in particular states and territories (or overridden if there are inconsistent provisions in such legislation).

In practice, a relatively small number of licenses has been granted and several of these have been for the purpose of improving fertility treatment procedures, rather than for human stem cell research. Information about the licenses that have been granted is available on a public database [17] and the Licensing Committee also reports to the Australian Parliament on the operation and outcomes of the licenses [18].

The First Review of the Legislation: The Lockhart Committee

The two 2002 Acts provided that they must be reviewed three years after they came into effect and, in 2005, in cooperation with the states and territories, the federal Minister responsible for the legislation appointed the Legislation Review Committee, whose categories of members and terms of reference were stated in the original Acts. This committee was known as the Lockhart Committee, named after its Chair, the late John Lockhart AO, and the present author was Deputy Chair. The Committee produced two reports, one for each Act [19], after wide-ranging community consultation throughout the country.

The Lockhart Committee made 54 recommendations. The principal recommendations were as follows:

- The legislative framework and prohibited practices in the 2002 legislation should remain in place, including the ban on reproductive cloning and creating an embryo for research by fertilising a human sperm and egg; and the prohibition on selling human eggs.
- Therapeutic cloning, also known as somatic cell nuclear transfer (SCNT), should be permitted for research, subject to the same requirements of licensing, reporting and ethical review as those for other types of human embryo research. SCNT is the 'Dolly technique' in which an embryo is formed from a somatic or body cell, not by a sperm fertilising an egg. This type of embryo has genetic material (DNA) almost entirely from the person whose body cell was used to form the embryo, with only a small amount of mitochondrial DNA coming from the egg that was used to 'incubate' the DNA from that person. Stem cells derived from that embryo would be 'matched' to the person so that if they were used for treatment in future, they would not rejected by the person's immuno-suppressive system, like material donated from another person.
- Research should be permitted on human embryos formed for fertility treatment that are not fit for implantation (formerly, these embryos were discarded, because only embryos that had been frozen for later implantation and then declared 'excess' were available for research, due to the waiting periods required in seeking consent for donation for embryos).
- Research should be allowed on an egg in the process of fertilisation, provided it stops at the point of syngamy when an embryo is formed. (This recommendation retained the ban on forming a human embryo for research by combining human sperm and eggs but would allow important research on an egg being fertilised which might help scientists to understand more about foetal abnormalities).

- Provisions regarding consent, administrative procedures and oversight should be reviewed, with relevant guidelines and procedures being amended as necessary.
- It should be lawful to create human-animal hybrid or chimeric embryos using animal eggs, for research, so that research on human embryonic stem cells can proceed despite the small number of human eggs donated for research. (These embryos formed using an enucleated animal egg would, like embryos formed by SCNT, contain almost entirely the DNA from the person whose DNA is used to create them.)
- It should be lawful to create embryos for research that contain DNA from more than two people (the people providing the sperm and egg). Adding DNA from a third person (an egg donor) might help avoid the transmission of mitochondrial disease to the next generation by replacing the cytoplasm of the woman with a family history of mitochondrial disease with cytoplasm from another woman's donated egg.
- Consideration should be given to a more flexible system of regulation, with the licensing committee being authorised to grant licences and make rulings within the tenor of the Acts and regulations, even if not expressly permitted by them, on condition that it reports immediately to the NHMRC and to Parliament.
- The legislation should be reviewed again three (or six) years after the amending legislation comes into effect.
- There should be ongoing community education and consultation.

Nearly all of the recommendations of the Lockhart Committee were implemented, either by the amending federal legislation in 2006 and later amending legislation in most states and territories [20] or by administrative changes made by the NHMRC and other bodies. The legislative structure and prohibited practices in the 2002 legislation remained mostly unaltered, including the ban on reproductive cloning and creating a 'sperm-egg embryo' for research, and the prohibition on selling human eggs. However, creating an embryo by SCNT (therapeutic cloning) was permitted for research, subject to licensing, reporting and ethical review. Research was also permitted on embryos formed for fertility treatment that were not fit for implantation, and on an egg in the process of fertilisation up to syngamy. The NHMRC updated the consent provisions in the National Statement, together with other guidelines concerning consent for egg and tissue donation and deciding when embryos are unsuitable for implantation.

On the other hand, there were some recommendations that the federal (and state or territory) Parliament did not accept, such as using animal eggs to 'incubate' human DNA to derive stem cells for research (which was legalised in the UK in 2008 after extensive debates in both Houses of Parliament); and creating human embryos for research that contain DNA from more than two people (which has recently been achieved in research in Newcastle, UK). The federal Parliament also rejected the Committee's proposal for a more flexible system of regulation.

As noted earlier, the amending federal legislation was required to be reviewed again as soon as possible after three years from when it became effective, which meant by the end of 2010 or as soon as possible after that date: see the current *Research Act* Part 5 [21]; and the current *Prohibition Act* Part 4. Both Acts had to be reviewed together by the same review committee, whose members had to be approved by the states as well as the federal Minister. The terms of reference for the second review, like those for the first review, were set out in the two Acts. Section 47A(4) of the current *Research Act*, for example, required the Review Committee to consider:

(a) Developments in assisted reproductive technology, including technological, medical and scientific developments, and the actual or potential clinical and therapeutic applications of such research;

(b) Developments in embryonic stem cell research, including technological, medical and scientific developments, and the actual or potential clinical and therapeutic applications of such research;

(c) Community standards;

(d) A brief analysis of international developments and legislation relating to the use of human embryos and related research;

(e) An analysis of research resulting from the licences granted;

(f) Any National Stem Cell Centre and any national register of donated excess ART embryos;

(g) An evaluation of the effectiveness of legislative provisions and NHMRC guidelines relating to proper consent;

(h) An evaluation of the range of matters for which the NHMRC Licensing Committee may issue a licence and any recommendations to increase, decrease or alter these arising from the evaluation;

(i) An analysis of any research or clinical practice which has been prevented as a result of legislative restrictions;

(j) The extent to which the NHMRC Licensing Committee has effectively used information and education tools to assist researchers working in the field, and any ongoing need for legally binding rulings;

(k) The extent of Commonwealth/State cooperation in the area of human embryo research and the requirement for further Commonwealth or State legislation on the matter.

The legislation required the Review Committee to report on 'recommendations about amendments that should be made to this Act' (s 47A(5)) and to consult (a) 'the Commonwealth and the States; and (b) a broad range of persons with expertise in or experience of relevant disciplines', setting out their views 'in the report to the extent that it is reasonably practicable to do so' (s 47A(6)).

The Second Review of the Legislation: The Heerey Committee

The members of the second legislation review committee were appointed in late 2010. They were the Hon Peter Heerey QC (Chair), a retired judge of the Federal Court of Australia; the present author; Professor Ian Frazer, a leading researcher in the field of immunology and cancer research at the University of Queensland and the Diamantin Institute; Dr Faye Thompson, an educator in nursing and midwifery; and Reverend Kevin McGovern, a Catholic priest and Director of the Caroline Chisholm Centre for Health Ethics [22]. The Committee's terms of reference (set out above) were similar to those of the Lockhart Committee.

The Heerey Committee's Report was tabled in the Federal Parliament on 7 July 2011 [23]. In broad terms, its recommendations were similar to those of the Lockhart Committee. As the Minister said when the Report was tabled, 'The 33 recommendations that were produced [by the Heerey Committee] relate largely to updating and clarifying the legislation, or keeping existing requirements' [24]. However, the Committee also considered important ethical issues.

The Need for Human Embryo Research to Continue

Perhaps the most contentious issue the Committee had to consider was whether human embryo research should be allowed to continue. Even people who initially accepted the need for human embryo research have questioned whether it can still be justified, especially the use of SCNT to create embryos for use in research, knowing that the embryos will then be destroyed. Induced pluripotent stem (iPS) cells have been obtained from human body cells without creating embryos to derive them [25] and these iPS cells seemed, at least at first, to have similar potential to embryonic stem cells for research and possible treatment. Also, direct 'reprogramming' of human body cells is now possible, enabling stem cells to be transplanted from one part of the body to another, or to be used to form new organs or tissue to replace those that have become diseased. To date, these techniques seem to offer the best hope of success in actual treatment. Also, the relatively few developments to date using human embryonic stem cells have all come from embryos donated from fertility treatment programs. No stem cells have been derived from embryos created by SCNT. Some people therefore say that, even if research is allowed on donated human embryos that would otherwise be discarded, research involving the deliberate creation of embryos for research cannot be justified.

However, as the Committee noted, 'Research on human ES [embryonic stem] cells has already produced substantial benefits' such as the derivation of 'four human ES cell lines from embryos identified as affected by Huntington's disease' (Report para 5.2). We do not yet know the potential of iPS cells. We do not know which type of stem cell research will ultimately be the most successful. At present, the most promising developments seem to be in the area of autologous cell transplants (using cells from the patient) and cellular reprogramming (again using the patient's own cells), but many scientists still see benefits in working with embryonic stem cells. Research on early embryos is necessary to understand how pluripotent stem cells develop and differentiate into other kinds of cells. Knowledge gained from embryo research has assisted scientists doing research on iPS cells. Research inevitably takes time but there has been some progress. In the US, an increasing number of new stem cell lines of human embryonic stem cells has been approved for federally funded research [26]. These cell lines are being used despite legal challenges [27]. Human trials involving cells derived from (donated) human embryonic stem cells have just started [28].

The Heerey Committee noted the reasons to continue human ES cell research including research involving SCNT (Report para 5.3). At the same time, the Committee said that 'An embryo must be the object of special consideration and respect, beyond that due to other human tissue' (Report para 5.2). Although it recommended (by majority) that 'Research involving embryos and ES cells should continue to be permitted subject to the statutory controls in the present legislation' (Recommendation 2), it also recommended caution in licensing embryo research involving SCNT embryos. The Committee specifically noted the lack of progress in SCNT research in animals and humans. It emphasised the statutory responsibility of the Licensing Committee to take into account, when it is considering any future application for a licence to undertake research involving SCNT, 'the likelihood of significant advance in knowledge or improvement in technologies for treatment as a result of the use of excess ART embryos or human eggs, or the creation or use of other embryos, proposed in the application, which could not reasonably be achieved by other means' (Report, Recommendation 3).

Creating Embryos with DNA from More than Two People

If an embryo is created using the sperm and an egg from prospective parents who are at risk of having a child with mitochondrial disease, and another egg donated by a second woman who has no such risk, it may be possible to avoid the transmission of mitochondrial disease to children. In Newcastle, UK, that procedure has been shown to work. There may therefore be an argument to ensure that this research is lawful in Australia. (It would seem, in fact, to be lawful at present, under licence, provided the 'three-parents' embryo is formed by SCNT and not by fertilisation of an egg and sperm, but that could be clarified. People may also say that this is another reason not to ban SCNT.) However, the Heerey Committee recommended that there should be no change in the current legislation to facilitate this research, given its early stage of developement (Report, Recommendation 7)

Payment for Donating Eggs

If human embryo research continues, large numbers of human eggs may be needed and there is a shortage of human eggs for use in research. This raises the issue whether monetary payments or other inducements should be permitted for donating human eggs for use in research. In the UK, Canada, and Australia, the tradition in medical research has been that all tissue used in research should be given gratuitously, including human eggs and embryos, and payments are not permitted beyond reasonable expenses, such as reimbursement of the donors' medical expenses and compensation for loss of earnings due to the donation. Similarly, European countries disapprove of commercialization, obtaining financial gains and other inducements from the donation of human reproductive materials. However, in the UK, the policy of the Human Fertilisation and Embryology Authority (HFEA) on 'egg sharing' modifies the general approach of altruism to some extent, as women who are prepared to donate some of their eggs for research may gain earlier access to fertility treatment programs. US law allows payment for human eggs in certain circumstances. There is no federal legislation governing the sale of human eggs and they may be sold for a 'fair price' for use in fertility programs, and in privately funded research. The meaning of 'a fair price' is open to interpretation [29] The state of New York has recently gone further by legislating to allow federally funded researchers to pay women for donating their eggs for research [30].

Many people are concerned that paying women to donate eggs would create a bad precedent for other types of human tissue donation. They say it is 'commodifying' human tissue. However, one might say that the women are being paid for going through the process of donating eggs – taking drugs to stimulate their ovaries and undergoing surgery to collect the eggs. If no eggs are collected, or the eggs are unsuitable to be used in the research, the women would still be paid for their participation.

However, the Heerey Committee recommended that payment should not be permitted beyond 'reasonable expenses' associated with the procedure of egg collection. (Report, Recommendation 6).

Human-Animal Hybrids

Even if the law permitted women to be paid for donating their eggs for research, it is unlikely that there would be a large number of eggs available for research. Egg donation is invasive and may have risks that we do not yet know. However, there may be an alternative. It may be possible to 'incubate' the nucleus of a human cell in an enucleated animal egg in order to produce embryonic stem cells that are almost entirely human. This is currently prohibited in Australia [31], Canada, and in many European and other countries. It is not currently banned in the US (the Human-Animal Hybrid Prohibition Bill was introduced in 2008 but has not been passed), but federal funding is not permitted for this research in the US. In the UK, on the other hand, it is lawful under licence. The first human-animal embryo was created in 2008 by scientists at Newcastle University under a licence from the Human Fertilisation and Embryology Authority and the validity of such a licence was confirmed in 2008 when the Human Fertilisation and Embryology Act 1990 was amended by Parliament. Since then, two more licences have been granted and the first human-animal embryo has been formed, though no stem cells have yet been derived from it. However, despite having licences to do this research, British scientists have not been able to get research grants to do it [32]. Perhaps one reason is that the funding bodies and their reviewers do not believe that the proposed research will be successful, or they consider that other research will be more productive.

The Heerey Committee considered issues arising in relation to the use of animal eggs to generate human ES cells by SCNT but recommended that this should not be permitted. One reason was that 'early studies have shown that using animal eggs brings its own set of molecular and genetic issues that need further research' (Report para 4.23)

RESEARCH ON HUMAN STEM CELLS THAT DO NOT INVOLVE HUMAN EMBRYOS

There are several types of human stem cell research and translational medicine that are not covered by the legislation that has been described above, because they do not involve assisted reproductive technology or the derivation or use of human embryos. These types of research include research on human stem cells derived from human somatic or body cells, human cell lines and human induced pluripotent stem cells (hiPS cells).

Stem Cells Derived from Human Somatic or Body Cells; hiPS Cells

The most familiar example of stem cells derived from human somatic cells being used in treatment is the use of donated human bone marrow to treat leukemia and other conditions. As noted earlier, this procedure is now outline and it has been undertaken for more than 30 years.

In recent research, scientists have been trying to 'reprogram' people's own body cells directly into other types of cells, which might be used to treat disease in that person without being rejected by the person's immunological system. This technology has been used to 'grow' new body parts by populating a plastic scaffold in the shape of that body part with stem cells derived from the patient's cells [33]. Patients who are treated with their own cells could avoid a life-time of immuno-suppressive drugs which often have adverse effects on the patient's body and quality of life.

Another recent development with stem cells derived from somatic cells is the derivation of hiPS cells which are also tissue-matched to the person. Research on animals suggests that it may be possible for these hiPS cells to be reprogrammed not only into other types of body cells but also into gametes (sperm and eggs). If that happened, the gametes could conceivably be fertilised, so that body cells could ultimately be used to form an embryo. At that point, the legislation on human cloning and stem cells would be relevant. However, at present, that type of research is not covered by the federal and state legislation on human cloning and embryo research. It could be done without obtaining a licence and periodic reporting to the licensing committee. However, there are other regulatory controls: ethical guidelines, review by an ethics committee, good laboratory practice and the like.

Human Cell Lines

In Australia, the use of human cell lines in research is comparatively unregulated. The NHMRC *National Statement* has various provisions stating when ethical review must be sought to use human cell lines in research, including those imported from other countries: see especially chapters 3.4 and 3.6, which are currently under review.

CLINICAL TRIALS AND TREATMENT FOR HUMAN PATIENTS

If stem cells or products derived from them are placed in human patients, either in clinical trials or treatment, ethical review must be undertaken, and researchers must comply with legislation on quality control (the *Therapeutic Goods Act 1989 (Cth)*). Doctors, other health professionals and their institutions are also bound by common law principles in the law of negligence and contract to take reasonable care in treating patients, who could sue for compensation if they suffer injury or loss as a result of treatment that is not performed with due care and skill. In an extreme case, a criminal prosecution is possible.

CONCLUSIONS AND PROSPECTS

It can be seen from the discussion in this chapter that Australian law rigorously controls human embryo research with a federal licensing system, mandated reporting and criminal penalties. Other types of human stem cell research are regulated relatively lightly and incidentally. Once potential treatments are used experimentally or in treatment procedures for human patients, that is governed by the same principles as any other clinical trial or treatment, however the stem cells are derived. That is entirely appropriate.

The recent review of the federal legislation on human cloning and stem cell research (the Heerey Review) has yet to be discussed in parliament and the wider community. However, the Committee recommended that the legislation should remain much the same, so that research would still be permitted on donated embryos and SCNT embryos, subject to licence and ethical oversight.

When the Heerey Committee's recommendations are more widely discussed, there may well be renewed calls to prohibit human embryo research, especially as there are now a range of other sources of human stem cells for use in research and possible future treatment. On the other hand, some scientists may want to extend even further the embryo research that is currently permitted, for example by creating 'fresh' embryos for research by fertilising a human egg with human sperm, which is currently prohibited in Australia and was not recommended by the Heerey Committee.

However, there is much to be said for minimal legislative change at present. Australian scientists do not seem to be clamouring to extend the research that they are currently undertaking. When I have informally asked groups of scientists at stem cell conferences, whether the current law is restricting their research activities, they have not been pressing for change. They have said that the law is generally operating well and they are not seeking amendments. There seems to be little embryonic or stem cell research that scientists want to do that the legislation prevents them from undertaking. Most seem to be happy to leave the current legislation much as it is and many members of the community are apprehensive about changes that have occurred in other countries, such as payment for human eggs and the creation of human-animal embryos.

REFERENCES

[1] Prohibition of Human Cloning for Reproduction Act 2002 (Cth). Australian Comm Law - - C2008C00694. PP 1-19; available at: http://www.comlaw.gov.au/Series/C2004A01081 (22 July 2011); Research Involving Human Embryos Act 2002 (Cth) Australian Comm Law- C2008C00689. PP 1-43available at http://www.comlaw.gov.au/Series/C2004A01082 (22 July 2011).

[2] National Statement on Ethical Conduct in Human Research. Australian Government, NHMRC, 2007, ARC Publication. E72; PP. 1-103 available at: http://www.anu.edu.au/ro/ORI/Human/NHMRC_Human_Ethics_National_Statement_2007.pdf (22 July 2011).

[3] Australian Code for the Responsible Conduct of Research. Australian Government, NHMRC Publication R 39. PP 1-12.1 available at: http://www.nhmrc.gov.au/_files_nhmrc/publications/attachments/r39.pdf (22 July 2011).

[4] Therapeutic Goods Act 1989. Australian Government Publication. Act No. 21. PP 1-267 available at: http://www.comlaw.gov.au/Series/C2004A03952 (22 July 2011).

[5] Skene L, Kerridge I, Marshall B, McCombe P, Schofield P. The Lockhart Committee: Developing Policy through Commitment to Moral Values, Community and Democratic Processes. J. Law Med 2008; 16: 132-138.

[6] See note 1 above.

[7] Prohibition of Human Cloning for Reproduction and the Regulation of Human Embryo Research Amendment. Act 2006. Australian Comm Law - No. 172. PP 2-29.

[8] Council of Australian Governments, Research involving Human Embryos and Prohibition of Human Cloning Inter-Governmental Agreement, 5 April 2002. Available at: http://www.coag.gov.au/intergov_agreements/docs/human_cloning.pdf (22 July 2011).

[9] NHMRC, Commonwealth and State Legislation available at: http://www.coag.gov.au/intergov_agreements/docs/human_cloning.pdf (22 July 2011).

[10] Nemes I. Therapeutic cloning in Australia: One small step from man, one giant leap for mankind. J Law Med 2008; 6: 139, 142-3.

[11] The Australian Health Ethics Committee of the NHMRC updated the earlier *Ethical Guidelines on the use of assisted reproductive technology in clinical practice and research* (ART guidelines 2001): available at http://www.nhmrc.gov.au/publications/synopses/e78syn.htm (22 July 2011).These guidelines include guidelines for egg donation. The NHMRC Licensing Committee also published objective criteria for deciding when an embryo is unsuitable for implantation in a woman (and so able to be used in research): *Contextual information for the Objective Criteria for determining embryos that are unsuitable for implantation issued by the National Health and Medical Research Council* (NHMRC): available at http://www.nhmrc.gov.au/_files_nhmrc/file/research/embryos/Contextual_information.pdf (22 July 2011).These Criteria were later prescribed in the *Research Involving Human Embryos Regulations 2003 (Cth)* Reg 2.2 : available at: http://www.comlaw.gov.au/ComLaw/Legislation/LegislativeInstrumentCompilation1.nsf/0/0B2FCC66A5BD3715CA25744E0006B3CD/$file/ResearchInvHumEmbryosReg2003.pdf (22 July 2011).

[12] NHMRC, Embryo Research Licensing Committee, Information Kit, 2008. Available at: http://www.nhmrc.gov.au/guidelines/publications/hc56 (22 July 2011).

[13] This Act is available at http://www.comlaw.gov.au/comlaw/Legislation/ActCompilation1.nsf/0/2AA59FCD74133106CA2577CA000E2551?OpenDocument (22 July 2011)

[14] Fertility Society of Australia, *About FSA*, available from: http://www.fsa.au.com/about/ (22 July 2011).

[15] Fertility Society of Australia, *Reproductive Technology Accreditation Committee, Certification Scheme (RTAC Scheme) (August 2008)*: available at http://www.fertilitysociety.com.au/wp-content/uploads/20080924-rtac-scheme-final2.pdf (at 22 July 2011).

[16] That is, services that are partly funded by the Health Insurance Commission: Saunders D. The Australian experience of self-accreditation. In Gunning J, Szoke, H. (Eds), *The* Regulation of Assisted Reproductive Technology 2003; 225: 230.

[17] NHMRC, *Database of Licences authorising the use of excess ART embryos*; available from http://www.nhmrc.gov.au/research/embryos/monitor/database/index.htm (at 22 July 2011).

[18] NHMRC, Licensing Committee Publications and Reports to Parliament: available from http://www.nhmrc.gov.au/research/embryos/information/reports/index.htm (at 22 July 2011).

[19] Legislation Review: Prohibition of Human Cloning Act 2002 and Research Involving Human Embryos Act 2002: available from http://www.nhmrc.gov.au/_files_nhmrc/file/research/embryos/review/legislation_review_reports_full_doc_19dec05.pdf (at 22 July 2011).

[20] The 2006 amending Commonwealth Act is noted in note 7 above. Victoria, New South Wales, Tasmania, Queensland, the ACT and South Australia all passed amending complementary legislation. Western Australia did not amend its legislation

[21] Note that the Research Act Part 5 is headed 'PART 5-SUNSET CLAUSE, REVIEW PROVISION ANDREGULATIONS' (emphasis added). A 'sunset clause' usually indicates that a regulatory instrument will become inoperative if a mandated review is not undertaken as prescribed in the Act. However, despite this heading, there is no provision in the Act providing that the Act will cease if there is not a review within the prescribed period and it is a principle of statutory interpretation that the words in a heading in an Act do not have statutory effect. Section 47A(1), requiring the review, has the words 'to be undertaken as soon as possible after the third anniversary of the day on which the amending Act received the Royal Assent' (emphasis added). However, s 47A (3) states that 'The persons undertaking the review must give the Council of Australian Governments and both Houses of the Parliament a written report of the review *before the fourth anniversary of the day on which the amending Act received the Royal Assent*' (emphasis added) – ie the words 'as soon as possible' are not there.

[22] 2010 Legislation Review website available at https://legislationreview.nhmrc.gov.au/2010-legislation-review-committee [22 July 2011).

[23] The Report of the Heerey Committee, the Independent Review of the *Prohibition of Human Cloning for Reproduction Act 2002* and the *Research Involving Human Embryos Act 2002,* is available at www.legislationreview.nhmrc.gov.au (22 July 2011).

[24] Minister for Mental Health and Ageing, Mark Butler, Press release, 7 July 2011, available at: http://www.health.gov.au/internet/ministers/publishing.nsf/Content/mr-yr11-mb-mb067.htm (22 July 2011).

[25] So seeming to avoid the ethical issue related to research involving embryos: Ensenat-Waser R *et al.* Reprogrammed induced pluripotent stem cells: how suitable could they be in reproductive medicine? (2009) 91 *Fertility and Sterility* 971-974; Deng W. Induced pluripotent stem cells: paths to new medicines. (2010) 11 *EMBO Reports* 161-165

[26] See, press release of National Institutes of Health. First Human Embryonic Stem Cell Lines Approved for Use Under New NIH Guidelines. 2 Dec, 2009, available at http://www.nih.gov/news/health/dec2009/od-02.htm (22 July 2011).

[27] Federally funded research involving human embryos was suddenly halted in the US by a preliminary injunction granted by a district judge in August 2010: Katsnelson A. US judge puts temporary block on human embryonic stem cell research. 23 August, 2010: available at http://blogs.nature.com/news/thegreatbeyond/2010/08/us_judge_puts_temporarly_block.html (22 July 2011). The US Justice Department appealed this decision and the US Court of Appeals in Washington lifted the injunction pending the final determination of the legal issues: Ertelt S. Appeals Court Allows Obama to Fund Embryonic Stem Cell Research During Suit. 9 Sept 2010, available at http://www.lifenews.com/2010/09/09/bio-3169/ (22 July 2011). (This could not happen in Australia. Australian law is different from American law and the Australian legislation clearly allows human embryo research to be undertaken under licence and ethical review.)

[28] The US based company Geron Corporation received federal approval in January 2009 from the US Food and Drug Administration (FDA), after extensive safety evaluation, to undertake clinical trials for human patients using neuronal precursor cells derived from human ES cells to treat spinal injury. The trials were deferred for further tests on animals but the FDA said in July 2010 they could proceed. In the trials, the cells will be inserted into 'the lesion site of the patient's injured spinal cord': News release of Geron Corporation, 'Geron to Proceed with First Human Clinical Trial of Embryonic Stem Cell-Based Therapy', 30 July 2010: available at http://www.geron.com/media/pressview.aspx?id=1229 (22 July 2011). Other trials involving patients with Stargardt's macular dystrophy, which causes blindness, have also started recently, using human embryonic stem cells: Stem cell trial on blindness. *The Australian*, 23 Nov 2010: available at http://www.theaustralian.com.au/news/world/stem-cell-trial-on-blindness/story-e6frg6so-1225958789009 (22 July 2011).

[29] Isasi RM, Knoppers BM. Monetary Payments for the Procurement of Oocytes for Stem Cell Research: In Search of Ethical and Political Consistency Stem Cell Res 2007; 37: 39, available at http://www.sciencedirect.com/science/article/pii/S1873506107000074 (22 July 2011).

[30] Siemaszko C. $10,000 is an Egg-cellent Price, Says Stem Cell Panel. *NY Daily News*, 25 June 2009: available at http://www.nydailynews.com/news/2009/06/26/2009-06-26_10000_is_an_eggcellent_price_says_stem_cell_panel.html (22 July 2011).

[31] The federal legislation has a limited exception; it is lawful to create a hybrid embryo by fertilising an animal egg by human sperm and to develop the resultant embryo up but not including the final point of fertilisation (the first mitotic division), but only for the purpose of an accredited fertility clinic testing sperm quality.

[32] Connor S. UK won't fund hybrid embryo studies. *The Canberra Times*, 6 Oct, 2009.

[33] See, for example, Windpipe transplant success in UK child. *Stem Cell Research News*, 19 March 2010, available at: http://stemcellresearchnews.net/news_frame.asp?article=7a3322c3-8086-40d7-8c8e-6e77d734c94c&title=Windpipe%20transplant%20success%20in%20UK%20child (22 July 2011).

CHAPTER 8

Pluripotent Stem Cells & Neurodegenerative and Ischaemic Brain Diseases: Basic Science to Clinical Applications

Kuldip S. Sidhu[1]* and Perminder S. Sachdev[2]

[1]Stem Cell Laboratory, Faculty of Medicine, University of New South Wales, NSW, Australia and [2]Neuropsychiatry Institute, Prince of Wales Hospital, University of New South Wales, NSW, Australia

Abstract: Degenerative disorders of the central nervous system (CNS), such as Alzheimer's disease (AD), Parkinson's disease (PD) and Huntington's disease (HD), represent a growing public health challenge worldwide. The ability to accurately diagnose and track neurodegenerative disorders is becoming increasingly important as we enter a new clinical era of disease-modifying therapies. Such treatments may be most effective in the earlier stages of the disease before neurodegeneration becomes too widespread, therefore, it may be important to identify subjects at early stages of the disease. Valid cellular models of these diseases are also needed to examine pathogenesis and test novel treatments at the preclinical stage. The recent advancement in stem cell technology, particularly in pluripotent stem cell biology, has opened up new vistas for neurodegenerative disease research. Both human embryonic stem cells and induced pluripotent stem cells possess the inherent ability to produce all cell types including neuronal tissues in large numbers. The greatest potential for induced pluripotent cells derived from affected individuals is likely to be their utility for modelling and understanding the mechanisms underlying neurodegenerative processes, and for searching for new treatments, including cell replacement therapies. However, much work remains to be done before pluripotent cells can be used for preclinical and clinical applications. In this chapter we discuss briefly, the clinical challenges in early diagnosis of these diseases, the progress so far in the use of stem cells for various neurodegenerative diseases at the levels of discovery, transition and translation, and the challenges of generating specific neural cell subtypes including the recent interest in the use of pluripotent stem cells to model these diseases in the dish. Progress in these areas will substantially accelerate effective application of pluripotent stem cells.

Keywords: Pluripotent stem cells, Cnetral nervous sytem, Alzheimer's disease, Parkinson's disease, Huntington's disease, ischaemic brain, reprogramming, clinical trials.

INTRODUCTION

Neurodegenerative disorders represent a large group of heterogeneous disorders in which specific neuronal subtypes degenerate and die. In most cases, these disorders arise for unknown reasons and progress relentlessly [1]. Currently these disorders are treated by conventional therapies that only alleviate the symptoms and patients live in a state of sustained and progressive disability. There are limited numbers of disease modifying therapies that are currently under investigation. According to the 2005 report from the World Health Organization (WHO) [2], AD, PD and other dementias made up more than 14% of the global burden of neurological illnesses, and their prevalence is expected to rise world-wide with the ageing of society. The search for novel treatments that will slow, stop or even reverse the process of neurodegeneration is therefore of utmost public health importance.

Cell transplantation into the brain as a strategy to treat neurodegenerative disorders has a long history. The transplantation of specific neuronal or glial tissue grafts from cadaver has been tried for many years, but the option is limited because of unavailability of the appropriate donor tissue, and the clinical outcome remaining questionable. Initial clinical trials aiming to replace lost neurons were performed in patients with

***Address correspondence to Kuldip S. Sidhu:** Director, Stem Cell Laboratory, Faculty of Medicine, The University of New South Wales, Level 2, Clinical Sciences Building, Prince of Wales Hospital, Randwick, NSW 2031, Australia; Tel: + 61 2 9382 2622; Fax: + 61 2 9382 2622; E-mail: k.sidhu@unsw.edu.au

Parkinson's disease or Huntington's disease. Similarly, allogenic fetal midbrain tissue alleviates the symptoms of these patients [3]. Ongoing trials use fetal striatal tissue to treat patients with Huntington's disease. Preliminary data indicate a mild disease modifying effect, but these trials are still at an early stage [4]. In this context, ethical questions regarding the use of tissue from aborted human fetuses, scarce tissue availability, as well as limited safety and quality control of such tissue have raised concerns regarding these experimental therapies. The potential of regenerative medicine lies in the possibility of therapies that offer flexibility, adaptive growth, and improved host acceptance mostly related to stem cells or reprogrammed cells. Realization of this potential will require a multitude of tasks, which include, in addition to organ and tissue replacements, new and advanced developments in the conventional methods of transplantation medicine and cell therapy.

Stem or progenitor cells can generate neural tissue, raising new possibilities for repairing the nervous system, and such cells can be generated from several sources (Fig. **1**). Embryonic stem cells (ESCs) are derived from blastocysts and have an almost unlimited capacity to self-renew. However, major ethical concerns exist regarding the use of human ESCs, and governmental restrictions continue to limit clinical applications in many jurisdictions [5]. More recently, induced pluripotent stem cells (iPS) were generated, which obviate the ethical concernsof ESCs. iPS cells also offer a source of autologous neural stem cells, eliminating the risk of immunological complications. However, these cells also may form tumors, and safety cannot yet be guaranteed. Last but not least, a very novel possibility exists to generate neurons from other adult cells such as fibroblasts via direct reprogramming without the intermediate iPS state [6]. If it becomes possible to generate specific mature neurons such as dopaminergic neurons, then the need for stem or progenitor cells may be eliminated, opening very novel avenues for restorative medicine. At present, human neural stem cells (NSCs) derived from fetal tissue seem closest to clinical applications. Unlike rodent fetal-tissue-derived NSCs, their human counterparts also display remarkable plasticity, expandability, and a variegate potential for differentiation. These cells potentially can overcome, in part, ethical issues of an insufficient supply of cells from fetal donor material. On the other hand, long-term expansion of NSCs seems to curb the argument that ESCs are the only source for well-characterized and/or standardized tissue for restoration. However, despite the enthusiasm generated by recent reports, the process by which NSCs differentiate into specific neurons is still poorly understood. Autologous bone-marrow-derived stem cells have been applied to patients with haematological disorders for many years. An ongoing discussion persists whether such cells can replace damaged neurons or can provide trophic support for neurons at risk in patients with neurodegenerative disorders. In this chapter, we will provide a short overview of current strategies of stem cell technology in alleviating CNS disease burden and concepts to translate stem cell research into clinical applications.

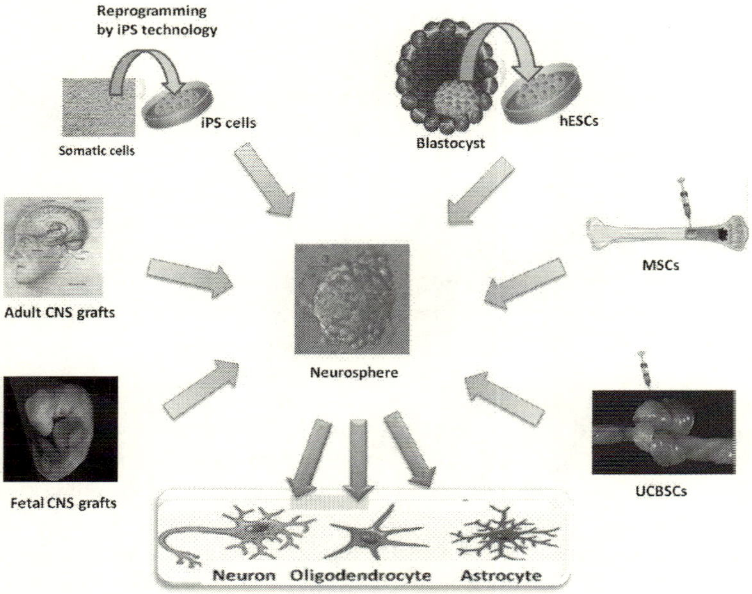

Fig. 1: Sources of tissues/cells for neuronal lineages required for transplantation strategies in disorders of CNS.

NEURODEGENERATIVE DISEASES, DISEASE PROCESS AND CLINICAL CHALLENGES

a. Alzheimer's Disease (AD)

Alzheimer's disease is the most common neurodegenerative disease, affecting 35 million patients worldwide. Clinically, it is characterized by progressive loss of short-term memory and other cognitive functions toward a state of profound dementia, and histopathologically, by the appearance of senile plaques and neurofibrillary tangles in affected brain regions [7]. The clinical diagnosis of AD is primarily based on medical history, cognitive tests and neurological investigations. The major risk factor for AD is advancing age, and other risk factors such as cerebrovascular disease, lifestyle, diet, apolipoprotein E4 gene, *etc.* have been identified. Majority of AD cases are sporadic; a few familial (FAD) cases have an early-onset autosomal dominant disorder with almost 100% penetration.

Alzheimer's disease is pathologically characterized by neuronal loss and aggregation of two proteins, beta-amyloid (Aβ) and hyperphosphorylated tau (pTau). Aβ is formed from the cleavage of APP into soluble monomers that then aggregate into fibrils and are eventually deposited as plaques in the extracellular space. Tau is a microtubule-associated protein that undergoes hyperphosphorylation and accumulates as intraneuronal inclusions or tangles in the brains of individuals with Alzheimer's disease [8-9]. Over the past 15 years, three genes have been identified that account for familial forms of Alzheimer's disease, and they have given important clues to the pathogenesis of Alzheimer's disease and to direct the basis for new therapies. The first mutations causing Mendelian Alzheimer's disease were identified in the *APP* gene [10], although mutations in two other genes, *presenilin 1* and *2* (*PSEN1* and *PSEN2*), that form the γ-secretase complex components, are more commonly found [11-12]. Mutations in these three genes act by either increasing the steady state level of Aβ, altering the Aβ42/40 ratio, or altering the amyloidogenic potential of Aβ. The point mutations cause different clinical phenotypes and a few of them are the "Dutch type", the "Flemish type", the "London type", the "Arctic type", the "Australian type", the "Belgian type", the "German type" and the "Swedish type" [13]. Thus, it was inferred that the aberrant processing of Aβ led to its aggregation and that this was the basis of FAD. A major genetic risk factor for the development of sporadic Alzheimer's disease is the *APOE* gene and the risk of late-onset alzheimer's disease increases with increased dosage of the ε4 allele. ApoE is synthesized in astrocytes and acts as a ligand for the receptor-mediated endocytosis of cholesterol-containing lipoprotein particles. It has also been shown to bind Aβ and is a component of amyloid plaques [14]. Whether ApoE affects Aβ clearance or operates through its function in lipid metabolism is not yet clear in Alzheimer's disease.

Neuropathology of AD

The histopathological hallmarks of AD include intracellular neurofibrillary tangles (NFTs) and extracellular deposit of β–amyloid (Aβ) in neuritic plaques and diffuse deposits, atrophy, synaptic reduction and neural loss [15-16].

The Aβ peptide consists of 38 to 42 amino acid residues and is a product generated from proteolytic processing (Fig. **2**) of the amyloid precursor protein (APP) by way of three enzymatic cleavages during normal cellular metabolism [16, 17]. The enzyme α-secretase cleaves APP and releases the non-amyloidogenic secreted form of APP (sAPPα). The amyloidogenic pathway involves β-secretase that cleaves APP at the N-terminal of Aβ and γ–secretase cleaves at its C-terminal site. Active γ–secretase comprises four membrane proteins, presenilin, nicastrin, Alpha-1, and Pen-2 [18], which not only cleave APP but also notch and other type 1 transmembrane receptor [19].

Although the form $A\beta_{1\text{-}40}$ is most common, only the $A\beta_{1\text{-}42}$ peptide has a strong tendency to self-aggregate and form long, insoluble amyloid fibrils, which accumulate and result in the formation of amyloid plaques [20-21]. Reactive astrocytes and activated microglia also initiate inflammatory responses in the brain in cases of AD by secreting cytokines and chemokines, contributing to the degeneration of neurons [22-24]. Many studies have shown that Aβ induces oxidative stress as manifested by lipid peroxidation, and DNA and protein oxidation, indicating that Aβ-induced oxidative stress is involved in the pathogenesis of AD [25-26]. The distribution patter of amyloid deposits differs from that of NFTs [27]. NFTs are intracellular lesions, found in the cytoplasm of degenerating neurons, which are composed of hyperphosphorylated aggregates of the microtubule-associated tau protein. The latter plays a role in stabilising the microtubules during their assembly, and abnormal hyperphosphorylation of tau protein is believed to lead to destabilizing of microtubules, dysfunction in axonal transport and progressive neuronal loss [28-29]. A selective reduction in the level of acetylcholine (ACh) and choline acetyltransferase (ChAT) activity, predominantly in the hippocampus and cerebral cortex, has also been reported in the brains of AD patients [30-31]. In severe AD cases, the level of acetyl cholinesterase (AChE) is reduced compared with that in normal subjects, whereas the level of butyrylcholinesterase (BuChE is increased [32].

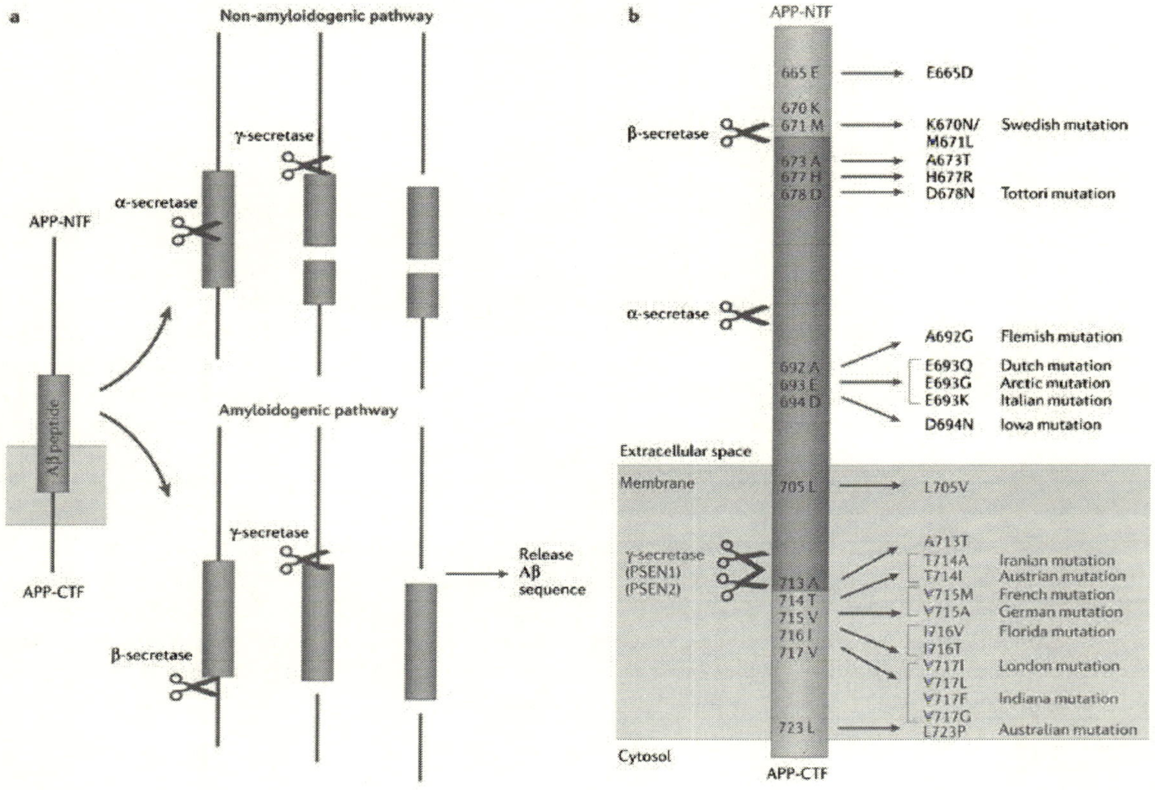

Fig. 2: APP processing and APP mutations associated with early-onset alzheimer's disease. a | Amyloid precursor protein (APP) processing involves proteolytic cleavage by several secretases. The non-amyloidogenic pathway is initiated by α-secretase cleavage, which occurs in the middle of the amyloid (Aβ) sequence (red), and results in the release of several soluble APP fragments. The amyloidogenic pathway releases Aβ peptides through cleavage by β- and γ-secretases. **b** | Part of the APP amino-acid sequence; mutations associated with early-onset alzheimer's disease have been highlighted. Most mutations are clustered in the close vicinity of secretase-cleavage sites, thereby influencing APP processing, and are named after the nationality or location of the first family in which that specific mutation was demonstrated. The Aβ sequence is indicated in red. A, alanine; Aβ, amyloid-β; APP-NTF, N-terminal fragment of the amyloid precursor protein; APP-CTF, C-terminal fragment of the amyloid precursor protein; D, aspartic acid; E, glutamic acid; F, phenylalanine; G, glycine; H, histidine; I, isoleucine; K, lysine; L, leucine; M, methionine; N, asparagine; P, proline; PSEN, presenilin; Q, glutamine; R, arginine; S, serine; T, threonine; V, valine; Y, tyrosine. [from 17].

In vivo measurements by means of positron emission tomography have provided data on early changes in the course of the disease, demonstrating that the cognitive decline in AD patients correlates with a loss of neuronal nicotinic acetylcholine receptors (nAChRs) [33-34]. nAChRs are associated with various transmission activities and receptor activation modulates the release of several transmitters including Ach, glutamate, γ-aminobutyric acid (GABA), noradrenalin, and dopamine. They belong to a gene superfamily of ligand-gated ion channels, which also includes glycine-, 5-hydroxytryptamine- and $GABA_A$ receptors [35]. Autographic studies in human *post mortem* brain tissues revealed a significant loss of nAChRs in AD patients compared with age-matched control subjects [34]. The protein levels of the α3 and α4 subunits were reduced in the temporal cortex, and the α7 subtype was reduced in the hippocampus of AD brain [36]. Consistently, reduction in the number of [^{3}H]-epibatidine and [^{3}H]-nicotine binding sites, which have high affinity for the α3 and α4 nAChR ubunits, was also observed in cortical regions of the AD patients who carried the APP $_{swe}$ double mutation [37]. Most studies have revealed that the muscarinic acetylcholine receptors (mAChRs) are preserved or increased in AD brain [38-39], although a decrease number of M2 mAChRs has been observed [40]. The mAChRs belong to the superfamily of g-protein-coupled receptors that include receptors for a variety of neurotransmitters. Five different subtypes of mAChRs (M1-M5), each encoded by a different gene, have been identified in eth CNS (Caulfield, 1993). Degeneration of cholinergic neurons in eth brain is a principal feature of AD and the population of neurons in the nucleus basalis and the hippocampus are severely affected. These cholinergic deficits in AD are positively correlated with cognitive impairment [41-44] as well as non-cognitive disturbances.

There is also strong evidence emerging that there are neuroinflammatory changes in AD patients, and these changes are independent of tau and Aβ pathogenesis and can cause cognitive, neurodegenerative, or oxidative damage [45-46]. One of the most well characterized neuroinflammatory responses in AD has been increases in interleukin-1. For example, IL-1β can exacerbate Aβ pathogenesis *via* increasing β-secretase (BACE) [19, 47].

Stem cells have been proposed as an important source of cells for several CNS disorders including AD, HD, PD, multiple sclerosis and spinal cord injury [48-52]. A considerable numbers of studies have been performed to investigate whether transplantation of stem cells into the injured or diseased CNS has the potential to repair circuitries and thereby restore neurological functions. The ideal donor cells for neural transplantation should be expandable *in vitro*, exhibit an immature but committed phenotype, and be able to migrate to the desired region of the brain or spinal cord upon transplantation.

Stem cell transplantation studies in animal models of neurodgeneration have indicated that these cells are able to migrate to sites of injury and differentiate into neurons and glial cells in the brain [53]. The microenvironment surrounding the implant is populated by glial and endothelial cells which are important regulators of stem cell migration and differentiation. Disturbance in the microenvironment due to brain injury or pathology could affect the environmental equilibrium and expose the cells to factors that are different from those in their normal conditions.

Novel therapeutic strategies are being developed taking advantage of the ability of NPCs to proliferate *in vitro* and to survive following transplantation into the mammalian brain, where they may integrate and express foreign genes or repopulate the damaged or diseased CNS. Recent studies indicate that human NPCs may hold great promise to replace degenerative cholinergic neurons, ameliorating the behavioural deficits of AD [54-56]. It has been demonstrated that human NPCs, injected into the lateral ventricle of 24-month old rats, improve cognition four weeks after transplantation [56]. These findings indicate that the aged rat brain is not only capable of providing the necessary environment conditions for the neural precursors to retain their pluripotency, but also the potential for investigation of replacement therapies to improve feature of memory and cognition in age-associated disorders. Transplantation of neurons derived from hESCs has also been successfully integrated *in vivo* [57]. Together, these important findings indicate the great advantage and opportunities that these cells may offer as a source of specialised human cells for future therapeutic applications.

b. Parkinson's Disease (PD)

Parkinson's disease is the second most common neurodegenerative disorder, afflicting over 6 million people worldwide. Clinically, there are progressive motor dysfunctions comprising bradykinesia, rigidity and tremor, as well as non motor features of cognitive decline, neuropsychiatric symptoms, sleep disturbance and autonomic dysfunction. Pathologically, PD is identified by intracellular inclusions known as Lewy bodies and dopaminergic neuronal loss that initiates in the substantia nigra. As with Alzheimer's disease, the incidence of Parkinson's disease increases with age and Parkinson's disease is largely a late-onset sporadic neurodegenerative condition. However, 5–10% cases are familial, transmitted in either an autosomal-dominant or autosomal recessive fashion. A number of genes have been linked to our understanding of pathogenesis [58] including missense mutations and duplications and triplications in the *SNCA* gene that are rare cause of autosomal dominant PD [59-60]. However, the gene product α-synuclein is the major component of the Lewy body in sporadic disease and therefore appears to be central to Parkinson's disease pathophysiology. Mutations in the *LRRK2* gene account for a substantial proportion of autosomal-dominant Parkinson's disease cases and one missense mutation, the G2019S mutation, occurs in 5% of familial cases and 1–2% of sporadic cases of Parkinson's disease [61]. Notably, this mutation varies in different populations occurring in 30–40% of North African familial and sporadic Parkinson's disease cases. Mutations in *PARK2*, *PINK1* and *PARK7* (also known as *DJ1*) cause autosomal-recessive, early-onset Parkinsonism [62-64]. These genetic discoveries have highlighted the importance of the ubiquitin proteasome system, mitochondrial dysfunction and oxidative stress in Parkinson's disease pathogenesis. Other genes, such as *UCHL1*, *NR4A2* (also known as *NURR1*) and *HtrA2/OMI*, have also been associated with Parkinson's disease, although their role remains unclear. The most common genetic risk factor for Parkinson's disease appears to be heterozygous mutations in the *glucocerebrosidase* gene (*GBA*) [65]. Homozygous mutations in *GBA* cause a recessive lysosomal storage disease known as Gaucher's disease. However, the frequency of heterozygous mutations in *GBA* reaches ~4% in sporadic Parkinson's disease populations, compared with ~1% in controls. Notably, this represents an example of multiple low-frequency risk variants contributing to disease susceptibility, which cannot be detected by current genome-wide association studies (GWAS) methodology.

Because PD results from the loss of dopaminergic neurons, the prospect of utilizing cell replacement therapies has attracted substantial interest. The first attempt was to use fetal mesencephalic tissue for transplantation, and the results were successful in the earliest reports [66-68]. However, not all trials showed beneficial outcomes. The sham surgery-controlled study also demonstrated some clinical benefits in younger but not in older patients [69]. Another study showed no significant treatment effects [70]. Moreover, fetal mesencephalic transplantation is associated with several problems. First, off-medication dyskinesia increased 6-12 months after the transplantation in 15-56% of patients [70-71]. Second, graft-induced inflammatory responses might influence the longevity of transplanted cells [72]. Third, tissue availability limits the clinical use [67]. As a result, fetal mesencephalic transplantation is not recommended as a conventional therapy for PD. Graft-induced dyskinesia is thought to be caused by unfavourable composition of the fetal mesencephalic grafts. The fetal mesencephalic tissue includes not only dopaminergic but also non-dopaminergic neurons [67]. The exclusion of serotonin and GABA neurons and enrichment of substantia nigra dopamine neurons may decrease the occurrence of dyskinesia [72]. To achieve this goal, probable solutions include refinement of dissection methods for fetal tissue transplantation, isolation of desired cell types and/or removal of unwanted cellular populations using fluorescence and/or magnetic-activated cell sorting (FACS/MACS), and using stem cells (ES cells and iPS cells) as an alternative cell source [67]. Recent evidence has shown that dopamine neurons derived from ES cells and bone marrow-derived neural progenitors are functional when grafted into parkinsonian rats [67, 73-74]. Several methods are able to improve the effectiveness of midbrain dopamine neuron generation and/or retrieval from stem cells. These include manipulating transcription factors like Nurr1, Pitx3 or Lmx1a, co-culture with astrocytes and using fluorescence-activated cell sorting [72]. The replacement of fetal tissue by stem cells also solves the problem of availability and ethical issue [72]. The ability of deriving large quantities of correctly differentiated dopamine neurons makes stem cells promising cell sources for transplantation in PD.

c. Ischemic Stroke

While not a neurodegenerative disorder, ischemic stroke is a major cause of morbidity and mortality in older individuals and often co-occurs with AD. The only effective treatment for acute ischemic stroke is thrombolytic agents such as recombinant tissue plasminogen activator (rt-PA). For patients receiving thrombolytic therapy shortly after the stroke (3–4.5 h), only 31-50% of them obtained favourable outcomes, and 6.4% of patients developed brain hemorrhage. New therapeutic strategies with neuroprotection or neurorestoration are crucial for improving the prognosis of patients with stroke. Ischemia affects the behaviour and proliferation status of NSCs. For example, focal ischemia of brain enhances endogenous neurogenesis, angiogenesis, axonal sprouting and synaptogenesis [75-76]. However, the proportion of damaged or dead neurons replaced by the new neurons is small. Besides, neurogenesis does not occur in some ischemic regions. This is probably either due to an unfavourable microenvironment of the ischemic sites, or because these sites are distant from the SVZ and SGZ, which are most abundant in NSCs [76]. Pharmacological treatments aimed at enhancing neurogenesis, angiogenesis and axonal outgrowth were successful in animal studies. These included erythropoietin, statins, phosphodiesterase 5 inhibitors, granulocyte-colony stimulating factor, nicotinic acid and minocycline [75]. Limited clinical data have indicated beneficial therapeutic potential of these drugs in humans, but further clinical survey is required.

The difficulties of cell-replacement therapy are due to variable cell types involved in ischemic stroke, which include neurons, astrocytes, oligodendrocytes and endothelial cells of blood vessels [75]. Although transplantation of bone marrow-derived mesenchymal stem cells promoted functional recovery, the effect was caused by activation of endogenous restoration of injured brain rather than cell replacement [67, 75, 77]. ES cells have been demonstrated to have greater developmental potential and more significant survival rate than adult stem cells after transplantation [78]. Transplantation of ES cells also recovered behavioural dysfunction induced by middle cerebral arterial occlusion in an animal model [79]. However, the ethical consideration, the limited availability and the possibility of immune rejection after transplantation restrict the accessibility of ES cells.

STEM CELLS AND THERAPEUTIC POTENTIALS

a. Embryonic Stem Cells

Human embryonic stem (ES) cells are pluripotent and can produce the entire range of major somatic cell lineage of the central nervous system (CNS) and thus form an important source for cell-based therapy of various neurological diseases(see also chapter 1). They can be expanded for many years and differentiated into neural stem or precursor cells and subsequently into brain cells. More details about these cells can be found in chapter 22 of this eBook. Despite their potential use in regenerative medicine, the progress is hampered by difficulty in their use because of safety issue and lack of proper protocols to obtain purified populations of specified neuronal cells. Most neurological conditions such as spinal cord injury and Parkinson's disease involve damages to projection neurons. Similarly, certain cell populations may be depleted after repeated episodes of attacks such as the myelinating oligodendrocytes in multiple sclerosis. Motoneurons are the key effector cell type for control of motor function, and loss of motoneurons is associated with a number of debilitating diseases such as amyotrophic lateral sclerosis (ALS) and spinal muscular atrophy hence, repair of such neurological conditions may require transplantation with exogenous cells. Transplantation of neural progenitor cells in animal models of neurological disorders and patients in some clinical trial cases has shown survival of grafted cells and contribution to functional recovery. Recently, considerable progress has been made in understanding the biochemical, molecular, and developmental biology of stem cells. But translation of these *in vitro* studies into the clinic has been slow. Major hurdles are the lack of effective donor cells, their *in vivo* survival and difficulty in remodelling the non-neurogenic adult CNS environment. Several factors play a role in maintaining their functions as stem cells. It is becoming increasingly apparent that the role of developmental signalling molecules is not over when embryogenesis has been completed. In the adult, such molecules might function in the maintenance of stem cell proliferation, the regeneration of tissues and organs, and even in the maintenance of their differentiated state. A major challenge is how to teach the naïve ES cells to choose a neural fate, especially the subclasses of neurons and glial cells that are lost in neurological conditions. Thus selecting specific cell types from ESC-derived tissue could increase markedly the efficacy and safety of the resulting tissue. The first clinical application of ESC-derived tissue in the central nervous system is likely to consist not of dopaminergic neurons for PD but of oligodendrocytes for the treatment of spinal cord injury. Several groups have developed protocols for the relatively pure isolation of oligodendrocytes [80-81]. These cells are now under clinical trials in patients after Food and Drug Administration (FDA) final approval [82]. Similarly, another recent FDA approval has been provided for clinical trials with hESC-derived retina pigment epithelial cells for patients with macular degeneration.

b. Induced Pluripotent Stem Cells

The advent of induced pluripotent stem cell (iPSC) technology has opened up new vistas to generate patient-specific pluripotent stem cells from somatic cells. During the last 5 years, the iPSCs produced from a variety of somatic cell sources have been found to be very similar if not identical to embryonic stem cells. Until recently, these cells were produced by viral transduction of four transcriptional factors that rendered these cells unfit for therapeutic purposes. However, current emphasis is on the development of new methods for generating iPSCs with minimal or no genetic modifications *via* excisable lentiviral and transposon vectors or through repeated application of transient plasmid, episomal, and adenovirus vectors and very recently the use of small molecules, synthetic mRNA and microRNAs. The iPSC technology is emerging as an unprecedented opportunity in biomedical research and regenerative medicine. However, to harness the full potential of this technology, a number of issues that need to be resolved pertaining to iPSC safety, stability, culture variability, their comparison with ES cells, the reprogramming mechanisms, and better ways to direct a specific reprogramming process including lineage specifications. This expert opinion will bring out the current developments emphasising on the improved and new methods of generating iPSCs.

Reprogramming of somatic cells has received much interest lately in regenerative medicine as it can offer a potentially accessible source of cells that can be used to replace the defective or diseased cells in patients suffering diseases such as AD, PD, Type 1 diabetes and cardiomyopathy that involve specific cellular pathology. The two most remarkable techniques used in reprogramming are, somatic cell nuclear transfer (SCNT) and induced pluripotent stem (iPS) cell generation. Both techniques can reprogram a differentiated cell back to an embryonic state [83-84]. Nuclear transfer experiments date back to the 1950s when frogs (*Rana pipiens*) were cloned by replacing nuclei of eggs with cells from the late blastula [85], and since then SCNT has been successfully reported in a number of mammalian species following the birth of Dolly the sheep [86]. Although SCNT remains the method of choice for generating isogenic pluripotent stem cells, this procedure is technically challenging and inefficient with low success rates, and not applicable to humans because of ethical constraints. The discovery of the induced pluripotent stem cell (iPSC) technique using defined transcription factors by Yamanaka invigorated this field and changed the paradigm, shifting the debate away from SCNT program particularly for humans [87]. Over the past five years, the iPSC technology has advanced significantly. iPSCs have been generated from a broad range of somatic cell types [88] and species [89], differentiated into various seemingly functional cell types, and used to study basic biology and model human diseases. These studies have opened up new possibilities that iPSCs could be potentially developed as unlimited and patient-matched cell sources for cell-based therapy or drug discovery.

Thus far, protocols to generate dopaminergic neurons have been developed and these have been used to alleviate some deficits in an animal model of PD [20, 21]. One key characteristic of iPS also is tumor formation Many laboratories around the globe currently reprogram cells from patients with various diseases including neurodegenerative disorders to identify novel drug targets. It can be anticipated that such research will lead to the identification of small molecules that finally have the potential to halt or even cure some of these devastating illnesses. Most recently, elegant work demonstrated that it may be possible to reprogram adult somatic cells into mature neurons without the intermediate step of iPS. This strategy is ethically acceptable beyond any doubt and also may eliminate the risk for tumor formation. Wernig *et al.* [90] analysed a number of so-called helix-loop-helix transcription factors that are expressed during neuronal development. They found that ascl1, in combination with some other factors, could transform fibroblasts from adult mice into mature neurons. One can imagine that reprogramming into dopaminergic neurons could be possible *via* the addition of supplementary dopaminergic neuron specific transcription factors such as Nurr1, and so on. Future studies will show whether it is possible to generate such specific neurons that are sufficiently mature for transplantation and also lack the risk of tumor formation. Although this study is a very novel area of research, it may represent the most promising avenue, thus far. Although iPS cells are very promising, clinical transplantation of tissue derived from these cells may not happen within the near future.

c. Neural Stem Cells

NSCs in the adult brain are located in the subventricular zone (SVZ) of the lateral ventricle and the subgranular zone (SGZ) of the hippocampal dentate gyrus (Fig. **3**). These regions have common cellular niche components which include astroglia, ependymal cells, vascular cells, NSC progeny and mature neurons. A number of common extracellular niche signals like Wnt, Sonic Hedgehog, bone morphogenic protein antagonists, membrane-associated Notch signalling, leukemia inhibitory factor, transforming growth factor-alpha, fibroblast growth factors, neurotrophins and extracellular matrix regulate this niche [90]. These cellular and extracellular components regulate the behaviors of NSCs in a region-specific manner. For example, SVZ NSCs give rise to Dlx2+ Mash1+ intermediate progenitor cells which subsequently give rise to PSA-NCAM+ doublecortin+ (DCX+) neuroblasts and migrate towards the olfactory bulb (OB). In contrast, SGZ NSCs do not differentiate into interneuron-lineage cells like those in the OB, but give rise to local glutamatergic excitatory dentate granule cells [91]. The region-specific development of these NSCs is not only due to intrinsic characteristics of the NSCs themselves, but also due to the dictation of local microenvironment (*i.e.*, the niche).

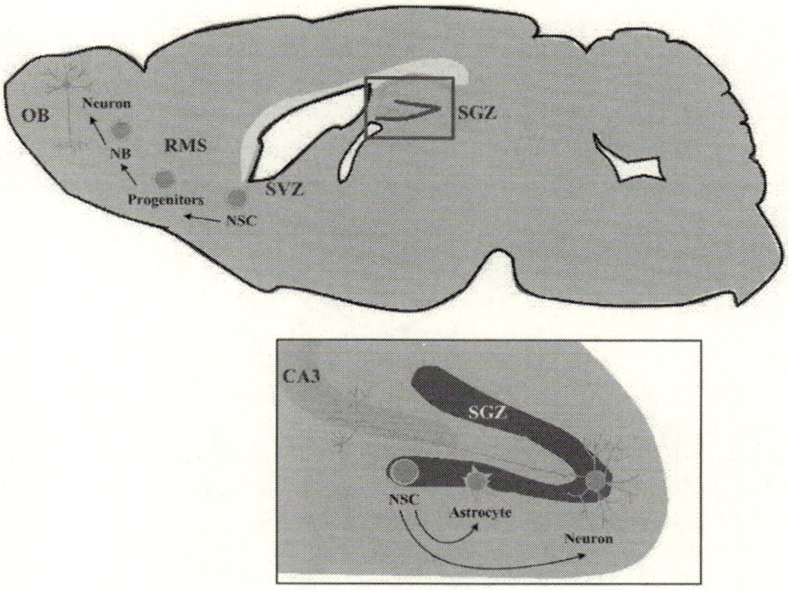

Fig. 3: The two niches of neural stem cells (NSCs). The subventricular zone (SVZ) of the lateral ventricle and the subgranular zone (SGZ) of the hippocampal dentate gyrus have common cellular niche components and extracellular niche signals. The development of NSCs of the two niches is different in a region-specific manner. SVZ NSCs give rise to Dlx2+ Mash1+ intermediate progenitor cells which subsequently give rise to PSA-NCAM+ doublecortin+ (DCX+) neuroblasts (NB) and migrate towards the olfactory bulb (OB). SGZ NSCs give rise to local glutamatergic excitatory dentate granule cells. RMS: rostro-migratory stream; GL: granular layer [91].

NSCs can be extracted directly from fetal or adult nervous tissue and can be maintained in several serum-free and serum containing growth media that allow the proliferation of such cells when supplemented with appropriate mitogens such as epidermal growth factor (EGF) and basic fibroblast growth factor (FGF-2). NSCs derived from embryonic or fetal central nervous tissue generate oligodendrocytes, neurons, and astrocytes in an approximate ratio of 1:5:25, respectively, when allowed to differentiate spontaneously after removal of growth factors or mitogens in serum-free media [92-93]. The possibility of stable expansion and *in vitro* differentiation into neurons make human NSCs an attractive cell source for reconstructive transplantation strategies in many brain diseases such as PD, stroke, head trauma, and spinal cord injury. Fetal-tissue-derived NSCs are the only source of stem- or progenitor-cell-derived brain cells that have entered the clinical arena. Preclinical studies showed that transplantation in the brain of mice suffering from a disorder resembling infantile neuronal ceroid lipofuscinosis (NCL, Batten's disease) integrate into the host brain, release the defective enzyme, and also protect affected neurons [94]. Accordingly, StemCells Inc. (Palo Alto, CA) completed the first FDA-approved clinical trial in patients with NCL in January 2009 (http://www.stemcellsinc.com/clinicaltrials/clinical trials.html). A second clinical trial in patients suffering from Pelizaeus–Merzbacher disease (PMD) characterized by defective myelination was approved by the FDA and currently is being conducted at the

University of California in San Francisco. However, the risk of tumor formation may not be as limited as has been deducted from animal experiments. After repeated transplantation in a centre in Russia, a young patient with telangiectasia ectatica developed several neuroblastoma that may have been derived from the transplants [95]. However, the positive data resulting from the well-documented patients transplanted in the context of the aforementioned approved clinical trials are encouraging. Other researchers have increased the proliferative potential of NSCs *via* immortalization. Also, such cells seem to be relatively safe, and clinical application in patients with a stable deficit after stroke was granted. These NSCs are modified genetically *via* ectopic expression of the oncogene c-myc. Regulatory and conditional ethical approval are in place to commence a phase I clinical trial in the United Kingdom. These immortalized stem-cell lines will be administered in patients who suffered stroke.

d. Mesenchymal Stem Cells

Bone-marrow-derived stem cells comprising mesenchymal stem cells (MSCs) have been applied to patients with hematological malignancies and other disorders for many years. Mesenchymal or bone-marrow-derived stem cells also provide the advantage of minimizing immune reactions because these can be derived from the respective patient. Many optimistic reports have been made about transdifferentiation or conversion of mesenchymal stem cell into NSCs or neurons, but the proof of functional neurons derived from MSCs was not provided [96]. Thus, the concept of transdifferentiation was jeopardized. However, as indicated, it now seems possible to convert fibroblasts directly into neurons. These findings again may change our view of the differentiation capacity of MSCs as well. Besides the question of transdifferentiation, it is also possible that these cells provide trophic support to damaged neurons. The transplantation of such cells into the brain or the cerebrospinal fluid, as performed in some centers, could result in bone marrow at the site of transplantation. Recently, Venkataramana *et al.* [97] reported the first open label clinical pilot study with autologous bone marrow-derived stem cells transplanted into the striatum of patients with advanced PD. The good news of this study is that no adverse events occurred during the observation period as well as some possible clinical improvement. Although this was trial carefully designed with a favorable outcome in respect to adverse effects, many questions remain and preclinical studies need to demonstrate whether such cells either can differentiate into exogenous functional neurons or provide trophic support for endogenous cells.

e. Differentiation to Neuronal Lineages

The production of neurons involves several sequential steps precisely orchestrated by signalling events [98]. The initial step is the specification of neuroepithelia from ectoderm cells, the process known as neural induction, which is accomplished by inductive interaction with nascent mesoderm and definitive endoderm. Despite being a topic of intensive study, there is still no consensus on the mechanisms and signals involved in neural induction. Bone morphogenetic protein (BMP) antagonism has been viewed as the central and initiating event in neural induction. According to this concept, neuroepithelial specification occurs as a default pathway [99]. However, recent findings challenge this neural default model and indicate some positive instructive factors, such as fibroblast growth factors (FGFs) and Wnt. Hence, a balanced view of neural induction most likely needs to include both instructive and inhibitory factors. FGF may induce a neural state at an early stage, and BMP antagonists may subsequently stabilize the neural identity. When the neuroectodermal fate is determined, the neural plate folds to form the neural tube, from which cells differentiate into various neurons and glia [100]. The neural tube is patterned along its rostrocaudal and dorsoventral axes to establish a grid-like set of positional cues [101]. Similarly, the neural plate acquires a rostral character, and is subsequently caudalized by exposure to Wnt, FGF, BMP, and RA signals [102] to establish the main subdivisions of the central nervous system (CNS): the forebrain, midbrain, hindbrain, and spinal cord. Therefore, the precursor cells in each subdivision along the rostro-caudal axes are fated to subtypes of neurons and glia depending on its exposure to unique sets of morphogens at specific concentrations.

Enrichment of neural progenitors from differentiating hESC has been achieved by exploiting the observation that cells of neural morphology form spontaneously within hESC colonies after prolonged culture. Reubinoff *et al.* [103] demonstrated the mechanical isolation of these neural progenitors and repeating in chemically defined medium supplemented with B27, human Epidermal Growth Factor (hEGF), and basic fibroblast growth factor-2 (bFGF2) resulted in the formation of spherical structures called neurospheres [103]. These neurospheres have highly enriched neural progenitor cells, with 99% of cells expressing N-CAM, 97% nestin, and 90.5% A2B5 [103]. According to Zhang *et al.* [104], hESC generated neuroectodermal cells usually do not form typical neurospheres. Instead, they from aggregates of columnar cells in the form of neural-tube like rosettes, where only after the long term expansion of the neural rosette clusters, will they form the morphology of neurospheres [104]. Therefore, the neurospheres formed in the spontaneous differentiation cultures may represent neural precursors at a much later developmental stage.

The major difference between Zhang's and Reubinoff's methods is that Zhang utilizes the EB pathway whereas Reubinoff spontaneously differentiates hESC colonies for a prolonged time of 3 weeks [103-104]. Both protocols require the isolation of neuroepithelial cells from other non-neural cells, and propagation of these neurospheres in culture. Isolation of these neural rosettes is done either by enzymatic treatment or mechanical dissection. Both groups utilize serum-free media (DMEM/F12) (1:1) supplemented with different types of nutrients catered for neural induction. The neurospheres are then plated on laminin or ornithin coated plates for further neural differentiation.

Other laboratories have created specific neuronal types with markers consistent with midbrain dopamingergic neurons, spinal motor neurons and striatal neurons [105-109]. Recognising that all such populations, although enriched, remain heterogeneous, there is a need for additional selection methods to further purify neuronal sub-type lineages. Sub-type specific reporter lines are one such method. Whilst a key aim of positionally specifying hESC-NPCs is to work towards the generation of cell-based therapies for diseases that target a sub-population of cells, they will also provide a valuable resource for research into those very same diseases [110]. Future application of regional specification will be particularly powerful in attempting to understand disease specificity when applied to induced disease-specific pluripotent stem cells (iPS cells). The emergence of iPS cells, through manipulation of somatic cells by over expression of key transcription factors to generate embryonic stem (ES) like cells offers additional experimental and therapeutic possibilities. Combining somatic cell reprogramming technology to generate patient specific iPS cell lines carrying disease-causing mutations, with neural developmental insights, offers the unique opportunity to develop human *in vitro* models of disease. Indeed, recent studies have confirmed the utility of such an approach with resulting neuronal cells displaying a disease phenotype as well as being a unique assay for drug discovery and testing [111-112]. Neural precusors were generated from an iPS cell line developed from fibroblasts taken from a spinal muscular atrophy patient, a disease characterised by selective loss of lower motor neurons caused by mutation in the survival motor neuron 1 gene (SMN1) gene. Compared with a wild-type iPS cell line, SMAiPS cells were equally amenable to differentiation into immature motor neurons, but numbers of motor neurons observed at later timepoints post-differentiation were much reduced in the SMA-iPS cell cultures, compared with wild type, indicative of either impaired production or selective degeneration of motor neurons [111]. In another recent example, iPS cells were generated from a patient with familial dysautonomia (FD), a fatal peripheral neuropathy, caused by a point mutation in the IKBKAP gene leading to depletion of autonomic and sensory neurons. FD-iPS cell derived neural precursors have particularly low levels of IKBKAP, mis-splicing of IKBKAP, and defects in neurogenic differentiation and migration behaviour [112]. Together these studies demonstrate that *in vitro* developmentally based human neuronal culture systems can reproduce a phenotype normally observed *in vivo* in juvenile or adult life.

CLINICAL TRIALS

The growing amount of preclinical data showing the potential benefit associated with stem cell therapy warrants the development of well-controlled clinical trials to investigate therapeutic safety and efficacy for central nervous system. There are only two clinical trials approved by FDA with human embryonic stem cell-derived cells, one for spinal cord injury and other for macular degeneration as indicated earlier. However, a number of clinical trials are in progress for various neurodegenerative diseases using autologous sources of stem cells such as MSCs, umbilical cord blood stem cells *etc.* Preliminary clinical trials investigating the role of cell therapeutics for ischemic stroke have been limited to date and powered only to evaluate safety (Table **1**). The majority of these studies are restricting enrolment to patients with MCA territory infarcts and do not assess the role of these cells in other areas of the brain.

A phase 1 trial (http://www.clinicaltrials.gov/ Identifier: NCT00859014) is evaluating the safety of bone marrow aspiration and infusion of bone marrow mononuclear cells (BMMCs) in adults within 24–72 hours of ischemic stroke. Primary outcome measures include a series of short- and long-term safety assessments with a secondary evaluation of neurological function as measured up to 90 days after injury. Similarly, a Phase 2 trial is recruiting patients to determine the potential efficacy of CD34+ cell implantation (http://www.clinicaltrials.gov/ Identifier: NCT00950521). A number of other centres such as the hospital universitario Central de Asturias, Asturias, Spain, Federal University of Rio de Janeiro, Rio de Janeiro, Brazil Imperial College London, London, England) are currently enrolling patients for a Phase 1, 2 trial investigating the safety and efficacy of the intra arterial delivery of autologous CD34+ progenitor cells into the MCA after ischemic stroke.

Table 1. Listing of location and details of current clinical trials being completed to investigate the potential role of bone marrow-derived progenitor cell therapeutics for the treatment of ischemic stroke.

Location of Study	**Study Design**	**Deliver Route**	**Sample Size**	**Inclusion Criteria**	**Outcomes**	**Cell Type**	**Time Window**
United States (The University of Texas in Houston)	Single arm	IV	10	Autologous BMMCs	- MCA stroke - 18–80 yo - NIHSS 6 to 20	Safety and feasibility	24 to 72 hrs
Taiwan (The China Medical University Hospital)	Randomized (cell infusion versus conventional treatment)	IC	30	Autologous peripheral blood CD34+ cells	- Stable deficits hemiplegia - 35–70 yo - NIHSS 9 to 20	Safety and efficacy	6 months to 5 years
Spain (Hospital Universitario Central de Asturias)	Single arm	IA	20	Autologous CD34+ bone marrow cells	- MCA stroke - 18–80 yo - NIHSS ≥ 8	Safety	5 to 9 days
France (University Hospital of Grenoble)	Randomized (Control versus 2 treatment groups)	IV	30	Autologous bone marrow derived progenitor cells	- Carotid territory stroke - 18–65 yo - NIHSS > 2	Feasibility and tolerability	6 weeks
United Kingdom (Imperial College London)	Single arm	IA	10	Autologous CD34+ bone marrow cells	- MCA stroke - 30–80 yo - Severe stroke conforming to the TACS phenotype (weakness, homonymous hemianopia and a focal cognitive deficit	Safety and tolerability	7 days
Brazil (Federal University of Rio de Janeiro)	2 arms (non randomized: 10 IA/5IV)	IV/IA	15	Autologous BMMCs	- MCA stroke - 18–75 yo - NIHSS 4 to 20	Safety	3hrs to 90 days

KEY: IV: intravenous; IA: intra-arterial; IC: intracerebral; BMMC: bone marrow mononuclear cells; MCA: middle cerebral artery; NIHSS: National Institutes of Health Stroke Scale; TACS: total anterior circulation stroke.

Using the search terms "stem cell" and "brain" in Medline identified nearly 300 studies, including studies evaluating the use of adult stem cells (administered *via* various routes, in various numbers, and using various cell types) to treat hypoxic/ischemic encephalopathy, cerebral palsy, multiple sclerosis, amyotrophic lateral sclerosis, neuronal ceroid lipofuscinosis, Parkinson's disease, and others. A single Phase I study using bone marrow-derived mononuclear cells in children after isolated TBI has recently been completed. In this study, 10 children age 5–14 years with a Glasgow coma scale score of 5–8 were treated with 6 × 106 bone marrow-derived mononuclear cells per kg body weight delivered intravenously within 48 hours of an isolated TBI. To determine the safety of administration, systemic and cerebral hemodynamics, laboratory parameters, chest radiographs, and serial clinical assessments were monitored. There were no identifiable adverse events with close monitoring of the neurologic, pulmonary, renal, hepatic, and hematologic systems. Functional and neuropsychological testing, including the Glasgow Outcome Scale, the Pediatric Injury Functional Outcome Scale, and the Wechsler Abbreviated Scale of Intelligence, revealed recovery consistent with (or improved from) expected baselines. Magnetic resonance imaging volumetric data revealed no significant change in grey matter, white matter, intracranial volume, or CSF space at 1 and 6 months as measured relative to expected norms [113]. This study should open the door for translation of cell therapies, particularly among patients with neurologic diseases and among pediatric patients. Given the apparent safety of this study, the development of larger, multicenter studies to further assess dosing and efficacy of autologous cell therapy for TBI is underway. Additionally, similar (more dispensable) progenitor cell populations, such as cord blood cells, may be safe and efficacious as well and warrant further study.

A number of studies have been completed with using an autologous source of BMMC that were infused either intravenously or by lumber puncture in SCI patients [114-117]. Neurologic improvements were greater in patients with the greatest leukocytosis following GM-CSF treatment. Neuropathic pain occurred in a third of the subacute and chronically treated patients but in only one of sixteen acutely treated patients. One control group patient developed neuropathic pain [117].

CONCLUSIONS AND PROSPECTS

Developing stem cell technology represents a prime research interest. Analyzing such cells *in vitro* and *in vivo* will help to understand neuronal development in many species including humans. The translation toward clinical transplantation of stem or progenitor cells currently is restricted to NSCs as far as well documented trials are concerned. The use of bone marrow-derived stem cells is applied broadly outside of clinical trials and is far away from having demonstrated safety or efficacy. Considerable hope has grown for several years that stem cell therapy could be useful in reversing the progressive downhill course that characterizes several neurodegenerative diseases. A few clinical trials have been carried out in Parkinson's disease. The results indicate that transplantation of dopaminergic neurons recovered from human fetuses can replace endogenous degenerating dopamine neurons and provide some improvement in symptoms. However, the availability of tissues, ethical issues, and concerns in regard to safety and quality control has limited the progress. It is well known that bone marrow-derived stem cells have been used in patients with hematologic disorders for many years, but a major challenge in this field is whether bone marrow-derived cells can transdifferentiate into functional neurons or whether they can provide trophic support for injured neurons in patients with chronic degenerative diseases. There are potential advantages and disadvantages of both embryonic stem cells as well as induced pluripotent stem cells. Recent work suggests that it might be possible to reprogram adult somatic cells into mature neurons without the intermediate step of induced pluripotent stem cells. Some work has been done with autologous bone marrow derived stem cells with the expectation that these cells would minimize immune reactions; however, a direct demonstration of functional neurons derived from the stem cells has not been provided.

Prior to large, multicenter clinical trials investigating the potential efficacy of stem cell therapies for CNS insults, a number of issues need to be addressed. Further research into optimal cell dosing, cell delivery method, and techniques for *in vivo* cell tracking need to be completed to ensure the safety of potential trials while affording them the best possible chance at success. Additional preclinical work to more clearly delineate the stem cell mechanism of action would also aid in the planning of quality-controlled clinical studies. Overall, while the very preliminary clinical trials offer novel data supporting the potential efficacy of cell therapeutics for CNS injury, a great deal of additional work is needed to ensure the safety and efficacy of stem cell therapy prior to widespread clinical trials.

REFERENCES

[1] Przedborski S, Vila M, Jackson-Lewis V. Neurodegeneration: what is it and where are we? J Clin Invest 2003; 111(1): 3-10.

[2] World Health Organization. Neurological disorders: public health challenges. Geneva: WHO Press, 2006.

[3] Olanow CW, Goetz CG, Kordower JH, *et al.* A double-blind controlled trial of bilateral fetal nigral transplantation in Parkinson's disease. Ann Neurol 2003; 54(3): 403-14.

[4] Bachoud-Levi AC, Gaura V, Brugieres P, *et al.* Effect of fetal neural transplants in patients with Huntington's disease 6 years after surgery: a long-term follow-up study. Lancet Neurol 2006; 5(4): 303-9.

[5] Sidhu KS. Motoneurons from human embryonic stem cells: present status and future strategies for their use in regenerative medicine. In: Maiese K, Ed. Neurovascular medicine: pursuing cellular longevity for healthy aging. New York: Oxford University Press 2009; 231-54.

[6] Abujarour R, Ding S. Induced pluripotent stem cells free of exogenous reprogramming factors. Genome Biol. 2009; 10(5): 220.

[7] Zhu X, Raina AK, Perry G, Smith MA. Apoptosis in Alzheimer disease: a mathematical improbability. Curr Alzheimer Res 2006; 3(4): 393-6.

[8] Hattiangady B, Shetty AK. Aging does not alter the number or phenotype of putative stem/progenitor cells in the neurogenic region of the hippocampus. Neurobiol Aging 2008; 29(1): 129-47.

[9] Hagg T. Endogenous regulators of adult CNS neurogenesis. Curr Pharm Des 2007; 13(18): 1829-40.

[10] Goate A, Chartier-Harlin MC, Mullan M, *et al.* Segregation of amissense mutation in the amyloid precursor protein gene with familial Alzheimer's disease. Nature 1991; 349: 704-6.

[11] Chartier-Harlin MC, Crawford F, Houlden H, *et al.* Early-onset Alzheimer's disease caused by mutations at codon 717 of the beta-amyloid precursor protein gene. Nature 1991; 353: 844-6.

[12] Mullan M, Crawford F, Axelman K, *et al.* A pathogenic mutation for probable Alzheimer's disease in the APP gene at the N-terminus of beta-amyloid. Nat Genet 1992; 1(5): 345-7.

[13] Felsenstein KM, Hunihan LW, Roberts SB. Altered cleavage and secretion of a recombinant beta-APP bearing the Swedish familial Alzheimer's disease mutation. Nat Genet 1994; 6(3): 251-255.

[14] Tanzi RE, Kovacs DM, Kim TW *et al.* The gene defects responsible for familial Alzheimer's disease. Neurobiol Dis 1996; 3: 159-168.

[15] Iwatsubo T, Saido TC, Mann DM, *et al.* Full length amyloid-beta (1-42(43) and amino-terminally modified and trancated amyloid-beta 42(43) deposit in diffuse plaques. Am J Pathol 1996; 149(6): 1823-30.

[16] Selkoe DJ. Alzheimer's disease is a synaptic failure. Science 2002; 298(5594): 789-91.

[17] Van Dam D, De Deyn PP. Drug discovery in dementia: the role of rodent models. Nat Rev Drug Discov 2006; **5:** 956-70.

[18] Kimberly WT, LaVoie MJ, Ostaszewski BL, *et al.* γ-secretase is a membrane protein complex comprised of presenilin, nicastrin, aph-1 and pen-2. Proc Natl Acad Sci U S A 2003a; 100(11): 6382-7.

[19] Kimberly WT, Esler WP, Ye W *et al.* Notch and amyloid precursor proteins are cleaved by similar gamma-secretase(s). Biochemistry 2003b; 42(1): 137-44.

[20] Lemere CA, Lopera F, Kosik KS, *et al.* The E280A presenilin 1 Alzheimer mutation produces increased A beta 42 deposition and severe cerebellar pathology. Nat Med 1996; 2(10): 1146-50.

[21] Iwatsubo T, Odaka A, Suzuki N, Mizusawa H, Nukina N, Ihara Y. Visualization of A beta 42(43) and A beta 40 in senile plaques with end-specific A beta monoclonals: evidence that an initially deposited species is A beta 42(43). Neuron 1994; 13(1): 45-53.

[22] Morgan D, Gordon MN, Tan J, Wilcock D, Rojiani AM. Dynamic complexity of the microglial activation response in transgenic models of amyloid deposition: implications for Alzheimer therapeutics. J Neuropathol Exp Neurol 2005; 64(9): 743-53. Review.

[23] Johnstone M, Gearing AJ, Miller KM. A central role for astrocytes in the inflammatory response to beta-amyloid; chemokines, cytokines and reactive oxygen species are produced. J Neuroimmunol 1999; 93(1-2): 182-93.

[24] McGeer PL, McGeer EG. Inflammation, autotoxicity and Alzheimer disease. Neurobiol Aging 2001; 22(6): 799-809.

[25] Butterfield DA, Drake J, Pocernich C, Castegna A. Evidence of oxidative damage in Alzheimer's disease brain: central role for amyloid beta-peptide. Trends Mol Med 2001; 7(12): 548-54.

[26] Christen Y. Oxidative stress and Alzheimer disease. Am J Clin Nutr 2000; 71(1): 621S-9S.

[27] Braak H, Braak E. Frequency of stages of Alzheimer-related lesions in different age categories. Neurobiol Aging 1997; 18(4): 351-7.

[28] Alonso AD, Grundke-Iqbal I, Barra HS, Iqbal K. Abnormal phosphorylation of tau and the mechanism of Alzheimer neurofibrillary degeneration: sequestration of microtubule-associated proteins 1 and 2 and the disassembly of microtubules by the abnormal tau. Proc Natl Acad Sci U S A 1997; 94(1): 298-303.

[29] Iqbal K, Alonso AC, Chen S, *et al.* Tau pathology in Alzheimer disease and other tauopathies. Biochim Biophys Acta 2005; 1739(2-3): 198-210.

[30] Nordberg A. Neuroreceptor changes in Alzheimer disease. Cerebrovasc Brain Metab Rev 1992; 4(4): 303-28.

[31] Perry EK, Johnson M, Kerwin JM, *et al.* Convergent cholinergic activities in aging and Alzheimer's disease. Neurobiol Aging 1992; 13(3): 393-400.

[32] Giacobini E. Cholinergic function and Alzheimer's disease. Int J Geriatr Psychiatry 2003; 18(Suppl 1): S1-5.

[33] Kadir A, Almkvist O, Wall A, Långström B, Nordberg A. PET imaging of cortical 11C-nicotine binding correlates with the cognitive function of attention in Alzheimer's disease. Psychopharmacology (Berl) 2006; 188(4): 509-20.

[34] Nordberg A. Nicotinic receptor abnormalities of Alzheimer's disease: therapeutic implications. Biol Psychiatry 2001; 49(3): 200-10.

[35] Bernard EA. Receptor classes and the transmitter-gated channels. Trends Biochem Sci 1992; 17(10): 368-74.

[36] Guan ZZ, Zhang X, Ravid R, Nordberg A. Decreased protein levels of nicotinic receptor subunits in the hippocampus and temporal cortex of patients with Alzheimer's disease. J Neurochem 2000; 74(1): 237-43.

[37] Marutle A, Warpman U, Bogdanovic N, Lannfelt L, Nordberg A. Neuronal nicotinic receptor deficits in Alzheimer patients with the Swedish amyloid precursor protein 670/671 mutation. J Neurochem. 1999; 72(3): 1161-9.

[38] Svensson AL, Alafuzoff I, Nordberg A. Characterization of muscarinic receptor subtypes in Alzheimer and control brain cortices by selective muscarinic antagonists. Brain Res 1992; 596(1-2): 142-8.

[39] Giacobini E. Cholinergic receptors in human brain: effects of aging and Alzheimer disease. J Neurosci Res 1990; 27(4): 548-60.

[40] Mash DC, Flynn DD, Potter LT. Loss of M2 muscarine receptors in the cerebral cortex in Alzheimer's disease and experimental cholinergic denervation. Science 1985; 228(4703): 1115-7.

[41] Coyle JT, Price DL, DeLong MR. Alzheimer's disease: a disorder of cortical cholinergic innervation. Science 1983; 219(4589): 1184-90.

[42] Whitehouse PJ, Price DL, Struble RG, Clark AW, Coyle JT, Delon MR. Alzheimer's disease and senile dementia: loss of neurons in the basal forebrain. Science 1982; 215(4537): 1237-9.

[43] Bartus RT, Dean RL, 3rd, Beer B, Lippa AS. The cholinergic hypothesis of geriatric memory dysfunction. Science 1982; 217(4558): 408-14.

[44] Collerton D. Cholinergic function and intellectual decline in Alzheimer's disease. Neuroscience 1986; 19(1): 1-28.

[45] Strittmatter WJ, Roses AD. Apolipoprotein E and Alzheimer disease. Proc Natl Acad Sci U S A. 1995; 92(11): 4725-7.

[46] Iwatsubo T, Saido TC, Mann DM, Lee VM, Trojanowski JQ. Full-length amyloid-beta (1-42(43)) and amino-terminally modified and truncated amyloid-beta 42(43) deposit in diffuse plaques. Am J Pathol 1996; 149(6): 1823-30.

[47] Parvathy S, Hussain I, Karran EH, Turner AJ, Hooper NM. Cleavage of Alzheimer's amyloid precursor protein by alpha-secretase occurs at the surface of neuronal cells. Biochemistry 1999; 38(30): 9728-34.

[48] Sugaya K. Possible use of autologous stem cell therapies for Alzheimer's disease. Curr Alzheimer Res 2005; 2(3): 367-6.

[49] Snyder EY, Macklis JD. Multipotent neural progenitor or stem-like cells may be uniquely suited for therapy for some neurodegenerative conditions. Clin Neurosci. 1995; 3(5): 310-6.

[50] Dunnett SB, Björklund A. Prospects for new restorative and neuroprotective treatments in Parkinson's disease. Nature 1999; 399(6738 Suppl): A32-9.

[51] Kordower JH, Chen EY, Winkler C, *et al.* Grafts of EGF-responsive neural stem cells derived from GFAP-hNGF transgenic mice: trophic and tropic effects in a rodent model of Huntington's disease. J Comp Neurol 1997; 387(1): 96-113.

[52] Rabchevsky AG, Smith GM. Therapeutic interventions following mammalian spinal cord injury. Arch Neurol 2001; 58(5): 721-6.
[53] Darsalia V, Kallur T, Kokaia Z. Survival, migration and neuronal differentiation of human fetal striatal and cortical neural stem cells grafted in stroke-damaged rat striatum. Eur J Neurosci 2007; 26(3); 605-14.
[54] Doering LC, Snyder EY. Cholinergic expression by a neural stem cell line grafted to the adult medial septum/diagonal band complex. J Neurosci Res 2000; 61(6): 597-604.
[55] Sugaya K, Brannen CL. Stem cell strategies for neuroreplacement therapy in Alzheimer's disease. Med Hypotheses 2001; 57(6): 697-700.
[56] Qu T, Brannen CL, Kim HM, Sugaya K. Human neural stem cells improve cognitive function of aged brain. Neuroreport 2001; 12(6): 1127-32.
[57] Muotri AR, Nakashima K, Toni N, Sandler VM, Gage FH. Development of functional human embryonic stem cell-derived neurons in mouse brain. Proc Natl Acad Sci U S A 2005; 102(51): 18644-8.
[58] Lesage S, Brice A. Parkinson's disease: from monogenic forms to genetic susceptibility factors. Hum Mol Genet 2009; 18(R1): R48-R59.
[59] Polymeropoulos MH, Lavedan C, Leroy E, *et al.* Mutation in the alpha-synuclein gene identified in families with Parkinson's disease. Science 1997; 276(5321): 2045-7.
[60] Singleton AB, Farrer M, Johnson J, *et al.* alpha-Synuclein locus triplication causes Parkinson's disease. Science 2003; 302(5646): 841.
[61] Gilks WP, Abou-Sleiman PM, Gandhi S, *et al.* A common LRRK2 mutation in idiopathic Parkinson's disease. Lancet 2005; 365(9457): 415-6.
[62] Kitada T, Asakawa S, Hattori N, *et al.* Mutations in the parkin gene cause autosomal recessive juvenile Parkinsonism. Nature 1998; 392(6676): 605-608
[63] Valente EM, Abou-Sleiman, Caputo V, *et al.* Hereditary early-onset Parkinson's disease caused by mutations in PINK1. Science 2004; 304(5674): 1158-60.
[64] Bonifati V, Rizzu P, van Baren MJ, *et al.* Mutations in the DJ-1 gene associated with autosomal recessive early-onset parkinsonism. Science 2003; 299(5604): 256-9.
[65] Neumann J, Bras J, Deas E, *et al.* Glucocerebrosidase mutations in clinical and pathologically proven Parkinson's disease. Brain 2009; 132(7): 1783-94.
[66] Lindvall O, Hagell P. Clinical observations after neural transplantation in Parkinson's disease. Prog Brain Res 2000; 127: 299-320.
[67] Lindvall O, Kokaia Z, Martinez-Serrano A. Stem cell therapy for human neurodegenerative disorders-how to make it work. Nat Med 2004; 10(Suppl): S42-S50.
[68] Kordower JH, Freeman TB, Snow BJ, *et al.* Neuropathological evidence of graft survival and striatal reinnervation after the transplantation of fetal mesencephalic tissue in a patient with Parkinson's disease. N Engl J Med 1995; 332(17): 1118-24.
[69] Freed CR, Greene PE, Breeze RE, *et al.* Transplantation of embryonic dopamine neurons for severe Parkinson's disease. N Engl J Med 2001; 344(10): 710-9.
[70] Olanow CW, Goetz CG, Kordower JH, *et al.* A double-blind controlled rial of bilateral fetal nigral transplantation in Parkinson's disease. Ann Neurol 2003; 54(3): 403-14.
[71] Hagell P, Piccini P, Bjorklund A, *et al.* Dyskinesias following neural transplantation in Parkinson's disease. Nat Neurosci 2002; 5: 627-8
[72] Hedlund E, Perlmann T. Neuronal cell replacement in Parkinson's disease. J. Intern Med 2008; 266(4): 358-71.
[73] Yang D, Zhang ZJ, Oldenburg M, Ayala M, Zhang SC. Human embryonic stem cell-derived dopaminergic neurons reverse functional deficit in parkinsonian rats. Stem Cells 2008; 26(1): 55-63.
[74] Glavaski-Joksimovic A, Virag T, Chang QA, et al. Reversal of dopaminergic degeneration in a parkinsonian rat following micrografting of human bone marrow-derived neural progenitors. Cell Transplant 2009; 18(7): 801-14.
[75] Zhang ZG, Chopp M. Neurorestorative therapies for stroke: Underlying mechanisms and translation to the clinic. Lancet Neurol 2009; 8(5): 491-500.
[76] Arvidsson A, Collin T, Kirik D, Kokaia Z, Lindvall O. Neuronal replacement from endogenous precursors in the adult brain after stroke. Nat Med 2002; 8(9): 963-970.
[77] Chopp M, Li, Y. Treatment of neural injury with marrow stromal cells. Lancet Neurol 2002; 1(2): 92-100.
[78] Takahashi K, Yasuhara T, Shingo T, *et al.* Embryonic neural stem cells transplanted in middle cerebral artery occlusion model of rats demonstrated potent therapeutic effects, compared to adult neural stem cells. Brain Res 2008; 1234: 172-82.

[79] Yanagisawa D, Qi M, Kim DH, *et al.* Improvement of focal ischemia-induced rat dopaminergic dysfunction by striatal transplantation of mouse embryonic stem cells. Neurosci Lett 2006; 407(1): 74-79.

[80] Keirstead HS, Nistor G, Bernal G, *et al.* Human embryonic stem cell-derived oligodendrocyte progenitor cell transplants remyelinate and restore locomotion after spinal cord injury. J Neurosci 2005; 25(19): 4694-705.

[81] Nistor GI, Totoiu MO, Haque N, Carpenter MK, Keirstead HS. Human embryonic stem cells differentiate into oligodendrocytes in high purity and myelinate after spinal cord transplantation. Glia 2005; 49(3): 385-96.

[82] Geron Corporation. World's first clinical trial of human embryonic stem cell therapy cleared. Regen Med 2009; 4(2): 161.

[83] Han J, Sachdev PS, Sidhu, KS. A Combined Epigenetic and Non-Genetic Approach for Reprogramming Human Somatic Cells. PLoS ONE 2010; 5(8): e12297.

[84] Han J, Chung H, Sachdev PS, Sidhu KS. Epigenetic modifications in reprogramming and lineage specifications from embryonic stem cells: current trends and relevance in developing regenerative medicine. In: Singh SR, Ed. Stem Cells, Regenerative Medicine and Cancer. New York: Nova Science Publishers 2011, pp. 3-30.

[85] Briggs R, King TJ. Transplantation of living nuclei from blastula cells into enucleated frogs' eggs. Pro Nat Acad Sci U S A 1952; 38(5): 455-63.

[86] Campbell KH, McWhir J, Ritchie WA, Wilmut I. Sheep cloned by nuclear transfer from a cultured cell line. Nature 1996; 380(6569): 64-66.

[87] Takahashi K, Yamanaka S. Induction of pluripotent stem cells from mouse embryonic and adult fibroblast cultures by defined factors. Cell 2006; 126(4): 663-6.

[88] Kim JB, Sebastiano V, Wu G, *et al.* Oct4-induced pluripotency in adult neural stem cells. Cell 2009; 136(3): 411-9.

[89] Li W, Wei W, Zhu S, *et al.* Generation of rat and human induced pluripotent stem cells by combining genetic reprogramming and chemical inhibitors. Cell Stem Cell 2009; 4(1): 16-9.

[90] Wernig M, Zhao J-P, Pruszak J *et al.* Neurons derived from reprogrammed fibroblasts functionally integrate into the fetal brain and improve symptoms of rats with Parkinson's disease. Proc Natl Acad Sci USA 2008; 105(15): 5856-61.

[91] Ma DK, Bonaguidi MA, Ming GL, Song H. Adult neural stem cells in the mammalian central nervous system. Cell Res 2009; 19: 672-82.

[92] Lee HJ, Kim JW, Yim SV, *et al.* Fluoxetine enhances cell proliferation and prevents apoptosis in dentate gyrus of maternally separated rats. Mol Psychiatry 2001; 6(6): 725-8.

[93] Chiou SH, Sheu BC, Chang WC, Huang SC, Hong-Nerng H. Current concepts of tumor-infiltrating lymphocytes in human malignancies. J Reprod Immunol 2005; 67(1-2): 35-50.

[94] Schulze-Bergkamen H, Brenner D, Krueger A, *et al.* Hepatocyte growth factor induces Mcl-1 in primary human hepatocytes and inhibits CD95-mediated apoptosis *via* Akt. Hepatology 2004; 39(3): 645-54.

[95] Xu H, Chen Z, He J, *et al.* Synergetic effects of quetiapine and venlafaxine in preventing the chronic restraint stress-induced decrease in cell proliferation and BDNF expression in rat hippocampus. Hippocampus 2006; 16(6): 551-9.

[96] Kim YJ, Park HJ, Lee G, *et al.* Neuroprotective effects of human mesenchymal stem cells on dopaminergic neurons through anti-inflammatory action. Glia 2009; 57(1): 13-23.

[97] Venkataramana NK, Satishkumar K, Balaraju S, *et al.* An open labeled study of unilateral autologous bone marrow derived mesenchymal stem cell transplantation in Parkinson's disease. Transl Res 2010; 155(2): 62-70.

[98] Wilson SI, Edlund T. Neural induction: toward a unifying mechanism. Nature Neurosci 2001; 4(Suppl): 1161-8.

[99] Munoz-Sanjuan I, Brivanlou AH. Neural induction, the default model and embryonic stem cells. Nature Rev Neurosci 2002; 3: 271-380.

[100] O'Rahilly R, Muller F. The development of the neural crest in the human. J Anat 2007; 211(3): 335-51

[101] Altmann CR, Brivanlou AH. Neural patterning in the vertebrate embryo. Int Rev Cytol 2001; 203: 447-482.

[102] Gunhaga L, Marklund M, Sjodal M, Hsieh JC, Jessell TM, Edlund T. Specification of dorsal telencephalic character by sequential Wnt and FGF signalling. Nat Neurosci 2003; 6(7): 701-7.

[103] Reubinoff BE, Itsykson P, Turetsky T, *et al.* Neural progenitors from human embryonic stem cells. Nature Biotechnol 2001; 19: 1134-40.

[104] Zhang SC, Wernig M, Duncan ID, Brustle O, Thomson JA. (2001). *In vitro* differentiation of transplantable neural precursors from human embryonic stem cells. Nature Biotechnol 2001; 19(12): 1129-33.

[105] Murry CE, Keller G. Differentiation of embryonic stem cells to clinically relevant populations: lessons from embryonic development. Cell 2008; 132(4): 661-80.

[106] Yan Y, Yang D, Zarnowska ED, *et al.* Directed differentiation of dopaminergic neuronal subtypes from human embryonic stem cells. Stem Cells 2005; 23(6): 781-790.

[107] Shin S, Dalton S, Stice SL. Human motor neuron differentiation from human embryonic stem cells. Stem Cells Dev 2005; 14(3): 266-269

[108] Li XJ, Hu BY, Jones SA, Zhang YS, Lavaute T, Du ZW, *et al.* Directed differentiation of ventral spinal progenitors and motor neurons from human embryonic stem cells by small molecules. Stem Cells 2008; 26(4): 886-93.

[109] Aubry L, Bugi A, Lefort N, Rousseau F, Peschanski M, Perrier AL. Striatal progenitors derived from human ES cells mature into DARPP32 neurons *in vitro* and in quinolinic acidlesioned rats. Proc Natl Acad Sci U S A 2008; 105(43):16707-12.

[110] Zeng X, Chen J, Deng X, Liu Y, Rao MS, Cadet JL, *et al.* An *in vitro* model of human dopaminergic neurons derived from embryonic stem cells: MPP+ toxicity and GDNF neuroprotection. Neuropsychopharmacology 2006; 31(12): 2708-15.

[111] Ebert AD, Yu J, Rose FF Jr, Mattis VB, Lorson CL, Thomson JA, *et al.* Induced pluripotent stem cells from a spinal muscular atrophy patient. Nature 2009; 457(7227): 277-80.

[112] Lee G, Papapetrou EP, Kim H, *et al.* Modelling pathogenesis and treatment of familial dysautonomia using patient-specific iPSCs. Nature 2009; 461(7262): 402-6.

[113] Cox CS Jr, Baumgartner JE, Harting MT, *et al.* (2011). Autologous bone marrow mononculear cell therapy for severe traumatic brain injury in children. Neurosurgery 2011; 68(3): 588-600.

[114] Callera F, do Nascimento RX. Delivery of autologous bone marrow precursor cells into the spinal cord *via* lumbar puncture technique in patients with spinal cord injury: a preliminary safety study. Exp Hematol 2066; 34(2): 130-1.

[115] Callera F, de Melo CMTP. (2007). Magnetic resonance tracking of magnetically labeled autologous bone marrow CD34+ cells transplanted into the spinal cord *via* lumbar puncture technique in patients with chronic spinal cord injury: CD34+ cells' migration into the injured site. Stem Cells Dev 2007; 16(3): 461-6.

[116] Deda H, Inci MC, Kurekci AE, *et al.* Treatment of chronic spinal cord injured patients with autologous bone marrow-derived hematopoietic stem cell transplantation: 1-year follow-up. Cytotherapy 2008; 10(6): 565-74.

[117] Yoon SH, Shim YS, Park YH, *et al.* Complete spinal cord injury treatment using autologous bone marrow cell transplantation and bone marrow stimulation with granulocyte macrophage-colony stimulating factor: phase I/II clinical trial. Stem Cells 2007; 25(8), 2066-73.

CHAPTER 9

Pluripotent Stem Cells - A Novel Source of Haematopoietic Cells for Transplantation and Transfusion Medicine

Catalina A. Palma[1], David Ma[1] and Robert Lindeman[2*]

[1]Blood Stem Cells and Cancer Research, St Vincent's Center for Applied Medical Research, St Vincent's Hospital, Darlinghurst NSW 2010 Australia and [2]Department of Haematology, Prince of Wales Hospital, Barker Street, Randwick NSW 2031 Australia

Abstract: Haematopoietic stem cells (HSCs) are commonly used to reconstitute the bone marrow and blood of patients suffering from diseases such as leukaemias, lymphomas and non-malignant conditions like severe bone marrow aplasia and thalassemia major. The mature progeny of HSCs, especially erythrocytes and platelets, are used regularly for rapid transfusion into patients with acute conditions like life-threatening bleeding. For HSC transplantation (HSCT), cells must originate from an immunologically matched donor, and the chronic lack of supply of suitable cells drives the search for alternate HSC sources. Human embryonic stem cells (hESCs) hold the promise for development into every tissue type. Multiple laboratory based studies have demonstrated the careful, ontological approach to HSC differentiation from hESCs, leading to the differentiation of cells which are similar to bone marrow HSCs. Significant challenges remain in achieving HSC replicas, primarily improvement to both efficiency of HSC generation and sustained engraftment *in vivo*. The differentiation of mature blood cells from hESCs, in order to generate large numbers of "off the shelf" cells for transfusions, has been successful with the production of seemingly fully mature and functional cells, albeit in low numbers. Concerns about the need for immunosuppression after hESC transplantation, has meant that the development of patient-specific induced pluripotent stem cells, which have also been differentiated into blood cells, also has enormous potential for use in the clinic. Presently, these cell products have significant barriers to overcome before clinical translation, such as efficient *in vivo* function, large-scale manufacture of consistent cell products and short- and long-term safety issues.

Keywords: Haematopoietic stem cells, embryonic stem cells, immunosuppression, transpalantion.

INTRODUCTION

Haematopoietic stem cell transplantation (HSCT) from immunologically matched donors (allogeneic) has been part of the treatment of haematologic disorders for over forty years [1-3]. Although most allogeneic transplants are performed for haematological malignancies such as acute myeloid or lymphoblastic leukaemias, lymphomas and multiple myeloma, transplantation is also potentially curative for non-malignant conditions such as aplastic anaemia. Reinfusion of autologous haematopoietic stem cells (HSCs) is also important in supporting high dose therapy for many cancers particularly haematological malignancies. Autologous HSCs may be harvested by apheresis from peripheral blood following mobilisation by chemotherapy and/or granulocyte-colony stimulating factor (G-CSF) [4]. Allogeneic HSCs are obtained from the peripheral blood of donors after G-CSF administration over a number of days or harvested from the bone marrow. Human leukocyte antigen (HLA) matched donors are either family members (most commonly siblings) or recruited from international bone marrow volunteer donor registries, while umbilical cord blood (UCB) stem cells also represent a source of haematopoietic progenitors and are now used in transplantation, particularly in children [3, 5].

There are a number of unmet needs in HSCT, which have prompted the exploration for novel HSC sources. Autologous HSCs used to support chemotherapy in the treatment of haematological malignancies are potentially contaminated with tumour cells. Early attempts at purging autologous HSCs of malignant cells by a variety of approaches (including in vitro chemotherapy, treatment with a variety of lectins or positive selection

***Address correspondence to Robert Lindeman:** Department of Haematology, Prince of Wales Hospital, Barker Street, Randwick NSW 2031 Australia; E-mail: Robert.Lindeman@SESIAHS.HEALTH.NSW.GOV.AU

Kuldip S. Sidhu (Ed)

of HSCs based on their expression of the CD34 antigen) led to compromised engraftment and gradually fell out of routine use. Decreasing family size has resulted in reduced availability of HLA-matched siblings in many populations. Although international donor registries have now grown to include several million individuals, identifying potential donors remains problematic for potential recipients belonging to ethnic minorities, while the time to harvest HSCs following the identification of a potentially suitable donor is often measured in months and presents a problem particularly for recipients with aggressive haematological malignancies. Although serological and molecular HLA matching has substantially improved, graft versus host disease is sometimes difficult to predict and modulate, and obtaining stem cells and lymphocytes from the same donor at some time after the donation of HSCs presents logistic difficulties. Cord blood banks have been established over the last 20 years, with considerable literature confirming reduced requirement for strict HLA matching when transplants are performed with cord blood. This, together with the ready availability of cord blood cryopreserved ready for use, has allowed allogeneic transplants to proceed when HLA-matched blood- or marrow-derived HSCs are unavailable. The observation that transplants could be performed using two separate UCB samples opened up the possibility of transplanting adults in whom the cell dose of UCB-derived HSCs from a single cord was otherwise limiting. Another option being explored is the possibility of expanding the number of HSCs in cord blood *in vitro* by exposure to haematopoietically relevant cytokines or to augment their capacity for proliferation by manipulation of the Wnt pathway. Finally, the neutropenic period that occurs with HSC transplants is a source of significant morbidity and mortality [6]. *In vitro* HSC manipulation to boost HSC numbers and enhance their partial differentiation prior to infusion has not been particularly successful, while co-infusion of bone marrow-derived mesenchymal stem cells (MSCs) to enhance early engraftment remains investigational. The advent of reduced intensity allogeneic transplants, in which the efficacy of the transplant in controlling the recipient's malignancy depends more on the immunological effect of the graft than on the intensity of chemotherapy or radiotherapy given prior to the infusion of the graft, has opened up the possibility of transplantation for far older individuals than was previously possible. HSCs have been conventionally used for their capacity to repopulate the marrow after transplantation and for their immunological properties in the allogeneic setting, but HSCs have also been used to enhance the repair of myocardial and other tissues and, after genetic manipulation, as a vehicle for enzyme replacement in individuals with rare inherited deficiencies such as Fanconi anaemia.

Ideal HSCs would circumvent the deficiencies outlined above. They would be abundant, either autologous or allogeneic (perhaps from HLA-matched donors) and able to be obtained from readily-accessible tissues and perhaps stored in banks for immediate use. These HSCs would be able to be differentiated down all haematopoietic lineages to support both short- and long term engraftment, and would also provide immune surveillance when a graft-versus-tumour effect was required to control malignancies. There has been some progress over the last decade towards the development of such a cell, and this chapter documents that progress.

DEVELOPMENT OF HSC TRANSPLANTATION

HSC Biology and their Niche

In adults, HSCs reside in the bone marrow and maintain all cell lineages found in the peripheral blood and immune system. They where identified in 1961 by Till and McCulloch when studying cells from the bone marrow and the spleen that could rescue and individual from radiation damage [7]. They first demonstrated the clonal nature of haematopoietic development by showing that single bone marrow-derived cell could give rise to colonies of differentiated blood cells [8] and reconstitute the lymphohaematopoietic system of a lethally irradiated mouse [9]. Since then, increasing numbers of cell surface markers have been identified allowing purification of cell populations and are progressively enriched for HSCs [10]. HSCs express the surface sialomucin CD34, an integral membrane glycoprotein which regulates adhesion to the stromal cells in the niche [11]. HSCs also express CD90, CD117 and CD133 but not lineage markers such as CD38, CD45, CD71 or HLA-DR [12].

HSCs are rare cells, with a frequency of 1 in 10^4 to 1 in 10^5 amongst bone marrow cells. Cord blood HSCs are morphologically small- to intermediate sized lymphoid cells, with high nuclear to cytoplasmic ratio and prominent nucleoli [13]. Most HSCs are preferentially quiescent *in vitro* (G_0 stage) although they do have a very high proliferative potential [14, 15]. Their proliferation (into daughter cells with equal potency) is dependent on a careful balance of cytokines, such as stem cell factor (SCF), thrombopoietin (TPO), FLK3 and interleukin (IL)- 6, as well as temperature and cell density. Nonetheless, HSCs are able to sustain the rapid turnover of blood and immune cells though the production of highly proliferative progeny, the haematopoietic progenitor cells (HPCs). These cells are able to undergo only a finite number of cell divisions, and therefore have a smaller proliferative potential than HSCs [16-18]. Their role is to rapidly increase the number of tissue cells while allowing stem cells to remain quiescent, a mechanism thought to have evolved as a defence against cancer progression: the reduced number of doublings protects stem cells from transmitting mutations to their progeny, while mutant differentiating cells are destined to being eliminated by rapid cell turnover [19, 20].

It was recognised in the 1970's that umbilical cord blood (UCB) also contained HSCs [21, 22] that constitute a valuable resource for transplantation [23]. Importantly, HSCs from UCB have higher expansion capacity than HSCs from peripheral blood [24] and are present at higher concentration [25, 26]. Requirements for HLA matching between donor and recipient are less stringent than for other HSC sources due to the more immature nature of cord blood cells.

Although HSCs are preferentially quiescent, they do have varying proliferative rates, with some being deeply quiescent reserve pool and others rapidly entering into cell cycle upon external stimuli. The crucial decision between self –renewal or differentiation and quiescence or proliferation is tightly balanced by the integration of intrinsic factors and extrinsic cues provided by the microenvironment, comprising of supporting cells and extracellular matrix (ECM) which together are called the niche (see also chapter 3). Schoefield *et al.* first postulated that the microenvironment plays a critical role in stem cell regulation, where it not only protects stem cells, but also contributes to maintaining their quiescence and regulates their number [27]. This multi-component model involves many factors, from other cells in the niche with direct cell-cell contact with HSCs, to cells secreting paracrine factors, ECM components, the chemical signature of the microenvironment and long range signals from the circulation [28].

The HSC niche contains a variety of non-haematopoietic cell populations including neural, vascular, adipocytic and fibroblastic elements, of which the osteoblasts is one of the major constituents identified to date. Osteoblasts are cells which produce bone, they are found to be present on the endosteal surface, in the same area as primitive HSCs [29, 30]. It was first shown in two distinct mouse models that increased osteoblasts number and activity would lead to an increase in HSC number [31, 32], while inducted ablation of osteoblasts would lead to haematopoietic failure [33]. It was also noted that osteoblasts have the potential to produce an array of cytokines, important for the expansion of haematopoietic myeloid progenitors, including G-CSF, granulocyte macrophage-colony stimulating factor (GM-CSF), IL-6, IL-1 and Transforming Growth Factor (TGF)-β. Early studies demonstrated that the co-transplantation of osteoblasts purified from long bones and Lineage negative bone marrow cells enhanced engraftment and haematopoietic reconstitution in lethally irradiated mice. HSCs are sensitive to proteins secreted by osteoblasts, such as osteopontin which acts to negatively regulate HSC proliferation, as demonstrated by an increase in HSC numbers, and BrdU uptake in bone marrow cells of Osteopontin deficient mice [34, 35]. Angiopoietin 1 (Ang1) is also secreted by osteoblasts and has been shown to maintain HSCs in a quiescent state through Ang1/Tie 2 signalling, by increasing their *in vivo* long term repopulating activity [36]. Recently, TPO was recognised as a mediator of osteoblast-HSC interaction in the niche. HSCs expressing the TPO receptor Mpl are quiescent bone marrow closely associated with the TPO producing osteoblasts, while exogenous TPO administration increased the entry of HSCs into G_0 prior to their proliferation. Conversely, inhibition of TPO signalling reduced the number of quiescent HSCs, releasing them from the niche [37]. Nonetheless, a direct correlation between osteoblasts and HSC number is however simplistic, are there are reports describing osteoblasts alteration that do not lead to HSC loss [38, 39].

Other cells in the bone marrow may also be involved in the regulation of the niche. One study demonstrated that adiponectin, an adipocyte-secreted molecule, increases proliferation of HSCs, while maintaining their mature state [40]. However a subsequent study demonstrated that the content of adipocytes in the marrow inversely correlated with haematopoietic activity [41]. The participation of osteoclasts has also been suggested, who demonstrated that upon activation, osteoclasts participate in the mobilisation of haematopoietic progenitors in the circulation through the cathepsi K-mediated cleavage of CXCL12 [42]. Furthermore, the osteoclastic bone resorption actively changes the calcium concentration in the bone marrow microenvironment, which has been previously implicated in HSC engraftment. HSCs deficient in the calcium-sensing receptor were unable to engraft bone marrow [43]. Despite these recent studies, much remains to be elucidated about the biology of the HSC niche.

HSC Transplantation

Allogeneic human HSCs from donor peripheral blood, bone marrow and UCB have been used to support bone marrow reconstitution and as a source of a graft immune response after conditioning chemotherapy with or without radiotherapy for a wide variety of haematological malignancies including acute myeloid and lymphoblastic leukaemia, lymphoma and multiple myeloma [44]. Autologous peripheral blood-derived HSCs are used to support reconstitution after high dose chemotherapy [4]. Allogeneic transplantation is also potentially curative for a variety of rare bone marrow failure syndromes such as Fanconi anaemia, dyskeratosis congenital, Blackfan Diamond disease and Kostmann Syndrome, and has been successfully used to correct congenital immunodeficiency syndromes and haemoglobinopathies such Sickle cell anaemia and β-Thalassaemia major (Table **1**) [45, 46]. Most transplants have been performed from an HLA matched sibling donor. Recent improvements in immunosuppression, HLA matching and supportive care have led to better outcome of transplants from HLA-matched unrelated donors. The high morbidity and mortality associated with transplantation for haematological malignancies make it critical to carefully weigh factors including alternative therapeutic approaches, predictors of the likelihood of their success, patient age and fitness for the rigorous process of a transplant, and graft factors such as the available cell dose (particularly with a cord blood source), and the extent of HLA matching, in determining whether or not to proceed with transplantation and in selecting between standard and low-intensity approaches to conditioning.

Although most transplants have been performed as treatment for haematological malignancies, there are a number of rare indications for transplantation where replacement of haematopoiesis is potentially curative, such as severe aplastic anaemia. Inherited metabolic diseases (IMD) are rare disorders caused by enzyme deficiencies and characterized by the accumulation of toxic metabolites in various tissues, and include the mucopolysaccharidoses and sphingolipidoses and leucodystrophies. Allogeneic HSC transplantation can halt progression of these diseases by providing a constant source of the missing enzyme through engrafted donor leucocytes [47].

Table 1. Common indications for HSCT in Australia and New Zealand in 2009 [48]

Haematological Malignancies	Multiple myeloma	489
	Non-Hodgkin's lymphoma	365
	Hodgkin's lymphoma	103
	Acute myeloid leukaemia	189
	Acute lymphoblastic leukaemia	54
	Chronic myeloid leukaemia	16
	Myelodysplastic syndrome	18
	Myeloproliferative disorders: polycytehmia, thrombocythemia	10
	Plasma cell leukaemia	3
	Chronic lymphocytic leukaemia	22
Bone Marrow Failure Syndromes	Diamond-Blackfan anaemia	1
	Severe aplastic anaemia	8
	X-linked lymphoproliferative syndrome	2
	Idiopathic chronic myelofibrosis	11

Primary immune deficiencies such as SCID, Wiskott-Aldrich Syndrome, leukocyte-adhesion deficiency (LAD), X-linked hyper IgM syndrome (X-HIM), haemophagocytic lymphohistiocytosis (HLD) and immunodysregulation and others, result from mutations in genes required for the production, function or survival of specific leukocytes. Curative therapy for these life-threatening diseases has been HSCT. The fact, however, that all that is required for their reversal is correction of a single gene defect in the lymphoid progeny of HSCs has made them ideal models for gene therapy approaches to modification of autologous HSCs in a transplant setting. The first syndromes for which gene therapy was attempted were x-linked severe combined immunodefiency (SCID) [49-51] and adenosine deaminase deficiency (ADA)- SCID [52]. These life-threatening diseases are typically treated by transplantation of HLA-identical HSCs, with survival rate of 90%, with slightly lower survival rates in a matched unrelated donor or haploidentical setting. Patients suffering from X-linked SCID were the first to undergo genetic manipulation of HSCs. Stem cells collected from bone marrow were transfected with the missing γ chain gene using γ retroviral vectors. Following reinfusion, functional T lymphocytes and NK cells were identified in 4 out of 5 children, with almost normal numbers and T cell phenotype and response to immunisation after 2 years. Among the patients who have subsequently undergone gene therapy with retrovirally transduced autologous stem cells there have, however, been five reports of leukaemia, presumably caused by insertional mutagenesis [53, 54]. A ten year follow up of the treated patients found T cells still carried the introduced gene, and seven out of 9 patients, including three ALL survivors, had sustained immune reconstitution [55].

Barriers to Transplantation

Remaining barriers to transplantation include donor availability, recipient vulnerability to infection following conditioning therapy, and difficulties achieving the optimum graft-versus-host reaction, which is beneficial in controlling malignancies but a cause of significant morbidity and mortality in patients following allogeneic transplantation.

International bone marrow donor registries have been established to provide HLA-matched unrelated donors for recipients without HLA-matched sibling donors and are an increasingly used HSC source (particularly in the context of smaller families and increasing recipient and therefore sibling donor age). An HLA-matched donor can usually be identified among the 13 million potential donors in the international registries, but the likelihood of finding a donor is lower for recipients among many ethnic minorities, and the delay to confirmation of the suitability and availability of a donor is still a significant factor in transplantation for haematological malignancies. Cord blood banks have filled some of the gaps in ready availability of donor material, particularly in view of the reduced requirement for HLA matching and the more rapid availability of "off the shelf" cryopreserved material, but the cell dose of HSCs in cord blood remains a limiting factor, particularly for adult recipients and engraftment occurs more slowly than after an adult HSC source [56-59]. Attempts to circumvent this problem by transplanting adult patients with two UCB samples has shown some promise, while attempts at expanding the CD34+ compartment *ex vivo* are ongoing [60].

Molecular HLA typing has resulted in improved HLA matching between unrelated donors and recipients, but extensive polymorphisms have also increased the complexity of selecting a suitable donor. There are also determinants outside the HLA system (minor HLA) which influence the likelihood of graft-versus-host disease following transplantation, and separation of a "graft-versus-leukaemia" for "graft-versus-host" effect remains an unmet target in transplant biology. All of these factors contribute to ongoing interest in engineering autologous or novel allogeneic cells to generate better HSCs for transplantation. In addition, the contribution of transplanted HSCs to non-haematopoietic tissues suggests the potential for engineered HSCs for treating other disorders.

Therefore there is a continual need not only for a renewable source of HSCs for transplantation for haematopoietic reconstitution, but also for mature cell types for the benefit of transfusion medicine. Human embryonic stem cells (hESCs) and induced pluripotent stem cells (iPSCs) are now known to be able to form mature cells of the haematopoietic lineage *in vitro*, including HSCs, neutrophils, NK cells, T cells, dendritic cells erythrocytes and platelets. This would provide an important adjunct population to that collected from the limited number of donors, and could also be a source of cells for immune therapies.

hESCs have been successfully differentiated into fully mature and functional subsets of each embryonic germ layer including cells from the haematopoietic system, and they remain one of the most promising sources for regenerative medicine. Phase 1 clinical trials using hESC-derived oligodendrocytes for spinal cord injury [61] were approved by the Food and Drug Administration in the USA in 2009 although the outcome remains to be published. Studies on the derivation and differentiation of iPSCs are also rapidly advancing [62-65]. Therefore the prospect of using hESC and iPSC derived haematopoietic products for diverse clinical therapies is not a distant prospect but a reasonable expectation.

One of the broad benefits of utilising hESC or iPSC derived cells is the reduced risk of transmissible infection agents and the decreased heterogeneity of transplantable tissue. A bank of hESCs, spanning all HLA subtypes, would provide limitless source of cells for transplantation and transfusion therapies for people of all races and ethnicities. Another advantage of hESC derived HSCs would be their immaturity compared to bone marrow derived HSCs, and advantage shared by UCB. Aged bone marrow accumulated defective elements and is clinically associated with lower long-term survival and greater incidence of graft vs host disease [66]. Future work may focus on producing drug resistant hESC-derived HSCs which may allow for more effective administration of chemotherapy to kill malignant tissue. Indeed, HSCT protocols are limited by the age of the patient, as the current myoablative conditioning regimens are characterised by high mortality rates among the older and weaker patients.

This chapter therefore focuses on the potentially clinically relevant use of haematopoietic cells derived from hESCs and iPSCs. The haematopoietic system is excellent candidate for hESCs or iPSCs reconstitution, particularly as standardised protocols are readily available for purification of uniform populations of cells. However, the ability to generate fully functional hESC-derived HSCs capable of long-term multilineage reconstitution of mouse models and ultimately patients remains a challenge and will depend upon further understanding of intrinsic gene regulation and extrinsic environmental cues.

PLURIPOTENT STEM CELL TYPES

Embryonic Stem Cells

The first hESC line was established from the inner cell mass of donated pre-implantation embryos created by artificial reproductive technology [67]. Typically the inner cell mass is separated from the trophectoderm using immunosurgery and mechanical dissection and are then plated onto feeder layer of cells that supply both soluble factors and extracellular support structure(see also chapter 1). These cells attach and grow over time and expand to form hESCs [67]. Worldwide over 300 hESC lines have been generated [68].

ESCs express a variety of markers which are common of pluripotent and undifferentiated cells including *Octamer-binding transcription factor* (*Oct*)- *4* and *Nanog*. They also express *Stage-specific embryonic antigen* (*SSEA*)*-3* and *-4* and *Tumour rejection antigen* (*TRA*) *1-60* and *1-81*. A core of transcription factors maintains their pluripotency, *OCT4, NANOG, SOX2* and *TCF3* [69]. Transplantation into immunodeficient mice form teratomas, demonstrating their innate pluripotency. hESCs when cultured under correct conditions, are capable of indefinite self renewal [70] and can be expanded to large numbers. This is thought to be possible due to their high levels of telomerase, which prevents telomere shortening and cell ageing, and underlies their ability for unlimited self-renewal [71-73]. Also hESCs do not appear to undergo replicative senescence, which normally limits adult stem cells to a maximum of approx 50 divisions. hESCs have been differentiated into cells of all three germ layers including neurons and glia, cardiomyocytes, endothelial cells, hepatocyte-like cells, insulin-producing cells and almost all blood cell lineages [74-76]. This provides us with a resource to potentially generate clinically relevant numbers of defined cell populations for tissue repair and regeneration.

Induced Pluripotent Stem Cells

iPSCs are produced by reprogramming somatic cells, such as fibroblasts, by transfer of transcription factors regulating pluripotency. Initially pioneered by Yamanaka *et al.*, who used *OCT4, SOX2, KLF4* and *cMYC* to derive mouse then human iPSCs [77], these cells are similar to hESCs in their morphology, gene expression, and the epigenetic status of their pluripotent-cell specific genes [77-79]. These cells were found to be able to differentiate into all three germ layers and form teratomas *in vivo* [79].

iPSCs were initially derived from somatic cells after lentiviral transduction of pluripotency genes which were randomly integrated into the genome of host cells. This caused concern as these might induce site induced mutagenic effects leading to abnormal cell phenotype. For this reason, along with safety concerns involved with the use of viral transduction, there have been efforts to establish a non-viral method of cell transduction. Cells have so far been successfully transfected using plasmids [80] or transposon systems [81], and with direct delivery of recombinant proteins [82] of the reprogramming transcription factors. Although the pluripotent transgenes become silenced during reprogramming, their re-activation during cell differentiation (especially of the proto-oncogene *c-MYC*) has been associated with tumour formation. This has lead to the view that *c-MYC* should be avoided for reprogramming of cells to be potentially used in clinical applications, however this affects the efficiency of iPSC generation which is known to better when *c-MYC* is used as this gene appears to "prime" cells for iPSC conversion [83, 84].

Somatic cells from ectoderm, endoderm and mesoderm have been used for iPSC geneeration, including murine and human fibroblasts [77, 78], B lymphocytes [85], human HSCs [86], keratinocytes [87], hepatocytes [88] as well as stem cells from haematopoietic [86], mesenchymal [79] and adipose [89] origin. Although a direct comparison between the efficiency of reprogramming of all this cell types has not been made, it appears that there is significant variability, with neural stem cells appearing to be the most amenable to reprogramming, with evidence that they can be transformed with a single transcription factor [90]. Furthermore it appears that individual iPSC clones may vary in their ability to differentiate, even within the same cell culture. This seems to be an artefact of the reprogramming process, where the cell's uptake of pluripotent transcription factors is not uniform, and therefore some colonies form which have aberrant differentiation potential. Therefore, before iPSCs can be used in a clinical setting, a standardised non-virally integrating method of reprogramming, and a method of purifying the iPSCs with true pluripotent potential will be required.

Different somatic cell iPSC origins have distinct differentiation properties, consistent with a genetic and epigenetic "memory" even after pluripotent reprogramming. In early passages after reprogramming, the transcriptional profile and epigenetic patterns of the original cell persist [91], and iPSCs from different sources do not have uniform differentiation capacity. For example, cells generated from murine adult B-cells or haematopoietic progenitors produce mixed haematopoietic colonies more frequently than iPSCs generated from adult fibroblasts [91, 92] because of residual methylation at gene loci required for haematopoietic development [92], while iPSC derived from hepatocytes have limited neuronal differentiation potential and are prone to teratoma formation [93]. Although these distinguishing transcriptional and epigenetic marks appear to gradually disappear with higher passage numbers, this "memory" may impact the clinical potential of iPSCs. Despite these difficulties, the possibility of generating autologous patient-specific iPSCs for replacement therapy, circumventing HLA matching considerations, continues to drive investigation in this area, particularly as their generation from terminally differentiated lineages avoids the need to use embryonic tissues and the ethical issues associated with their use. iPSCs have also been used in the investigation of disease biology and for pharmacological testing, as reprogrammed cells retain their genetic abnormalities. Human iPSCs disease lines now available include Down Syndrome, ADA-SCID, Guacher disease, Duchenne muscular dystrophy, Parkinson disease, Type 1 diabetes, Swachman-Dodian-diamond syndrome, Huntington disease and Lesch-Nyhan syndrome cell lines [94].

DIFFERENTIATION OF PLURIPOTENT STEM CELLS TO HSCS

Early Haematopoiesis- An Ontogenic Approach

When removed from the specific culture conditions that sustain their self renewal, ESCs and iPSCs spontaneously form embryoid bodies (EBs) that contain derivatives of three germ layers including cells of the haematopoietic lineage [95]. From this unguided approach, which proved the ability of ESCs to generate blood cells, there has now been a push towards directed differentiation of hESCs and iPSCs using combinations of growth factors, extracellular matrix proteins and/or genetic manipulation, which allows for differentiation along developmental pathways of normal ontogeny towards a mature cell [96, 97].

Analysis of *in vitro* differentiating pluripotent stem cells has revealed a close resemblance to *in vivo* embryonic development [98-100], particularly after the exposure of cultures to morphogens and inductive signals known from *in vivo* developmental models (such as the zebrafish, frog and chick), which enables directed differentiation [97]. From a developmental standpoint, ESCs are equivalent to early epiblast cells and can mimic aspects of early embryogenesis if guided by growth factors (eg. Wnt, TGF-β, Activin and BMP). ESCs progress through a primitive streak-like stage before forming the three germ layers, endoderm, ectoderm and mesoderm [101]. While early embryonic stages are able to be replicated [98], spontaneously differentiating EBs lack the anatomical structure of the embryo, which is both temporally and spatially defined, where posterior, mid and anterior regions of the primitive streak have differing gene expression patterns and developmental potential. Further, cells developing in EBs do not participate accurately in embryonic cell-to cell interactions, and miss exposure to physical stimuli that biologically occur at different stages of development (eg. the flow pressure and cellular movement occurring with onset of embryonic circulation [102]). It is therefore likely that *in vitro* differentiating ESCs are prone to generate immature, embryonic-type progenitor cells, which may not function properly upon transplantation in adults.

Cell fates during gastrulation are carefully regulated by the signalling environment, where members of the TGFβ family including BMP4 [103] and Nodal [104, 105] together with the Wnt family [106] are essential for the formation of the three germ layers. Together with regionalized inhibitors these factors form signalling domains that regulate germ layer specification in the embryo [107] and in ESCs. In many ESC differentiation studies, expression of *Brachyury* is used to monitor the formation of a primitive streak-like population and the onset of mesoderm induction. Gene expression analysis distinguishes between the formation of anterior primitive streak-like cells (expressing *FOXA2* and *Goosecoid* [108, 109]), developing mostly into endoderm, and posterior primitive streak-like cells that will differentiate into mesoderm and from there through activation of patterning genes (*e.g. Cdx/Hox, Evx1* [110, 111]), into haemangioblasts which are common precursors of HSCs and endothelial cells [101].

Blood develops in distinct waves consisting of primitive haematopoiesis and definitive haematopoiesis. In human primitive haematopoiesis, haemangioblasts are generated in the blood islands of the extra-embryonic yolk sac in the 21 day old human embryo [112]. Haemangioblasts have been identified in *in vitro* differentiating mESCs, hESCs and iPSCs after EB mediated differentiation with BMP4 and VEGF in particular [113-120]. Mouse haemangioblasts express the mesodermal marker *Brachyury* and can be identified by expression of *fetal liver kinase 1* (*FLK1*) [115] and *PDGFR* [121]. Human haemangioblasts are found in the CD34+CD45- cellular population and express KDR (VEGF receptor 2) [113] CD31 [122], and angiotensin-converting-enzyme (ACE/CD143). BMP4 signalling is required to induce the development of FLK-1 positive haemangioblasts from *Brachyury*-expressing populations [123, 124], while Wnt, Activin/Nodal and BMP signalling is required for haematopoietic mesoderm development from a Brachyury positive primitive streak population. Importantly it has also been shown that upregulation of Flk-1 correlates with an irreversible commitment to mesoderm fate [123].

Definitive haematopoiesis first occurs from a population of cells in the dorsal aortal region called the aorta-gonads-mesonephros (AGM) from day 27 in human embryos [125-127]. The AGM gives rise to multilineage myelo-lymphoid progenitors [128] and to definitive haematopoiesis characterised by enucleated erythrocytes synthesising fetal hemoglobin. In the embryo these regions harbouring haemangioblast development are believed to then seed intra-embryonic sites, such as the liver, spleen and bone marrow, which are the major sites of haematopoiesis until a few weeks after birth. The bone marrow then normally becomes then only source of haemopoiesis, producing HSCs capable of multilineage differentiation into all blood lineages [97, 129, 130]. In vitro, ESCs derived progenitors start expressing the haematopoietic antigen CD45 [119]. The emergence of CD34+CD45+ cells correlates with derivation of haematopoietic progenitor cells with colony forming potential in which can then give rise to haematopoietic colonies [122].

Two different protocols, or combinations of them, have typically been used for the generation of haematopoietic cells from pluripotent stem cells: the formation of EBs and/or culture on supportive stromal layers. Most of the protocols which are described below, highlight the need for careful sequential exposure of ESCs to growth factors as their temporal exposure, as well as concentrations, will alter cellular fate.

EB Formation

EB formation has been used by many groups to derive cells with haemangioblast potential [113, 119, 131-134]. Initial studies using EB-mediated haematopoietic differentiation from hESCs used a combination of FBS and defined cytokines (SCF, FLT3-L, IL-3, IL-6, G-CSF [135]). These studies described the spontaneous gain of haematopoietic markers (CD34, CD45) in small percentages of the cell population, and correlated it with the progressive acquisition of haematopoietic gene expression in an ontological progression: loss of *OCT4*, gain of *MIXL1, BRACHYURY, FLK1, RUNX1, KDR, GATA1, GATA2, P.U.1* and *CD34*. These cells could also produce haematopoietic colonies at low frequencies [135]. However this technique for differentiation remained inefficient and subject to the spontaneous differentiating ability of hESCs.

Zambidis *et al.* (2005) attempted to recapitulate the yolk-sack stages of embryonic hemato-endothelial development. EBs were cultured with FBS and hybridoma medium II PFHM II for upto 4 weeks. Cystic EBs became evident from 8 days after the initiation of culture. CD34 and CD31 expression peaked at day 12-15 coinciding with peak expression of key haematopoietic transcriptional regulators (*SCL, GATA1, GATA2, EKLF, PU.1*). Haematopoietic activity was shown to arise in EB cells from CD45 negative progenitors that give rise to haematopoietic blast colonies, followed by two sequential yolk sac-like primitive and definitive haematopoietic waves. The CFU potential of EBs differentiated in serum-free medium with haematopoietic growth factors showed that 6-9 day old EBs initiated yolk sac-like, primitive and definitive haematopoieses. Primitive macrophages and large nucleated erythroblasts expressing embryonic and fetal hemoglobins were observed during days 7-12 of EB differentiation. This was followed by some evidence for definitive erythropoiesis (that had limited amounts of adult hemoglobin), myelopoiesis, and megakaryopoiesis typical of a late yolk sac stage of development at days 12-20 of EB development. Progressive expression of *SCL/TAL1*, *RUNX1/AML1*, *GATA1*, and *GATA2* (genes known to play roles in the initiation of haematopoiesis from murine haemangioblasts) during this period indicated that this model could mimic early embryonic haematopoietic development [131].

Keller *et al.* (2007) presented a protocol where using a two step approach for EB mediated haematopoietic differentiation, the first driven by BMP4 and bFGF for primitive streak/mesoderm induction, the second by the addition of vascular endothelial growth factor (VEGF) for haematopoietic specification. The kinetics of EB-derived haematopoiesis demonstrated expression of *SCL, GATA-1* and *RUNX1* from day 3, consistent with the progression to CD34 expression detectable from day 4. The first haematopoietic colonies were developed from day 5 cells, and consisted predominantly of primitive erythroid progenitors, followed by erythroid colonies and macrophages that emerged at day 8 of differentiation [113]. Further development of this protocol into a 3 step process includes the addition of IL-6, IL-3, IL-11 and SCF to induce further haematopoietic maturation. This was assessed by the formation of erythroid, myeloid and macrophage colonies and the expression of CD45 which reached 72% by day 22 [136].

To try and assess the *in vivo* developmental potential of hESC-HSCs, Park *et al.* (2009) transplanted CD34+ cells from EBs differentiated in semi solid media into the yolk sac of the chick embryo. It was found that there was a 60% engraftment rate, and the cells engrafted haematopoietic organs, as assessed by their human CD45+ expression. These cells yielded human glycophirin A + erythroid, CD13+ myeloid and CD19+ lymphoid cells in the spleen and bursa of Fabricius. Cells from the spleen of chimeric embryos were also able to form CFU-GM in a methylcellulose assay. This study demonstrated that hESC-HSCs can home to intraembryonic blood forming organs in a developmental model [137].

Wang *et al.* (2005) performed one of the first *in vivo* studies and detailed transcriptome analysis of hESC derived CD34+ cells (hESC-CD34). They induced haematopoietic differentiation by EB formation and culture with SCF, GCSF, FTL3-L, IL3, IL6 and BMP-4, isolated CD45+ cells of which 9% were CD34+. Direct comparison to UCB derived CD34+ cells demonstrated that the hESC derived cells had similar CFU generating capacity, however these cells were approximately 3 times more likely to form CFU-G colonies and lacked BFRU-E and CFU-M potential of somatic CD34 cells. Transplantation into NOD/SCID mice demonstrated very low levels of engraftment (1%). Further analysis by microarray to detect gene expression differences between the hESC-CD34 and UCB-CD34 cells revealed differences in broad families of genes associated with cell-cell contact and migration, cell replication, and transcriptional regulation (including HOX clusters A and B). hESC-CD34 cells also expressed genes involved in a accelerated cell-cycle progression and loss of stem cell serf renewal ability, which may explain hESC-derived cell's limited proliferative capacity and low levels of human chimerism. Previous studies have demonstrated that haematopoietic progenitors derived from mESCs could be endowed the ability for *in vivo* reconstitution by ectopic expression of HoXB4 [138]. Transduction of hESCs with HOXB4 retrovirus led to increased cell expansion of hESC-CD34 cells by 2.5 fold, indicating an effect on proliferation capacity, however in contrast with previous studies it did not affect colony formation of mouse engraftment [139].

Further studies have attempted to define the differences between hESC-CD34 cells and human fetal blood CD34+CD45+ cells. Functionally hESCs had a much lower CFU capacity of 2.1% compared to fetal blood CD34+ cells which had a 37% capacity. Microarray demonstrated that these two cell types are genetically very different, with only 79 out of over 3000 genes being commonly expressed. Studies demonstrate that hESC-CD34 cells typically fail to reconstitute the mouse because of cellular aggregation causing fatal emboli formation. Gene analysis then demonstrated that several adhesion molecules were overexpressed, such as *NINJ2, CD47, CD62* and *ICAM1*. Further, they have low expression of the homing molecules *CXCR4* and *N-cadherin*, higher expression of pro-apoptotic genes and a dysregulation of genes controlling cell cycle [140]. The Wnt and Notch pathways are known to be important in fate specification for the embryo and for hESCs. It was found that in hESC-CD34 cells several members of the Wnt and Notch pathway, including ligands and FZD receptors and NOTCH4/HERP1/HEY1/HEY2, were expressed at very low levels. This would need to be therefore supplemented in culture. This need for a wide variety of cytokines in order to support the correct development of hESCs supports a role for a stromal layer in the specification of definitive haematopoietic cells from hESCs.

Notably, two groups have tried to develop methods to make EB-mediated haematopoietic differentiation of hESCs more consistent and efficient such as the forced aggregation of EBs by centrifugation [133, 141]. Bauwens *et al.* demonstrated that the differentiation lineage of hESCs was affected by colony size and EB size, where larger colonies were found to favour mesoderm development and smaller colonies favouring neural development [141]. This study highlighted the inherent difficulties associated with differentiation dependent on EB formation. Spin EBs are EBs produced from a define number of cells that are centrifuged to produce tight cell compaction, thereby improving experimental reproducibility compared to standard EB derivation methods. The number of cells used for EB formation and various augmenting factors were found to be critical: optimum haematopoietic differentiation was achieved when spin EBs were formed from at least 1000 hESCs and different cytokines were required at each of two distinct differentiation stages for efficient differentiation. This protocol yielded CD34 expression at day 6, concurrent with the expression of *RUNX.* CD34 expression peaked at day 10 and was maintained for a further 15 days, when upto 30% of the cells expressed CD45. More primitive blast forming units erythroid colonies (BFU-E) emerged on day 6, whereas more mature colony forming units (CFU-Es) and myeloid colonies arose in later points of the differentiation [133].

In an clinically important study (2011) hESC and iPSC derived CD34+ cells and progenitors were generated in a serum-, feeder- and animal- free system, making this a potentially clinically relevant protocol. In this two step protocol, EBs were first exposed too BMP-4 and bFGF for 3 days, followed by culture with VEGF, bFGF, SCF, FTL3-L and TPO for 4 days. This yielded 10% hESC-CD34 cells and 4% iPSC/CD34 cells [142], which is low compared to other studies, but this study has the advantage of being the first protocol which has promise for clinical translation.

Stromal Cells

In studies that first demonstrated haematopoietic development from hESCs, CD34+ haematopoietic precursor cells expression haematopoietic transcription factors were derived using a co-culture method with the fetal murine bone marrow cell line S17 or the yolk sac endothelial cell line C166 in the presence of FBS, without any other cytokines [134]. These cells were capable of forming characteristic myeloid, erythroid and megakaryocytic colonies after 14, 18 and 21 days in culture in this system. After 17 days in culture 2% of the cells had acquired CD34, which occurred in two stages. The first peak of expression occurred in conjunction with CD31+ expression which was postulated to mark the hemogenic endothelium state, while the second peak coincided with CD45 expression which marked the beginning of the cell's ability to form haematopoietic colonies in semi solid medium. These CD34+ cells also expressed *SCL* and *GATA-2*, and were able to form erythroid, myeloid and megakaryocyte colonies. It is important to note that these early type of differentiation protocols relied on the spontaneous differentiating ability of hESCs, and this protocol not only yielded haematopoietic cells, but also a variety of other cell types which are mentioned but not characterised as part of the study. Transplant of these cells into sublethally irradiated NOD/SCID mice (which were treated with an antibody against natural killer cells to avoid immune rejection of the transplanted cells) achieved long term engraftment of 1.8% of human CD45 cells upto 6 months post transplantation. Further, secondary engraftment was achieved following intravenous transplantation of bone marrow from primary recipients, but the level of engraftment was very low at less than 0.5% [122].

To further assess the ability of hESC-HSCs to engraft *in vivo*, cells differentiated on S17 stromal cell line were transplanted into a fetal sheep model, a more developmentally matched model than transplantation into adult mice. Animals received CD34+ purified cells and 3 to 29 months later were tested for human chimerism in the bone marrow and peripheral blood. Rates were low at approximately 0.1% [143].

A similar approach was used to co-culture hESCs with OP9 cells. OP9 stromal cell are derived form the calvarie of newborn op/op mice which lack M-CSF [144] and which has been demonstrated to be supportive of haematopoiesis and lymphogenesis from mESCs [145-147] and hESCs [120, 148]. This method was found to generate a higher level of CD34+ cells than those grown on S17s, 15-20% by day 7 of culture [148, 149]. The kinetics of hESC differentiation on OP9s demonstrate CD34 expression from days 3 to 7, followed by a gain in CD43 and CD41 expression at day 5, and CD45 expression by day 8. This study also demonstrated the presence of haematopoietic transcription factors *GATA-1* and *GATA-2*, as well as *SCL* and *FLK1*. These cells had haematopoietic clonogenic potential, and when further cultured on MS-5 stromal cells with SCF, FLT3-L, IL-7 and IL-3 the isolated CD34+ cells differentiated into B, NK cells as well as macrophages and granulocytes [148].

Other cell lines used include those derived form the AGM region (AM20.1B4), urogenital ridge (UG26.1B6) and fetal liver (EL08.1D2) to induce HSC differentiation. It was found that CD34 expression peaked at day 18-21 of cell culture with upto 16% of cells demonstrating CD34+ expression, which correlated with CD31, CD45 and KDR/VEGFR2 expression. Cells cultured on AGM stromal cells produced the higher level of CD34+CD45+ cells. These cells could all produce colonies on methylcellulose, both myeloid, mixed, and erythroid, they expressed mesodermal transcription factors (*BRACHYURY, MIXL1, SCL1, KDR, PECAM, HOXB4, PU1*). Most importantly, upon transplantation into NOD/LtsZ-Scid ILRy null mice, CD34+ cells from AGM stromal cells were found to engraft, with upto 28% of CD45 cells being of human origin [150]. Co-culture with stromal cells lines derived from the fetal liver (FHb-HTERT cells) promotes maturation and expansion of CD34+ and CD45+ cells beyond that produced by the S17 cell line but at still quite low levels at 4-5% [151].

In an attempt to further define to role that OP9 cells play in inducing haematopoietic differentiation from hESCs, cells were transduced with *HOXB4* and co-cultured with OP9 cells along with G-CSF, FLT-3L, IL3 and IL6. It was found that OP9 cells promote the proliferation and survival of CD34+ cells. CFU assays however demonstrated that OP9 cells did not increase primitive haematopoietic subsets with multilineage differentiation potential as there was no increase in the number of CFU mixed colonies. Furthermore, transplantation into irradiated mice found that culture on OP9 cells did not increase the engraftment potential of hESC-CD34 cells. Consistent with previous studies on hESCs, ectopic *HOXB4* expression did not support the generation of repopulating HSCs from hESCs, indicating that its effects are on hESC derived progenitors, not the repopulating cells [152].

A number of research groups have derived HSCs from hESCs, and transplanted them into the gold-standard model of haematological reconstitution, the non-obese diabetic/severe combined immune deficient (NOD/SCID) mouse. These mice do not reject human cells as they lack B and T lymphocytes and HSC can migrate to the bone marrow and lodge there, generating mature human blood cells that can be detected in the peripheral blood of the mouse. However, the capacity of HSC from hESC to contribute to haemopoiesis in this model appears limited. Transplantation of HSC from UCB results in an average of 10% human cells in the blood, whereas the maximum observed from hESC-derived HSC is 1%. Also, hESC- derived HSCs express comparatively low levels of major histocompatibility complex (MHC) class I, and it is thought that they are consequently removed from the NOD/SCID model by residual natural killer cell activity [122]. Thus, although it is clear that hESC can be used to derive HSCs that are capable of partially reconstituting haematopoiesis in a murine model, there is a need for better models in order to understand the clinical potential of hESC as a source of HSC.

Induced Pluripotent Stem Cells

Although many issues need to be resolved before iPSCs may be considered for clinical use, patient-specific iPSCs would be ideally suited for cellular therapy, since they are derived from the individual patient to be treated, therefore minimizing the risk of immune rejection and therefore increasing survival rates. The potential of iPC-based cell replacement therapy has been established in proof of concept studies conducted in humanized sickle cell anaemic mice in which iPSC were first created following reprogramming of tail-tip fibroblasts [153]. The mutant gene was then corrected with the wild type beta-globin gene *via* homologous recombination. The genetically modified iPSCs were then differentiated into haematopoietic progenitors and transplanted back into sublethally irradiated autologous mice. This resulted in successful engraftment and amelioration of sickle cell disease symptoms. Another study found significant increase in the plasma Factor VII levels and correction of excessive bleeding was observed in haemophilia A mice injected with functionally normal iPSC-derived endothelial cells from allogenic sources [154]

Studies using human cells have found that iPSCs can also be derived from patients with certain diseases, which can then be corrected *ex vivo*. iPSCs have been derived from patients with beta-thalassaemia [155] which may then have the HBB mutation corrected. In an important study, Raya *et al.* (2009) isolated fibroblasts and keratinocytes from patients suffering from Fanconi anaemia. The severe genetic instability of these cells did not allow for generation of iPSCs, however, upon correction of the defect by transduction with lentiviruses containing *FANCA* or *FANCD2* genes, iPSCs could be generated. These genetically corrected cells could then be differentiated into haematopoietic progenitors, as well as cells from the myeloid and erythroid lineages, with the same efficiency as iPSCs derived from normal patients [156].

In many transplant protocols, using iPSCs, or genetically corrected iPSCs, would be advantageous as the derived cells would be isogenic and would not face immune barriers in the host. Furthermore, since iPSCs are derived from terminally differentiated, mature cell such as fibroblasts and lymphocytes, there is no need for the use of human embryos or oocytes as with hESCs or SCNT respectively and are therefore not subject to the same ethical scrutiny. However long term safety of iPSCs needs to be established before clinical use.

It is important to demonstrate that iPSCs have comparable differentiation potential to hESCs. iPSCs generated from human fetal, neonatal and adult fibroblasts, reprogrammed with *POU5F1, SOX2, NANOG* and *LIN28* lentiviruses were differentiated on an OP9 co-culture along with the hESC lines H1, H7, H9, H13 and H14. Similarly to hESCs, iPSCs generated CD34+CD43+ haematopoietic progenitors and CD31+CD43- endothelial cells [64].

POTENTIAL USES OF "OFF-THE-SHELF" HAEMATOPOIETIC CELLS: CELLS FOR TRANSFUSION MEDICINE

Blood cells derived from pluripotent stem cells are a potentially useful replacement for donation-based blood for transfusion in clinical settings. A critical issue for clinical utilisation of hESCs is whether they can generate terminally mature progenies with normal function. Robust generation of murine erythroid cells and self-renewing immature erythroid progenitors [157, 158], megakaryocytes [159, 160], granulocytes [161, 162], eosinophils [163], T and B lymphocytes [164, 165] [166], macrophages [167], dendritic cells [168, 169], NK cells [170] has been reported from mESC. This work in mESC has laid the foundation for interrogating detailed aspects of blood cell biology, and provided a basis for the development of similar protocols in hESC and iPSCs, opening up prospects for use in cell replacement therapies. We will review here recent progress that has been made in blood cell production from hESCs.

Erythroid Cells

Erythrocytes are the most abundant cells in the blood which are present at $5x10^{12}$ cells per litre and function to transport oxygen and carbon dioxide through the body, and are involved in the regulation of vascular tone and the immune response. They circulate for approximately 120 days before they are engulfed by macrophages and recycled. The transfusion of erythrocytes is a well established therapy, with its origins over 200 years old. However, this treatment remains problematic with chronically insufficient supply world wide, the possibility of transmission of infections (from HIV to bacterial) and the requirement for immunological matching. Therefore, the possibility of generating large numbers of infection free, O RhD negative red cells remains a therapeutic goal.

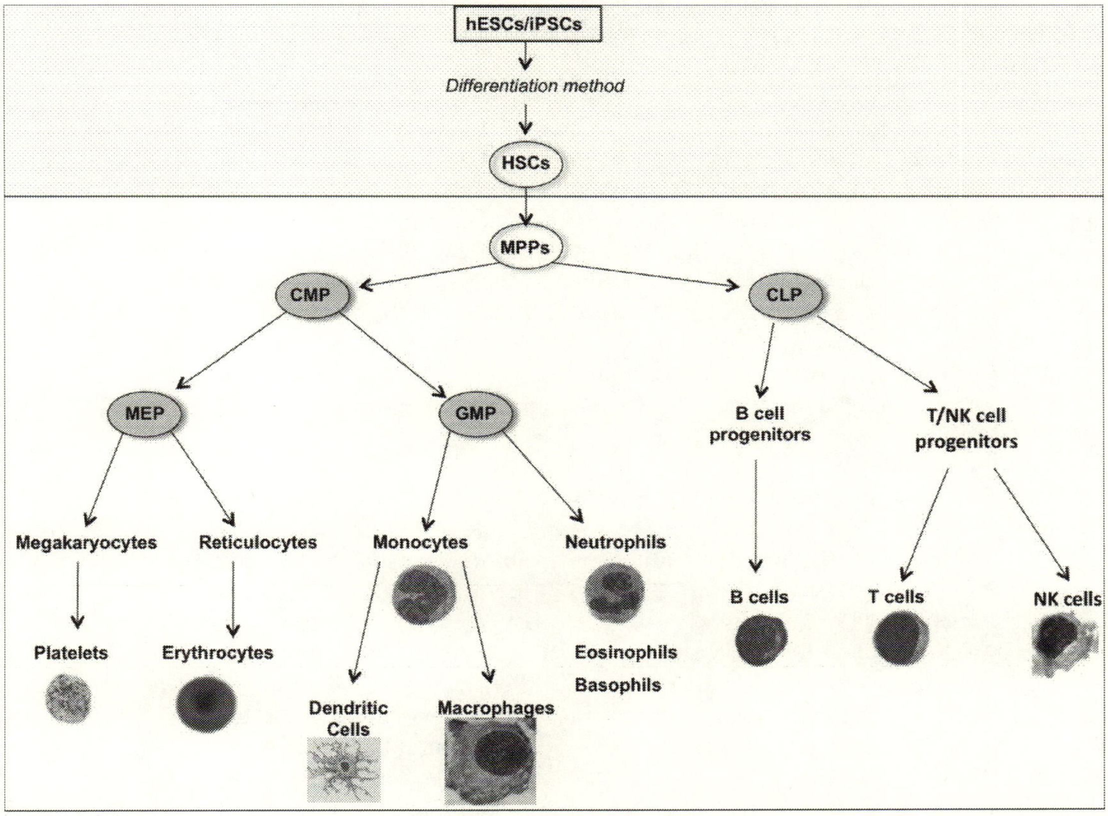

Fig. 1: Simple schematic of human haematopoietic differentiation. *In vitro* (top) differentiation of hESCs and/or iPSCs by EB or stromal cell methods can yield cells with an HSC phenotype. Further differentiation of these cells *in vitro* are thought to mimic the *in vivo* progression through stages of differentiation (bottom) that give rise to progenitor cells in both the myeloid lineage and lymphoid lineage. These progenitors may then undergo further differentiation to give rise to mature blood cells. MPP, Multipotent Progenitor; CMP, common myeloid progenitor; CLP, common lymphoid progenitor; MEP, megakaryocyte-erythoid progenitor; GMP, granulocyte-monocyte progenitor.

In the adult bone marrow definitive erythropoiesis is a multistep process regulated by EPO. It begins when an HSC-derived CMP passes through the megakaryocyte erythrocyte progenitor (MEP) stage and commits to the erythroid lineage (Fig. 1). The appearance of the proerythroblast marks the first stage of differentiation and is subsequently followed by early, intermediate, and late erythroblast stages, at which time the nucleus is expelled and the cell becomes a reticulocyte. Reticulocytes exit the bone marrow and become fully mature erythrocytes within the circulation, expressing adult forms of hemoglobin ($\alpha 2\beta 2$).

Haematopoietic differentiation of hESCs reproduces many aspects of embryonic haematopoietis and provides and *in vitro* model to elucidate mechanisms of early commitment. Phenotyping work has demonstrated that haematopoietic progenitors can be identified by the surface expression of CD43. This reliably separated CD34+ cells from CD34+CD43-KDR+CD31+ endothelial cells. This CD43 marker was also found to be expressed on erythroid progenitors lacking CD34 expression. Meanwhile, acquisition of CD41a was found to be associated with cells already committed to erytro-megakaryocytopoiesis [120].

Like with HSCs, two methods have been used to differentiate hESCs to erythrocytes, EB culture and culture on OP9 stromal cells. Early work by Chang *et al.* (2006) demonstrated that primitive erythrocytes expressing embryonic and fetal globin, but not mature globin, would be seen in developing EBs from 14-56 days in culture [171]. Similarly Zambidis *et al.* demonstrated the progression of spontaneously differentiating EBs through mesodermal-hematoendothelial, primary and definitive stages of haematopoietis [131]. Further work by Kennedy *et al.*, found that after 72-96 hours of EB differentiation with BMP-4 differentiated cells exhibited the phenotype of a human haemangioblast with both endothelial and haematopoietic markers (KDR or FLK1 and CD117) before the onset of primitive erythopoisis was reported [113].

Olivier *et al* [172] reported the large scale production of red blood cells from human ESC line H1. High numbers of erythrocytes (5000 fold increase in cell number) were obtained by a 14-day co-culture protocol with immortalised human fetal liver cells in the presence of serum containing medium with subsequent expansion of CD34+ cells in serum-free liquid cultures supplemented with cytokines and defined factors (hydrocortisone, iL-3, BMP4, Flt-3L, SCF, EPO, IGF-1, hemin). However, the erythrocytes obtained were mostly found to be nucleated primitive erythroblasts, which expressed a mixture of embryonic and fetal globins but not beta-globin characteristic of mature cells [172]. Further work prolonging the differentiation culture on fetal liver cells to 35 days was found to promote the differentiation of fetal liver-like erythroblasts which were smaller in size, expressed fetal hemoglobin and were able to enucleate, suggesting the hESC-derived erythropoiesis mimics human development and recapitulated accordingly the globin gene switch [173].

One of the more clinically relevant published studies is that of Lu *et al.* [174], who have demonstrated the differentiation of hESC into functional erythrocytes on a large scale using a 4-step procedure. The first step involved the formation of EBs by culturing hESC in serum-free medium, BMP4, VEGF and bFGF with the addition of SCF, TPO and FLT3-L after 48 h. Step 2 (haemangioblast induction) involved dissociation of the EBs and culture of the single cell suspension in blast colony growth medium, bFGF and recombinant tPTD-HoxB4 fusion protein for 10 days. The third step (erythoid cell differentiation) consisted of further culture with EPO for 5 days and thereafter with SCF, EPO and methylcellulose. Finally the erythroid cells were enriched through a plastic adherence step. Starting with one 6-well plate of hESC, they generated 10^{11}–10^{12} nucleated erythroid cells with comparable oxygen dissociation curves, response to pH (Bohr effect) and diphosphoglycerate (DPG) activity to normal adult erythrocytes, despite expressing mainly embryonic and fetal globins. However, after 28 d of culture only 16% of the cells expressed adult β globin by PCR and the majority maintained expression of fetal-globin, and just 10–30% of the cells became enucleated, although a large proportion of cells did acquire an erythroid phenotype with a decrease in size, increase in glycophorin A expression and chromatin and nuclear condensation. Also, in order to achieve this large expansion in cell numbers during differentiation, it was necessary to treat the cells with a recombinant HoxB4 protein [174].

The co-culture of hESCs with mouse fetal liver-derived stromal cells has also been used to generate erythroid cells [175]. This method allows for the production of a million cells from a starting population of 10 000 after 14 days in culture. These cells expressed fetal and embryonic globins with 8% expression of beta globin. They also expressed haematopoiesis related genes, such as *CD34, SCL, vWF, KCR, GATA1/2* and *EKLF*. Clonogenic assays demonstrated the development of E bursts units with high expression of βn globin, and with oxygen dissociation curves similar to those cells generated from cord blood [175].

iPSCs have also been used to generate erythocytes. Lapillone *et al.* [176] developed iPSC lines from human fetal fibroblasts and adult fibroblasts and compared their potential to hESC differentiation. Their iPSC lines were reprogrammed with an *OCT4, SOX2, LIN28* and *NANOG* viral cocktail, and erythroid differentiation initiated from EBs. These were cultured with SCF, TPO, FLT3-L, BMP4, VEGF, IL3, IL6 and EPO for 20 days to generate CD34+ cells. These were then dissociated and incubated with SCF, IL3 and EPO for 11 days. This protocol generated small numbers of erythocytes which expressed the markers CD71, CD36 and CD235a and expressed mostly fetal globin [176].

One of the major issues which needs to be addressed with iPSCs is the permanent genomic integration of transgenes. Despite the silencing of these pluripotent transgenes after reprogramming, background expression is still usually detected. To overcome these issues, there has been a shift to produce iPSCs which are virus free, produced by EBV-based episomal vectors. These iPSC lines are produced by reprogramming human neonatal fibroblasts with episomal vectors expressing *OCT4, SOX2, NANOG, LIN28, MYC, KLF4* and *LT* [177]. Dias *et al.*, recently produced erythrocyte progenitors from hESCs and iPSCs using co-culture with the murine MS5 bone marrow cell line. The first haematopoietic colony-forming cells which arise from hESCs and iPSCs have a CD34+ CD43+ phenotype, coexpress CD41a and CD235a and possess erythroid and megakaryocytic potentials. The lin-CD34+CD43+CD45+/- multipotent progenitors with a broad lymphomyeloid potential emerge 1-2 days later. These cells have GEMM colony forming potential but lack Erythroid colony forming potential. It has been found that over 70% of cells express CD235a after 8 days of expansion with MS5 bone marrow stromal cells, and in the presence of EPO, SCF, TPO, IL3, IL6 and Dexamethasone. These cells continued to expand for upto 90 days. However, even though these hESCs and iPSCs expressed erythroid markers, enucleation only occurred at a rate of 2-10% and cells expressed only small amounts of the β globin [177] which remains a major hurdle towards clinical translation.

Megakaryocytes and Platelets

Platelets place a central role in haemostasis and vascular repair. Their concentration of $3x10^{11}$ per litre makes them the second most abundant cell type in the peripheral blood and because of their short lifespan (7-10 days), platelets constantly need to be replaced in order to maintain normal levels. They do not possess a nucleus and are known to be cell fragments generated through the shearing of megakaryocyte precursors.

Megakaryocytes share a common precursor with erythrocytes, the MEP (Fig. **1**). Progressive commitment of MEPs to the megakaryocyte lineage is mostly regulated by TPO and typically involves an increase in characteristic cell surface markers such as glycoprotein (GP) IIb/IIIa (CD41, also known as integrin αIIbβ3, a receptor for fibrinogen), GPIb (CD42b/c), GPIX (CD42a), and GPV (CD42d), several rounds of endomitosis, and the accumulation of components necessary for platelet biogenesis such as α- and dense-granules, and proteins such as von Willebrand factor (vWF) and platelet factor 4 (PF4), as well as components of the GPIb/V/IX surface complex. Megakaryocyte development also involves an increase in size cause by the cytosolic accumulated of platelet-associated proteins and nuclear polyploidization.

In vitro megakaryopoiesis was first reported in 1995 using CD34+ HSCs as a starting cell source [178], and several other studies have confirmed that HSCs isolated from peripheral blood, bone marrow, and UCB are capable of producing megakaryocytes and functional platelets using standard cell culture methods [179]. Despite bioengineering advances [180], the limited *in vitro* expansion capabilities of primary CD34+ cells make these cells unable to replace donation as the principle source of platelets. hESCs may therefore be a better starting cell population for large-scale *in vitro* production.

The first study to report the *in vitro* production of megakaryocytes from hESCs was published in 2006 using an OP9 co-culture method [181], yet the megakaryocytes produced rarely generated any proplatelet-like structures even though they were of high ploidy. Since then, Takayama *et al.* has reported the successful generation of both megakaryocytes and functional platelets from hESCs and, more recently, from iPSCs [182, 183]. They co-cultured hESCs or iPSCs on C3H10T1/2 stromal cells for 14-15 days, handpicked saclike structures containing CD31+ haematopoietic progenitors and replated single cell suspensions onto fresh stroma in medium containing TPO, SCF, and heparin for 9–23 days. From these 50-60% of cells became polyploid, CD41a/CD42b/CD42a positive mature megakaryocytes, emerged from these cultures and produced platelets containing characteristic morphology, as assessed by electron micrography [182]. A variety of *in vitro* tests confirmed platelet functionality, and a laser-induced vascular injury model in the NOG mouse was used to show that their iPSC-derived platelets readily incorporate into newly developing thrombi *in vivo* [183].

Using differentiation on OP9 stromal cells and supplementation with VEGF and TPO, multipotent haematopoietic progenitors were regenerated along with efficient production of mature megakaryocytes expressing specific surface antigens such as CD41a, CD42a and CD42b [182]. Some megakaryoctes appeared to be shedding their cytoplasmic membranes and platelet-like particles were detected in culture supernatants. Electron microscopy confirmed normal microtubule formation, similar to plasma-derived human platelets, although ESC-derived platelets displayed fewer granules [182]. Stimulation of these hESC derived platelets with thrombin and ADP induced GPIIb/IIIa activation, and filopodia formed upon adherence to fibrogen-coated dishes, indicating functionality of hESC-derived platelets. On average, after one day in culture, approx 5 million platelets were generated from an initial $1x10^5$ hESCs. hESC derived megakaryocytes yielded fewer platelets *in vitro* than their adult counterparts *in vivo*, possibly due to some microenvironmental stimulus lacking in the *in vitro* differentiation system.

The use of both serum and animal feeder layers throughout these studies hinders the ability of this method to be adapted for clinical use and handpicking ES sacs adds considerable time and labor to the process as well. Alternative methods will likely have to be developed for clinical grade, large-scale production. An attempt to develop a serum- and feeder-free method for the generation of megakaryocytes [184] found, similarly to Takayama's studies, that efficient platelet generation from megakaryocytes still required conventional stroma co-culture as well as exposure to TPO, SCF, and IL11. The platelets developed by Lu *et al.*, [184] demonstrated all the ultrastructural and morphological criteria that are typical of blood platelets, and possessed characteristic properties of functional platelets, such as activation by thrombin, an ability to adhere to and spread on fibrinogen, von Willebrandt Factor, and type I collagen-coated surfaces, to aggregate when stimulated with physiological agonists, and to form/retract fibrin clots *in vitro*. A laser-induced murine thrombosis model also confirmed that hESC-platelets were capable of contributing to newly forming thrombi in the injured arteriolar wall *in vivo* [184].

In addition, the yield of megakaryocytes in these systems is low. Gaur *et al.* [14] reported the generation of less than one (0.1 to 0.4) megakaryocyte from each hESC. Although Takayama *et al.* [15] improved on this yield (generating two to five megakaryocyte per hESC), the use of handpicking hESC- sacs significantly hinders the ability to translate this to a clinically relevant scale. The protocol by Lu *et al.*, which uses haemangioblast/BC intermediates, is exponentially more efficient in generating CD41a+ megakaryocytes than previous methods, and can be carried out under defined serum- and animal feeder-free conditions [184]. However, if *in vitro* differentiated hESCs are to become a major source of platelets for transfusion, future work will be needed in order to determine a way to eliminate the need for stroma during the megakaryocytes to platelet step and to increase the efficiency of *in vitro* thrombopoiesis as well.

Dendritic Cells

Human dendritic cells area a type of immune cell which form an important part of the human immune system. Their main function is to present antigens on its surface to other immune cells. They originate from HSCs and can develop through both myeloid and lymphoid lineage differentiation pathways. Slukvin *et al.* [185] were the first group to produce functional dendritic cells from hESCs and did so by using a three step differentiation protocol adapted from a mESC protocol [185]. They co-cultured hESCs on OP9 stromal cells for 9-10 days to promote initial haematopoietic differentiation and then transferred cells to suspension culture with GM-CSF for the next 8–10 days. Live cells were purified and cultured in medium containing

GM-CSF and IL4 for an additional 7–9 days. This protocol produced cells with typical dendritic morphology and which expressed high levels of MHC Class I and II molecules, CD1a, CD11c, CD80, CD86, DC-SIGN and CD40 and were capable of antigen processing, triggering naïve T cells and presenting antigens to specific T cell clones through the MHC class I pathway. Further incubation of these dendritic cells with A23187 calcium ionophore induced expression of the mature markers CD83 and fascin [185].

Two other groups have since reported the generation of hESC-derived myeloid-lineage dendritic cells using EB formation and have done so in a serum-free or serum- and feeder-free manner [186, 187]. Tseng *et al.* utilized a serum- and feeder- free system together with BMP-4, GM-CSF, SCF, VEGF, IL4 TNF-α, IL1b, IFN-γ in a three step process. These hESC-derived dendritic cells had characteristic large eccentric nuclei, spiny dendritic processes and expressed dendritic cell surface markers CD11c, CD40, CD45, CD86, HLA class I, and HLA class II to varying degrees. Yields ranged from two dendritic cells per hESC [186] to 3–5 dendritic cells per hESCs [187]. Despite subtle differences compared to monocyte-derived dendritic cells, hESC-dendritic cells appear to be functional upon maturation in assays measuring IL12p70 secretion, chemotaxis, antigen uptake and proteolysis, induction of T-cell proliferation, and stimulation of antigen-specific cytotoxic CD8+ T-cell responses [186, 187] and have also been found to be susceptible to HIV infection like standard dendritic cells [188].

A major advantage of being able to derive and clinically use allogeneic, hESC– derived dendritic cell vaccines for active immunotherapy is the ability to generate large numbers of well-defined antigen-bearing cells that are amendable to a degree of standardization. Secondly, generation of dendritic cells from hESC lines for human immunotherapy circumvents the possibility that tumor-derived factors elicit dendritic cell dysfunction, as is often the case when autologous dendritic cells are used for tumor immunotherapy

Natural Killer Cells

Human natural killer (NK) cells are generated from HSC- derived CLPs (Fig. **1**) and have a half-life of 7–10 days in the peripheral blood. Their concentration of 1×10^8 cells/L comprises 1-2% of leukocytes, or 0.01-0.02% of all cells in the peripheral blood. NK cells belong to the innate immune system and provide rapid, nonspecific responses against various microbial infections and contribute to tumor cell detection and elimination. The two main populations of NK cells, immature and mature, are functionally different and can be discerned based upon expression of various cell surface markers. Immature NK cells have high cytokine production capacity and low cytotoxicity potential and are typically CD56+/CD16low/KIRlow/CD94+ [189]. Secretion of cytokines by immature NK cells activates macrophages and helps initiate a broad immunological response. In contrast, mature NK cells display low cytokine production capacity and high cytolytic potential and are CD56dim/CD16+/KIR+/CD94low. Their cytolytic functions depend upon the release of granzyme and perforin enzymes which are responsible for lysing and inducing apoptosis in target cells. The transition from immature to mature NK cells is thought to arise in secondary lymphoid tissues, yet the vast majority of NK cells in the PB are the mature CD56dim CD16+ cytolytic type [190].

Studies to derive NK cells were the first to show development of functional lymphocytes from hESCs [191, 192]. A two step culture method was used to support development of CD56+CD45+ hESC-derived lymphocytes that function like mature NK cells. The protocol begins by coculturing undifferentiated hESCs on M210-B4 (a mouse BM-derived stroma cell line that expresses laminin and collagen IV) for 17–20 days. CD34+/CD45+ double positive cells (<5% of the total cell population), are then isolated and transferred onto AFT024 stroma in medium containing SCF, FLT3-L, IL7 and IL15. After 30-35 days, this culture method appears to produce a highly enriched homogenous population of CD45+ CD56+ CD94+ CD117low NK cells. *In vitro* assays demonstrated that hESC-NK cells produced could secrete IFNγ in response to IL12/IL18 stimulation and displayed potent natural cytotoxicity against K562 erythroleukemia cells and ADCC against Raji cells, being able to lyse these tumour cells by both direct cell mediated cytotoxitity and antibody dependent cellular cytotoxicity [191]. These hESC-NK cells were subsequently shown to harbor natural cytotoxicity against other types of cancer cells and displayed *in vivo* antitumor activity in a xenograft mouse model [192]. More recently, this same group has been able to successfully produce functional NK cells from iPSCs and showed that they harbor anti-HIV activity [193]. Despite these exciting findings, the requirement for two different types of stroma co-culture as well as the need to isolate rare CD34/CD45 double positive cells limits the utility of this approach for large-scale, cost-effective, clinical grade production of hESC/iPSC-generated NK cells. Further optimization will need to be performed before such hESC/iPSC-derived NK therapies can move into clinical trials.

Lymphocytes: T and B Cells

As part of the adaptive immune system, T cells develop in the thymus and can be stimulated to mount antigen- specific immune responses against a variety of pathogens and cancer cells. They are present at a concentration of 1×10^9 cells per liter of peripheral blood, representing approximately 10% of leukocytes or 0.1% of all circulating blood cells. T cells can be divided into five main subtypes based on function and cell surface marker expression: effector or memory helper CD4+ T cells; effector or memory cytotoxic CD8+ T cells [194]; immunosuppressive regulatory T cells (Tregs); skin, gut, or lung-resident γδ T cells; CD1d-restricted NK T cells [195].

The first report that hESCs could be differentiated into T cells was published by [196]. hESCs were cultured OP9 cells for 7–14 days, whereupon CD34+ cells were isolated and injected into a Thy/Liv implant within SCID or RAG2−/− mice. These implants, which are small pieces of human fetal liver and thymus under the renal capsule, were sued to provide a more biologically relevant microenvironment for the hESC-CD34 cells. The hESC-derived cells were found to differentiate into CD4+ and CD8+ T cells within the thymus-like Thy/Liv environment at a rate of upto 70% of human CD45+ cells [196]. Galić *et al.* switched to an EB-based method three years later and noted improvements in the Thy/Liv-generate hESC-derived T cells [197]. Their EBs were exposed to BMP4, SCF, FLT-3L before *in vitro* and *in vivo* analysis. CD4+ /CD8+ double positive cells began to appear within 4 weeks while CD4+ single positive and CD8+ single positive cells that had undergone TCR rearrangements emerged within 8 weeks. In order to assess the ability of the hESC-T cells to respond to TLR mediated signals, the cells were stimulated with CD3 and CD28 *in vitro*, which induced the expression of the high affinity interlukin-2 receptor CD25, demonstrating that these cells are function and able to respond to TLR mediated signals [197]. These studies demonstrated the development of T cells, the engraftment of these cells into the SCID-Hu model is typically 1% or less, as measured by GFP+ labelling. In contract, previous studies that used SCID-hum mice to derived T cells from human BM, UCB or fetal liver have significantly higher levels of engraftment. Therefore this model is inefficient.

Extrapolating from results in various mouse studies, the use of immobilized Notch ligand, such as Delta1, has proven to be an effective strategy for lymphopoietic differentiation of human pluripotent cells [198]. It was shown that Delta1-expressing OP9 stroma allowed CD34+ cord blood cells to differentiate into CD7+ T-cell progenitors. These pro-T cells could engraft into the thymuses of NOD/SCID/γc−/− (NSG) mice and continue differentiating down the T-cell lineage [198]. Another study published in 2009 showed that mature T cells could be obtained from hESCs using a completely *in vitro* culture system [199]. Timmermans *et al.* co-cultured hESCs on OP9 cells in the presence of serum for 12 days and observed the appearance of "haematopoietic zones," which appear to be similar to the ES sacs described for megakaryocyte generation. They isolated CD34+CD43low cells from these zones and replated them onto Delta1-expressing OP9 cells for 5 to 7 weeks in the presence of FLT3-L, SCF, and IL7. CD45+ CD7+ CD117+ cyCD3e+ T-cell progenitors emerged within 6 days while CD4+/CD8+ double positive cells emerged from a larger CD3e+CD5+ population within 14 days. After 30 days, CD3+ T cells that had undergone TCR rearrangements were present and found to be functional in assays examining their proliferation in response to phytohemaglutinin (2500-fold proliferation rate) and IFNγ production when stimulated, and exhibited the mature phenotype of CD27+CD1a- [199]. In total, these differentiation protocols for T-cell generation have the potential to be further developed for use with iPSCs, and one day will hopefully be applicable to human immunotherapy.

B cells are part of the body's humoral immune response, responsible for antibody production, perform the role of antigen presenting cells (APCs) and modulate the adaptive immune system. To date there has been little work dedicated to the *in vitro* production of these cells and their subtypes from pluripotent stem cells.

One study by Vodyanik *et al.* (2005), demonstrated that culture of hESCs on OP9s followed by CD34+ purification then further culture on MS-5 stromal cells with SCF, FLT3-L, IL-7 and IL3 cause the production of CD19+ B-like cells amongst other lymphoid and myeloid subtypes [148]. Another study was able to show under OP9-co-culture conditions small percentages of CD19+ cells from hESC, indicating their potential to form B-cells [139]. Taken together, these studies corroborate the hypothesis that ESC differentiation protocols drive embryonic haematopoiesis and formation of progenitor cells resembling human yolk sac derived CD34+ cells [125], but identifying specific environmental cues will be essential to induce formation of adult-type cells.

Monocytes and Macrophages

The mononuclear phagocyte system consists of cells derived from progenitor cells in the bone marrow. These myeloid progenitor cells differentiation into monocytes which enter in the circulation and migrate into various tissues where they differentiate into macrophages [200]. Mononuclear phagocytes as they appear in tissues share certain features, including morphology, expression of enzymes and receptors, endocytic and phagocytic capability and secretion of cytokines in response to pathogen stimuli [201]. However, mononuclear phagocytes are phenotypically and functionally heterogeneous as a result of cellular differentiation, tissue distribution and activation by exogenous stimuli such as the T-cell cytokines IFN-γ and IL-4.

Previous work has shown that monocytes and mononuclear phagocytes, including macrophages, can be generated from both mESCs [169, 202, 203] and hESCs [132]. However, current methods using hES cells are complicated by the requirements for co-cultures with cell lines, use of complex cytokine cocktails and additional purification steps, which limits their use in studies requiring high numbers of homogenous cells.

In an initial report about producing more purified macrophages from hESCs, CD34+ cells derived by S17 stromal cell co-culture were isolated. These hESC-derived CD34+ cells were sorted and cultured in semi-solid medium to allow for development of myeloid colonies, which were then picked and cultured with GM-CSF and M-CSF for a further 5 days to generate macrophages. The hESC-Macrophages could be cultured in bulk and expressed a macrophage phenotype with CD14+ HLADR+. These cells also displayed functions including phagocytosis, up-regulation of the costimulatory molecule B7.1, IL-1 and TNFa cytokine secretion in response to lipopolysaccharide stimulation at comparable levels to somatic CD34+ cells, and maintenance of productive HIV infection [204].

Karlsson *et al.*, utilised HUES-2, KCL001 and KCL002 hESC lines to generate monocytes. Using a feeder free system, they mechanically dissected colonies and induced EB formation in low adherence plates for 3 days. These cells were then treated the M-CSF and IL-3 for differentiation into monocytes in serum containing media. Further differentiation into macrophages was induced by culture with MCSF and FCS [132]. Monocytes were found to migrate from EBs and were collected from the supernatant of culture for analysis. They have a high cytoplasm to cellular ratio and indented nucleus, and demonstrated to be a homogenous population. They expressed CD14 (>90% of cells), CD45, CD44, CD11a, CD11b, CD11c and CD54 by flow cytometry, and did not express CD83 or CD103 (dendritic markers), nor CD34 (haematopoietic progenitor marker). Cells were produced slowly, with no production for approximately 18 days (lag phase) followed by a rapid production phase and a subsequent period of continuous lower level production that continued for several months. Average monocyte yield was 8.7 million cells (calculated to be 3.4×10^5 monocytes per EB) [132].

The ability of these hESC derived monocytes to differentiate to macrophages was tested by inducing the adhesion of the cells to tissue culture coated plastic combined with M-CSF exposure for 6 days. Morphologically the cells became adherent and elongated, and uniformly expressed CD68 (a pan macrophage marker) and CD163. They also had downregulated CD14 expression, and expressed CD64, CD32 and CCR5. Transcriptome analysis verified that hESC-derived monocyte gene expression correlated significantly (r>0.9) with that of fresh blood monocytes. Confirmation of myeloid gene expression included expression of: PU1, CEBPα, membrane receptors EMR1 and EMR2, MPEG1, CD1c, CD4, CD18, CD32, CD33, CD68 and enzymes including cathepsis and serine carboxypeptiases [132].

Functional macrophage tests determined that they could endocytose acLDL with half the efficiency of blood macrophages cultured under the same conditions, in a scavenger receptor dependent manner. hESC-derived macrophages were found to be able to phagocytose yeast particles at least as well as blood monocytes derived macrophages. Furthermore, upon stimulation with LPS (a bacterial ligand for toll-like receptors) macrophages were found to produce high levels of TNF-α, IL10, IL-6 and IL1ra, monocyte chemotactic protein -1 and interferon inducible protein 10, however they did not produce IL-1b, eotaxin or IL-12 [132].

Granulocytes: Neutrophils

Neutrophils play an important role during host defence against invading microorganisms and number at 1.5-7 x10^9 cells per liter in the blood. They characteristically have a multilobed nucleus. The decreased number (neutropenia) or attenuated function of neutrophils results in serious infections. Demonstration of granulocytes from hESCs are commonly seen with CFU assay conditions when demonstrating the development of hESC-HSCs [134], however apart from these, two studies produced a highly enriched population of neutrophils by a two-step culture system [205, 206], where the cells where characterized according to surface antigen, gene expression as well as functional assays (phagocytosis and *in vivo* chemotactic activity).

In a feeder-free system hESCs were cultured with IGF-II, VEGF, SCF, FLT3-L, TPO and GCSF. After three days these cells formed sacs which contained cells which remained non-adherent while further differentiated for another 40 days in the same cytokine cocktail. Phenotyping these cells found expression of CD45, low CD34, but upto 90% expressed the myeloid markers CD11b and CD33, while 20-70% expressed the neutrophil makers CD66b, CD16b and GPI-80. Functionally these neutrophils had chemotactic capacity in response to the bacterial chemotactic peptide formyl-methionyl-leucyl-phenylalanine and neutrophil specific chemokine IL8. Using these neutrophils which migrated to IL8, phagocytic and respiratory burst activities were demonstrated. Furthermore, these neutrophils were found to have chomotactic activity *in vivo* in a murine air-pouch inflammatory model [205].

Yokoyama *et al* demonstrated that functional neutrophils can be produced form hESCs by using EB-differentiation in serum based medium followed by co-culture on OP9 cells for 7 to 14 days in medium supplemented with BMP4, SCF, Flt-3L, IL-6, FP6, TPO and G-CSF. During culture on OP9, hESC derived progenitors matures from mostly myeloblasts and promyelocytes on day 7 to banded neutrophils on Day 10 and mature, terminally differentiated neutrophils on day 14. The majority of OP-9-co-culture progenitors expressed CD45, while expression of primitive cells markers (CD133, CD34 and CD117) were lost at later timepoints, indicating progressive maturation. Functional comparisons of hESC and human PB derived neutrophils showed similar phagocytosis, chemotaxis, superoxide production, bactericidal [148] activity and oxidative burst capacity, despite slight differences in cellular phenotypes compared to normal neutrophils such as decreased CD16 expression and aberrant CD64 and CD14 expressions [206].

Using an EB formation strategy but different combinations of cytokines (M-CSF, IL-3) for 18 days, Karlson *et al.*, reported the homogenous production of functional CD14+ monocytes and macrophages from hESC which could endocytose acetylated low density lipoprotein (LDL) and opsonized yeast particles at similar levels to somatic macrophages. Furthermore these hESC-macrophages could upregulate MCH Calss II and mannose receptors after activation with IL-4, and were found to be able to produce chemokines and cytokines in response to lipopolysaccharide (TNFα, IL1b, IL10, IL6 and IL1a) [132].

Neutrophils have also been derived from human iPSCs, which mirrored previous hESC protocols. The culture system used consisted of the transfer of iPSCs onto OP9 cells and their culture with VEGF. After 10 days, TRA 1-85 CD34 VEGF receptor-2 (VEGFR-2) cells were sorted and co-cultured with OP9 cells in the presence of IL3, SCF, TPO, G-CSF for 30 days. Non adherent cells were analysed and found to contain 40% of cells with stab and segmented neutrophils which were positive for MPO. Upon phenotypic analysis these cells were found to be positive for CD45, CD11b, CD13, CD33 and CD16. However other cells contaminated this population (immature myeloid cells, macrophages). These iPSC-derived neutrophils were similar to peripheral blood mature neutrophils in morphology, contained functional neutrophil specific granules, and were equipped with the basic functions such as phagocytosis, superoxide production, and chemotaxis. These cells also expressed similar patterns of surface antigen, transcription factors (*PU.1, C/EPBa, MPO*, Lectoferrin, Gelatinase), and granule proteins to those of granulopoiesis in normal bone marrow [207].

Although in some of these studies there is strong evidence for the generation of mature functional neutrophils from pluripotent stem cells, in some cases even following normal ontogeny of myeloblasts, promyelocytes, myelocytes, metamyelocytes, mature stab and segmented neutrophils, there is a need for protocols which can provide large scale numbers of these cells before they can be considered a viable therapeutic alternative.

BARRIERS TO ADOPTION OF NOVEL TECHNOLOGIES

Future therapeutic applications for hESC derived haematopoietic cells will require significant improvement in a number of areas before the cells could be considered ready for entry into a clinical setting. Firstly, current protocols for blood cell differentiation involve animal products (such as FBS or the use of a stromal cell line) or viral transduction (such as with iPSCs). This needs to be replaced by the use of human products only and by strategies involving protein transduction or small molecules. Once this has been addressed, the efficiency of the differentiation protocols must be improved. For clinical translation, differentiation protocols are required to be highly scalable, with the ability to produce billions of cells reproducibly, efficiently and cheaply.

One very basic hurdle which to date is the biggest barrier to clinical translation, is the lack of protocols which produce high numbers of fully mature cells, with equivalent scope of function to their somatic equivalents. To date, although it would appear that hESC-HSCs has the same phenotype as a bone marrow HSC, transcriptome analysis has demonstrated that the cells exhibit important differences, which is reflected in the typically low engraftment levels by hESC-HSCs in murine models. This low engraftment may also result from multiple factors, including possible rejection by the host immune system, dose failures, the developmental stage of the cells derived from hESC differentiation and the quality or viability of the transplanted cells.

A safety concern which is common to all pluripotent stem cell potential therapies is tumour formation due to teratoma growth which can occur if a single pluripotent stem cell is transplanted along with their differentiated counterparts. The only way to ensure avoidance of teratoma formation is to differentiate hESCs in advance, enrich for the desired cell type (by FACS, MACS column separation, tagging of differentiated cells with an inducible genetic marker) and rigorously screen for the presence of undifferentiated cells. Stringent following of such procedures allowed for 200 mice to be transplanted with ESC-derived cardiomyocytes [208] without teratoma formation. However, tumours can also form from more differentiated cells. Transplantation of a population of hESC neural derivatives enriched for dopamine neurons into Parkinsonian rats led to the uncontrolled growth of a primitive population of neuroepithelial cells which remained proliferative for 70 days [209]. This may be due to the accidental transplant of a pluripotent progenitor, or a differentiation defect in the entire cell population. Cell dose is important and needs to be carefully monitored (too high a cell dose is thought to have caused dyskinesias in Parkinson's patients who received fetal neural tissue transplant [210].

The immune system presents a formidable barrier to allogeneic cell transplantation. hESCs express few MHC Class I antigens and have virtually no expression of class II molecules [211-213], which initially raised hopes that ESC derivatives might not induce an immune response when transplanted. However, once transplanted, ESC derivatives show increased expression of histocompatibility antigens, with further increases in response to cytokine stimulation, which leads to their rejection [212]. To circumvent this issue traditional immunosupression can be used, although its serious side effects make this an unattractive option. Banking of ESCs can provide cells with a repertoire of tissue types, but large numbers of cell lines (likely thousands) would be needed to cover a genetically diverse population such as that in the United States. The recent development of iPSCs solves this issue. The ability to generate pluripotent cells from readily available cells, such as skin fibroblasts, means that there is the potential to generate patient-specific cells that would be recognized as ''self'' by the immune system, preventing rejection of cell transplants. In addition, iPSCs could be generated from patients with specific diseases having genetic components, these could then be corrected *ex vivo* and transplanted back to the patient [153]. There are, however, drawbacks to the clinical use of iPSCs. The integrated proviruses are silenced during iPSC generation, but there is potential for re-activation of these viral transgenes, which include the protoongene *c-MYC*, and there is evidence that reactivation of *c-MYC* can result in tumour formation in chimeric mice generated from iPSCs [77]. Furthermore, continued expression of these pluripotent transgenes may inhibit complete iPSC differentiation and maturation, leading to an increased risk of teratoma formation. Recent analysis of copy number variations [214], point mutations [215] and DNA methylation [216] iPSCs and hESCs, determined that iPSCs display more genetic and epigenetic abnormalities than hESCs. It is unclear how these mutations affect the growth, differentiation, tumorougenicity or functional potential of iPSCs, however it does point to potential long term safety issues which need to be assessed before clinical translation. The second is that iPSC are not an ''off-the-shelf'' product and would likely only be produced after the patient becomes ill, precluding their use in the acute phase of the disease.

Therefore, despite a myriad of recent advances, therapeutic use of blood cells derived from pluripotent stem cells is not imminent. The urgent need for alternate supplies of HSCs and their mature derivatives has driven research in other areas, such as the induction of proliferation of somatic HSCs, which as recently met with great success with the finding that stimulation of the aryl hydrocarbon receptor induces the proliferation of repopulating stem cells with maintenance of their engraftment capabilities [60].

CONCLUSIONS AND PROSPECTS

HSCs have been used for transplantation therapy for the treatment of both malignant and non-malignant diseases for decades, while the use of blood products for transfusion medicine is a routine procedure. These cells are essential for patient treatment in modern medicine but are subject to availability problems, due to difficulties in finding an immunogenically similar donor or sourcing sufficient cells from volunteer donors to comply with medical demand. In order to circumvent these issues, numerous laboratory studies have been carried out to assess the ability of pluripotent stem cells to become novel sources of these cells. In recent reports hESCs have been able to produce functional, seemingly functional HSCs and mature cells of the erythroid, myeloid and lymphocyte lineage. Combined with the prevailing ethical dilemmas associated with the use of human embryonic tissue for research and clinical uses, it is important to investigate alternate potential sources of stem cells such as iPSCs and adult stem cells. However there are continued unresolved problems which prevent their imminent translation to the clinic.

REFERENCES

[1] Thomas ED, Blume KG. Historical markers in the development of allogeneic hematopoietic cell transplantation. Biol Blood Marrow Transplant 1999; 5(6): 341-6.

[2] Gatti RA, Meuwissen HJ, Allen HD, Hong R, Good RA. Immunological reconstitution of sex-linked lymphopenic immunological deficiency. Lancet 1968 2(7583): 1366-9.

[3] Ma D. Hematopoietic reconstitution following bone marrow transplantation. Bone Marrow Transplantation in Children: Raven Press Ltd; 1990. pp. 111-39.

[4] Ma DD, Biggs JC. Autologous bone marrow transplant in the treatment of malignant disease. Med J Aust 1981; 1(2): 57-8.

[5] Brunstein CG, Weisdorf DJ. Future of cord blood for oncology uses. Bone Marrow Transplant 2009; 44(10): 699-707.

[6] Ma DD, Varga DE, Biggs JC. Haemopoietic reconstitution after allogeneic bone marrow transplantation in man: recovery of haemopoietic progenitors (CFU-Mix, BFU-E and CFU-GM). Br J Haematol 1987; 65(1): 5-10.

[7] Till JE, Mc CE. A direct measurement of the radiation sensitivity of normal mouse bone marrow cells. Radiat Res 1961; 14: 213-22.

[8] Becker AJ, Mc CE, Till JE. Cytological demonstration of the clonal nature of spleen colonies derived from transplanted mouse marrow cells. Nature 1963; 197: 452-4.

[9] Osawa M, Hanada K, Hamada H, Nakauchi H. Long-term lymphohematopoietic reconstitution by a single CD34-low/negative hematopoietic stem cell. Science 1996; 273(5272): 242-5.

[10] Bryder D, Rossi DJ, Weissman IL. Hematopoietic stem cells: the paradigmatic tissue-specific stem cell. Am J Pathol 2006; 169(2): 338-46.

[11] Healy L, May G, Gale K, Grosveld F, Greaves M, Enver T. The stem cell antigen CD34 functions as a regulator of hemopoietic cell adhesion. Proc Natl Acad Sci USA 1995; 92(26): 12240-4.

[12] Yin AH, Miraglia S, Zanjani ED, Almeida-Porada G, Ogawa M, Leary AG, *et al.* AC133, a novel marker for human hematopoietic stem and progenitor cells. Blood 1997; 90(12): 5002-12.

[13] Mayani H, Lansdorp PM. Biology of human umbilical cord blood-derived hematopoietic stem/progenitor cells. Stem Cells. 1998; 16(3): 153-65.

[14] Goodell MA, Brose K, Paradis G, Conner AS, Mulligan RC. Isolation and functional properties of murine hematopoietic stem cells that are replicating *in vivo*. J Exp Med 1996; 183(4): 1797-806.

[15] Morrison SJ, Weissman IL. The long-term repopulating subset of hematopoietic stem cells is deterministic and isolatable by phenotype. Immunity 1994; 1(8): 661-73.

[16] Bradford GB, Williams B, Rossi R, Bertoncello I. Quiescence, cycling, and turnover in the primitive hematopoietic stem cell compartment. Exp Hematol 1997; 25(5): 445-53.

[17] Cheshier SH, Morrison SJ, Liao X, Weissman IL. *In vivo* proliferation and cell cycle kinetics of long-term self-renewing hematopoietic stem cells. Proc Natl Acad Sci USA 1999; 96(6): 3120-5.

[18] Mahmud N, Devine SM, Weller KP, *et al.* The relative quiescence of hematopoietic stem cells in nonhuman primates. Blood 2001; 97(10): 3061-8.

[19] Wilson A, Laurenti E, Oser G, *et al.* Hematopoietic stem cells reversibly switch from dormancy to self-renewal during homeostasis and repair. Cell 2008; 135(6): 1118-29.

[20] Park Y, Gerson SL. DNA repair defects in stem cell function and aging. Annu Rev Med 2005; 56: 495-508.

[21] Fauser AA, Messner HA. Granuloerythropoietic colonies in human bone marrow, peripheral blood, and cord blood. Blood 1978; 52(6): 1243-8.

[22] Knudtzon S. *In vitro* growth of granulocytic colonies from circulating cells in human cord blood. Blood 1974; 43(3): 357-61.

[23] O'Brien TA, Tiedemann K, Vowels MR. No longer a biological waste product: umbilical cord blood. Med J Aust 2006; 184(8): 407-10.

[24] Tanavde VM, Malehorn MT, Lumkul R, *et al.* Human stem-progenitor cells from neonatal cord blood have greater hematopoietic expansion capacity than those from mobilized adult blood. Exp Hematol 2002; 30(7): 816-23.

[25] Pettengell R, Luft T, Henschler R, *et al.* Direct comparison by limiting dilution analysis of long-term culture-initiating cells in human bone marrow, umbilical cord blood, and blood stem cells. Blood 1994; 84(11): 3653-9.

[26] Steen R, Tjonnfjord GE, Egeland T. Comparison of the phenotype and clonogenicity of normal CD34+ cells from umbilical cord blood, granulocyte colony-stimulating factor-mobilized peripheral blood, and adult human bone marrow. J Hematother 1994; 3(4): 253-62.

[27] Schofield R. The relationship between the spleen colony-forming cell and the haemopoietic stem cell. Blood Cells 1978; 4(1-2): 7-25.

[28] Scadden DT. The stem-cell niche as an entity of action. Nature 2006; 441(7097): 1075-9.

[29] Taichman RS, Reilly MJ, Emerson SG. Human osteoblasts support human hematopoietic progenitor cells *in vitro* bone marrow cultures. Blood 1996; 87(2): 518-24.

[30] Lord BI, Testa NG, Hendry JH. The relative spatial distributions of CFUs and CFUc in the normal mouse femur. Blood 1975; 46(1): 65-72.

[31] Calvi LM, Adams GB, Weibrecht KW, *et al.* Osteoblastic cells regulate the haematopoietic stem cell niche. Nature 2003; 425(6960): 841-6.

[32] Zhang J, Niu C, Ye L, *et al.* Identification of the haematopoietic stem cell niche and control of the niche size. Nature 2003; 425(6960): 836-41.

[33] Visnjic D, Kalajzic Z, Rowe DW, Katavic V, Lorenzo J, Aguila HL. Hematopoiesis is severely altered in mice with an induced osteoblast deficiency. Blood 2004; 103(9): 3258-64.

[34] Stier S, Ko Y, Forkert R, *et al.* Osteopontin is a hematopoietic stem cell niche component that negatively regulates stem cell pool size. J Exp Med 2005; 201(11): 1781-91.

[35] Nilsson SK, Johnston HM, Whitty GA, *et al.* Osteopontin, a key component of the hematopoietic stem cell niche and regulator of primitive hematopoietic progenitor cells. Blood 2005; 106(4): 1232-9.

[36] Arai F, Hirao A, Ohmura M, *et al.* Tie2/angiopoietin-1 signalling regulates hematopoietic stem cell quiescence in the bone marrow niche. Cell 2004; 118(2): 149-61.

[37] Yoshihara H, Arai F, Hosokawa K, *et al.* Thrombopoietin/MPL signalling regulates hematopoietic stem cell quiescence and interaction with the osteoblastic niche. Cell Stem Cell 2007; 1(6): 685-97.

[38] Kiel MJ, Radice GL, Morrison SJ. Lack of evidence that hematopoietic stem cells depend on N-cadherin-mediated adhesion to osteoblasts for their maintenance. Cell Stem Cell 2007; 1(2): 204-17.

[39] Lymperi S, Horwood N, Marley S, Gordon MY, Cope AP, Dazzi F. Strontium can increase some osteoblasts without increasing hematopoietic stem cells. Blood 2008; 111(3): 1173-81.

[40] DiMascio L, Voermans C, Uqoezwa M, *et al.* Identification of adiponectin as a novel hemopoietic stem cell growth factor. J Immunol 2007178(6): 3511-20.

[41] Naveiras O, Nardi V, Wenzel PL, Hauschka PV, Fahey F, Daley GQ. Bone-marrow adipocytes as negative regulators of the haematopoietic microenvironment. Nature 2009; 460(7252): 259-63.

[42] Kollet O, Dar A, Shivtiel S, *et al.* Osteoclasts degrade endosteal components and promote mobilization of hematopoietic progenitor cells. Nat Med 2006; 12(6): 657-64.

[43] Adams GB, Chabner KT, Alley IR, Olson DP, Szczepiorkowski ZM, Poznansky MC, *et al.* Stem cell engraftment at the endosteal niche is specified by the calcium-sensing receptor. Nature 2006; 439(7076): 599-603.

[44] Nivison-Smith I, Bradstock KF, Dodds AJ, *et al.* Hematopoietic stem cell transplantation in Australia and New Zealand, 1992-2004. Biol Blood Marrow Transplant 2007; 13(8): 905-12.

[45] Gaziev J, Lucarelli G. Stem cell transplantation for hemoglobinopathies. Curr Opin Pediatr 2003; 15(1): 24-31.

[46] Galanello R, Origa R. Beta-thalassemia. Orphanet J Rare Dis 2010; 5: 11.

[47] Prasad VK, Kurtzberg J. Emerging trends in transplantation of inherited metabolic diseases. Bone Marrow Transplant 2008; 41(2): 99-108.

[48] Australasian Bone Marrow Transplant Registry Annual Data Summary 2009. Sydney, Australia2009.

[49] Hacein-Bey-Abina S, Le Deist F, Carlier F, *et al.* Sustained correction of X-linked severe combined immunodeficiency by *ex vivo* gene therapy. N Engl J Med 2002; 346(16): 1185-93.

[50] Cavazzana-Calvo M, Hacein-Bey S, de Saint Basile G, *et al.* Gene therapy of human severe combined immunodeficiency (SCID)-X1 disease. Science 2000; 288(5466): 669-72.

[51] Gaspar HB, Parsley KL, Howe S, *et al.* Gene therapy of X-linked severe combined immunodeficiency by use of a pseudotyped gammaretroviral vector. Lancet 2004 31; 364(9452): 2181-7.

[52] Aiuti A, Slavin S, Aker M, *et al.* Correction of ADA-SCID by stem cell gene therapy combined with nonmyeloablative conditioning. Science 2002; 296(5577): 2410-3.

[53] Hacein-Bey-Abina S, Garrigue A, Wang GP, *et al.* Insertional oncogenesis in 4 patients after retrovirus-mediated gene therapy of SCID-X1. J Clin Invest 2008; 118(9): 3132-42.

[54] Howe SJ, Mansour MR, Schwarzwaelder K, *et al.* Insertional mutagenesis combined with acquired somatic mutations causes leukemogenesis following gene therapy of SCID-X1 patients. J Clin Invest 2008; 118(9): 3143-50.

[55] Hacein-Bey-Abina S, Hauer J, Lim A, *et al.* Efficacy of gene therapy for X-linked severe combined immunodeficiency. N Engl J Med 2010; 363(4): 355-64.

[56] Laughlin MJ, Eapen M, Rubinstein P, *et al.* Outcomes after transplantation of cord blood or bone marrow from unrelated donors in adults with leukemia. N Engl J Med 2004; 351(22): 2265-75.

[57] Rocha V, Labopin M, Sanz G, *et al.* Transplants of umbilical-cord blood or bone marrow from unrelated donors in adults with acute leukemia. N Engl J Med 2004; 351(22): 2276-85.

[58] Rubinstein P, Carrier C, Scaradavou A, *et al.* Outcomes among 562 recipients of placental-blood transplants from unrelated donors. N Engl J Med 1998; 339(22): 1565-77.

[59] Wagner JE, Barker JN, DeFor TE, *et al.* Transplantation of unrelated donor umbilical cord blood in 102 patients with malignant and nonmalignant diseases: influence of CD34 cell dose and HLA disparity on treatment-related mortality and survival. Blood 2002; 100(5): 1611-8.

[60] Boitano AE, Wang J, Romeo R, *et al.* Aryl hydrocarbon receptor antagonists promote the expansion of human hematopoietic stem cells. Science 2010; 329(5997): 1345-8.

[61] Keirstead HS, Nistor G, Bernal G, *et al.* Human embryonic stem cell-derived oligodendrocyte progenitor cell transplants remyelinate and restore locomotion after spinal cord injury. J Neurosci 2005; 25(19): 4694-705.

[62] Niwa A, Umeda K, Chang H, *et al.* Orderly hematopoietic development of induced pluripotent stem cells *via* Flk-1(+) hemoangiogenic progenitors. J Cell Physiol 2009; 221(2): 367-77.

[63] Ye Z, Zhan H, Mali P, Dowey S, *et al.* Human-induced pluripotent stem cells from blood cells of healthy donors and patients with acquired blood disorders. Blood 2009; 114(27): 5473-80.

[64] Choi KD, Yu J, Smuga-Otto K, *et al.* Hematopoietic and endothelial differentiation of human induced pluripotent stem cells. Stem Cells 2009; 27(3): 559-67.

[65] Park IH, Lerou PH, Zhao R, Huo H, Daley GQ. Generation of human-induced pluripotent stem cells. Nat Protoc 2008; 3(7): 1180-6.

[66] Kollman C, Howe CW, Anasetti C, *et al.* Donor characteristics as risk factors in recipients after transplantation of bone marrow from unrelated donors: the effect of donor age. Blood 2001; 98(7): 2043-51.

[67] Thomson JA, Itskovitz-Eldor J, Shapiro SS, *et al.* Embryonic stem cell lines derived from human blastocysts. Science 1998; 282(5391): 1145-7.

[68] Guhr A, Kurtz A, Friedgen K, Loser P. Current state of human embryonic stem cell research: an overview of cell lines and their use in experimental work. Stem Cells 2006; 24(10): 2187-91.

[69] Boyer LA, Lee TI, Cole MF, *et al.* Core transcriptional regulatory circuitry in human embryonic stem cells. Cell 2005; 122(6): 947-56.

[70] Avery S, Inniss K, Moore H. The regulation of self-renewal in human embryonic stem cells. Stem Cells Dev 2006; 15(5): 729-40.

[71] Riazi AM, Kwon SY, Stanford WL. Stem cell sources for regenerative medicine. Methods Mol Biol 2009; 482: 55-90.

[72] Loh YH, Wu Q, Chew JL, *et al.* The Oct4 and Nanog transcription network regulates pluripotency in mouse embryonic stem cells. Nat Genet 2006; 38(4): 431-40.

[73] Zeng X, Rao MS. Human embryonic stem cells: long term stability, absence of senescence and a potential cell source for neural replacement. Neuroscience 2007; 145(4): 1348-58.

[74] Hoffman LM, Carpenter MK. Characterization and culture of human embryonic stem cells. Nat Biotechnol 2005; 23(6): 699-708.

[75] Kyba M, Daley GQ. Hematopoiesis from embryonic stem cells: lessons from and for ontogeny. Exp Hematol 2003; 31(11): 994-1006.

[76] Orkin SH, Zon LI. Hematopoiesis: an evolving paradigm for stem cell biology. Cell 2008; 132(4): 631-44.

[77] Takahashi K, Yamanaka S. Induction of pluripotent stem cells from mouse embryonic and adult fibroblast cultures by defined factors. Cell 2006; 126(4): 663-76.

[78] Yu J, Vodyanik MA, Smuga-Otto K, *et al.* Induced pluripotent stem cell lines derived from human somatic cells. Science 2007; 318(5858): 1917-20.

[79] Park IH, Zhao R, West JA, *et al.* Reprogramming of human somatic cells to pluripotency with defined factors. Nature 2008; 451(7175): 141-6.

[80] Okita K, Nakagawa M, Hyenjong H, Ichisaka T, Yamanaka S. Generation of mouse induced pluripotent stem cells without viral vectors. Science 2008; 322(5903): 949-53.

[81] Kaji K, Norrby K, Paca A, Mileikovsky M, Mohseni P, Woltjen K. Virus-free induction of pluripotency and subsequent excision of reprogramming factors. Nature 2009; 458(7239): 771-5.

[82] Zhou H, Wu S, Joo JY, *et al.* Generation of induced pluripotent stem cells using recombinant proteins. Cell Stem Cell 2009; 4(5): 381-4.

[83] Okita K, Ichisaka T, Yamanaka S. Generation of germline-competent induced pluripotent stem cells. Nature 2007; 448(7151): 313-7.

[84] Wernig M, Meissner A, Cassady JP, Jaenisch R. c-Myc is dispensable for direct reprogramming of mouse fibroblasts. Cell Stem Cell 2008; 2(1): 10-2.

[85] Hanna J, Markoulaki S, Schorderet P, *et al.* Direct reprogramming of terminally differentiated mature B lymphocytes to pluripotency. Cell 2008; 133(2): 250-64.

[86] Loh YH, Agarwal S, Park IH, *et al.* Generation of induced pluripotent stem cells from human blood. Blood 2009; 113(22): 5476-9.

[87] Aasen T, Raya A, Barrero MJ, Garreta E, Consiglio A, Gonzalez F, *et al.* Efficient and rapid generation of induced pluripotent stem cells from human keratinocytes. Nat Biotechnol 2008; 26(11): 1276-84.

[88] Aoi T, Yae K, Nakagawa M, *et al.* Generation of pluripotent stem cells from adult mouse liver and stomach cells. Science 2008; 321(5889): 699-702.

[89] Sun N, Panetta NJ, Gupta DM, *et al.* Feeder-free derivation of induced pluripotent stem cells from adult human adipose stem cells. Proc Natl Acad Sci USA 2009; 106(37): 15720-5.

[90] Kim JB, Sebastiano V, Wu G, *et al.* Oct4-induced pluripotency in adult neural stem cells. Cell 2009; 136(3): 411-9.

[91] Polo JM, Liu S, Figueroa ME, *et al.* Cell type of origin influences the molecular and functional properties of mouse induced pluripotent stem cells. Nat Biotechnol 2010; 28(8): 848-55.

[92] Kim K, Doi A, Wen B, *et al.* Epigenetic memory in induced pluripotent stem cells. Nature 2010; 467(7313): 285-90.

[93] Miura K, Okada Y, Aoi T, *et al.* Variation in the safety of induced pluripotent stem cell lines. Nat Biotechnol 2009; 27(8): 743-5.

[94] Park IH, Arora N, Huo H, *et al.* Disease-specific induced pluripotent stem cells. Cell 2008; 134(5): 877-86.

[95] Doetschman TC, Eistetter H, Katz M, Schmidt W, Kemler R. The *in vitro* development of blastocyst-derived embryonic stem cell lines: formation of visceral yolk sac, blood islands and myocardium. J Embryol Exp Morphol 1985; 87: 27-45.

[96] Murry CE, Keller G. Differentiation of embryonic stem cells to clinically relevant populations: lessons from embryonic development. Cell 2008; 132(4): 661-80.

[97] Lengerke C, Daley GQ. Patterning definitive hematopoietic stem cells from embryonic stem cells. Exp Hematol 2005; 33(9): 971-9.

[98] Lengerke C, Schmitt S, Bowman TV, *et al.* BMP and Wnt specify hematopoietic fate by activation of the Cdx-Hox pathway. Cell Stem Cell 2008; 2(1): 72-82.

[99] North TE, Goessling W, Walkley CR, *et al.* Prostaglandin E2 regulates vertebrate haematopoietic stem cell homeostasis. Nature 2007; 447(7147): 1007-11.

[100] Davidson AJ, Ernst P, Wang Y, *et al.* cdx4 mutants fail to specify blood progenitors and can be rescued by multiple hox genes. Nature 2003; 425(6955): 300-6.

[101] Gadue P, Huber TL, Paddison PJ, Keller GM. Wnt and TGF-beta signalling are required for the induction of an *in vitro* model of primitive streak formation using embryonic stem cells. Proc Natl Acad Sci USA 2006; 103(45): 16806-11.

[102] Adamo L, Naveiras O, Wenzel PL, *et al.* Biomechanical forces promote embryonic haematopoiesis. Nature 2009; 459(7250): 1131-5.

[103] Hogan BL. Bone morphogenetic proteins in development. Curr Opin Genet Dev 1996; 6(4): 432-8.

[104] Schier AF. Nodal signalling in vertebrate development. Annu Rev Cell Dev Biol 2003; 19: 589-621.

[105] Conlon FL, Lyons KM, Takaesu N, *et al.* A primary requirement for nodal in the formation and maintenance of the primitive streak in the mouse. Development 1994 ; 120(7): 1919-28.

[106] Yamaguchi TP. Heads or tails: Wnts and anterior-posterior patterning. Curr Biol 2001; 11(17): R713-24.

[107] Gadue P, Huber TL, Nostro MC, Kattman S, Keller GM. Germ layer induction from embryonic stem cells. Exp Hematol 2005; 33(9): 955-64.

[108] Kinder SJ, Tsang TE, Ang SL, Behringer RR, Tam PP. Defects of the body plan of mutant embryos lacking Lim1, Otx2 or Hnf3beta activity. Int J Dev Biol 2001; 45(1): 347-55.

[109] Sasaki H, Hogan BL. Differential expression of multiple fork head related genes during gastrulation and axial pattern formation in the mouse embryo. Development 1993; 118(1): 47-59.

[110] Dush MK, Martin GR. Analysis of mouse Evx genes: Evx-1 displays graded expression in the primitive streak. Dev Biol 1992; 151(1): 273-87.

[111] Forlani S, Lawson KA, Deschamps J. Acquisition of Hox codes during gastrulation and axial elongation in the mouse embryo. Development 2003; 130(16): 3807-19.

[112] Jaffredo T, Nottingham W, Liddiard K, Bollerot K, Pouget C, de Bruijn M. From hemangioblast to hematopoietic stem cell: an endothelial connection? Exp Hematol 2005; 33(9): 1029-40.

[113] Kennedy M, D'Souza SL, Lynch-Kattman M, Schwantz S, Keller G. Development of the hemangioblast defines the onset of hematopoiesis in human ES cell differentiation cultures. Blood 2007; 109(7): 2679-87.

[114] Kennedy M, Firpo M, Choi K, *et al.* A common precursor for primitive erythropoiesis and definitive haematopoiesis. Nature 1997; 386(6624): 488-93.

[115] Choi K, Kennedy M, Kazarov A, Papadimitriou JC, Keller G. A common precursor for hematopoietic and endothelial cells. Development 1998; 125(4): 725-32.

[116] Lu SJ, Luo C, Holton K, Feng Q, Ivanova Y, Lanza R. Robust generation of hemangioblastic progenitors from human embryonic stem cells. Regen Med 2008; 3(5): 693-704.

[117] Feng Q, Lu SJ, Klimanskaya I, *et al.* Hemangioblastic derivatives from human induced pluripotent stem cells exhibit limited expansion and early senescence. Stem Cells 2010; 28(4): 704-12.

[118] Woll PS, Morris JK, Painschab MS, *et al.* Wnt signalling promotes hematoendothelial cell development from human embryonic stem cells. Blood 2008; 111(1): 122-31.

[119] Zambidis ET, Park TS, Yu W, *et al.* Expression of angiotensin-converting enzyme (CD143) identifies and regulates primitive hemangioblasts derived from human pluripotent stem cells. Blood 2008; 112(9): 3601-14.

[120] Vodyanik MA, Thomson JA, Slukvin, II. Leukosialin (CD43) defines hematopoietic progenitors in human embryonic stem cell differentiation cultures. Blood 2006; 108(6): 2095-105.

[121] Ema M, Takahashi S, Rossant J. Deletion of the selection cassette, but not cis-acting elements, in targeted Flk1-lacZ allele reveals Flk1 expression in multipotent mesodermal progenitors. Blood 2006; 107(1): 111-7.

[122] Tian X, Woll PS, Morris JK, Linehan JL, Kaufman DS. Hematopoietic engraftment of human embryonic stem cell-derived cells is regulated by recipient innate immunity. Stem Cells 2006; 24(5): 1370-80.

[123] Nostro MC, Cheng X, Keller GM, Gadue P. Wnt, activin, and BMP signalling regulate distinct stages in the developmental pathway from embryonic stem cells to blood. Cell Stem Cell 2008; 2(1): 60-71.

[124] Wiles MV, Johansson BM. Embryonic stem cell development in a chemically defined medium. Exp Cell Res 1999; 247(1): 241-8.

[125] Tavian M, Robin C, Coulombel L, Peault B. The human embryo, but not its yolk sac, generates lympho-myeloid stem cells: mapping multipotent hematopoietic cell fate in intraembryonic mesoderm. Immunity 2001; 15(3): 487-95.

[126] Tavian M, Cortes F, Charbord P, Labastie MC, Peault B. Emergence of the haematopoietic system in the human embryo and foetus. Haematologica 1999; 84 Suppl EHA-4: 1-3.

[127] Taoudi S, Gonneau C, Moore K, *et al.* Extensive hematopoietic stem cell generation in the AGM region *via* maturation of VE-cadherin+CD45+ pre-definitive HSCs. Cell Stem Cell 2008; 3(1): 99-108.

[128] Oberlin E, Tavian M, Blazsek I, Peault B. Blood-forming potential of vascular endothelium in the human embryo. Development 2002; 129(17): 4147-57.

[129] Galloway JL, Zon LI. Ontogeny of hematopoiesis: examining the emergence of hematopoietic cells in the vertebrate embryo. Curr Top Dev Biol 2003; 53: 139-58.

[130] Dzierzak E, Speck NA. Of lineage and legacy: the development of mammalian hematopoietic stem cells. Nat Immunol 2008; 9(2): 129-36.

[131] Zambidis ET, Peault B, Park TS, Bunz F, Civin CI. Hematopoietic differentiation of human embryonic stem cells progresses through sequential hematoendothelial, primitive, and definitive stages resembling human yolk sac development. Blood 2005; 106(3): 860-70.

[132] Karlsson KR, Cowley S, Martinez FO, Shaw M, Minger SL, James W. Homogeneous monocytes and macrophages from human embryonic stem cells following coculture-free differentiation in M-CSF and IL-3. Exp Hematol 2008; 36(9): 1167-75.

[133] Ng ES, Davis RP, Azzola L, Stanley EG, Elefanty AG. Forced aggregation of defined numbers of human embryonic stem cells into embryoid bodies fosters robust, reproducible hematopoietic differentiation. Blood 2005; 106(5): 1601-3.

[134] Kaufman DS, Hanson ET, Lewis RL, Auerbach R, Thomson JA. Hematopoietic colony-forming cells derived from human embryonic stem cells. Proc Natl Acad Sci USA 2001; 98(19): 10716-21.

[135] Chadwick K, Wang L, Li L, Menendez P, *et al.* Cytokines and BMP-4 promote hematopoietic differentiation of human embryonic stem cells. Blood 2003; 102(3): 906-15.

[136] Grigoriadis AE, Kennedy M, Bozec A, *et al.* Directed differentiation of hematopoietic precursors and functional osteoclasts from human ES and iPS cells. Blood 2010; 115(14): 2769-76.

[137] Park TS, Zambidis ET, Lucitti JL, Logar A, Keller BB, Peault B. Human embryonic stem cell-derived hematoendothelial progenitors engraft chicken embryos. Exp Hematol 2009; 37(1): 31-41.

[138] Kyba M, Perlingeiro RC, Daley GQ. HoxB4 confers definitive lymphoid-myeloid engraftment potential on embryonic stem cell and yolk sac hematopoietic progenitors. Cell 2002; 109(1): 29-37.

[139] Wang L, Menendez P, Shojaei F, *et al.* Generation of hematopoietic repopulating cells from human embryonic stem cells independent of ectopic HOXB4 expression. J Exp Med 2005; 201(10): 1603-14.

[140] Shojaei F, Menendez P. Molecular profiling of candidate human hematopoietic stem cells derived from human embryonic stem cells. Exp Hematol 2008; 36(11): 1436-48.

[141] Bauwens CL, Peerani R, Niebruegge S, *et al.* Control of human embryonic stem cell colony and aggregate size heterogeneity influences differentiation trajectories. Stem Cells 2008; 26(9): 2300-10.

[142] Chicha L, Feki A, Boni A, Irion O, Hovatta O, Jaconi M. Human pluripotent stem cells differentiated in fully defined medium generate hematopoietic CD34- and CD34+ progenitors with distinct characteristics. PLoS One 2011; 6(2): e14733.
[143] Narayan AD, Chase JL, Lewis RL, *et al.* Human embryonic stem cell-derived hematopoietic cells are capable of engrafting primary as well as secondary fetal sheep recipients. Blood 2006; 107(5): 2180-3.
[144] Ikehara S. A novel strategy for allogeneic stem cell transplantation: perfusion method plus intra-bone marrow injection of stem cells. Exp Hematol 2003; 31(12): 1142-6.
[145] Nakano T, Kodama H, Honjo T. Generation of lymphohematopoietic cells from embryonic stem cells in culture. Science 1994; 265(5175): 1098-101.
[146] Kitajima K, Tanaka M, Zheng J, Sakai-Ogawa E, Nakano T. *In vitro* differentiation of mouse embryonic stem cells to hematopoietic cells on an OP9 stromal cell monolayer. Methods Enzymol 2003; 365: 72-83.
[147] Fraser ST, Yamashita J, Jakt LM, Okada M, Ogawa M, Nishikawa S. *In vitro* differentiation of mouse embryonic stem cells: hematopoietic and vascular cell types. Methods Enzymol 2003; 365: 59-72.
[148] Vodyanik MA, Bork JA, Thomson JA, Slukvin, II. Human embryonic stem cell-derived CD34+ cells: efficient production in the coculture with OP9 stromal cells and analysis of lymphohematopoietic potential. Blood 2005; 105(2): 617-26.
[149] Trivedi P, Hematti P. Simultaneous generation of CD34+ primitive hematopoietic cells and CD73+ mesenchymal stem cells from human embryonic stem cells cocultured with murine OP9 stromal cells. Exp Hematol 2007; 35(1): 146-54.
[150] Ledran MH, Krassowska A, Armstrong L, *et al.* Efficient hematopoietic differentiation of human embryonic stem cells on stromal cells derived from hematopoietic niches. Cell Stem Cell 2008; 3(1): 85-98.
[151] Qiu C, Hanson E, Olivier E, *et al.* Differentiation of human embryonic stem cells into hematopoietic cells by coculture with human fetal liver cells recapitulates the globin switch that occurs early in development. Exp Hematol 2005; 33(12): 1450-8.
[152] Ji J, Vijayaragavan K, Bosse M, Menendez P, Weisel K, Bhatia M. OP9 stroma augments survival of hematopoietic precursors and progenitors during hematopoietic differentiation from human embryonic stem cells. Stem Cells 2008; 26(10): 2485-95.
[153] Hanna J, Wernig M, Markoulaki S, *et al.* Treatment of sickle cell anemia mouse model with iPS cells generated from autologous skin. Science 2007; 318(5858): 1920-3.
[154] Xu D, Alipio Z, Fink LM, *et al.* Phenotypic correction of murine hemophilia A using an iPS cell-based therapy. Proc Natl Acad Sci USA 2009; 106(3): 808-13.
[155] Wang Y, Jiang Y, Liu S, Sun X, Gao S. Generation of induced pluripotent stem cells from human beta-thalassemia fibroblast cells. Cell Res 2009; 19(9): 1120-3.
[156] Raya A, Rodriguez-Piza I, Guenechea G, *et al.* Disease-corrected haematopoietic progenitors from Fanconi anaemia induced pluripotent stem cells. Nature 2009; 460(7251): 53-9.
[157] Carotta S, Pilat S, Mairhofer A, *et al.* Directed differentiation and mass cultivation of pure erythroid progenitors from mouse embryonic stem cells. Blood 2004; 104(6): 1873-80.
[158] Wiles MV, Keller G. Multiple hematopoietic lineages develop from embryonic stem (ES) cells in culture. Development 1991; 111(2): 259-67.
[159] Fujimoto TT, Kohata S, Suzuki H, Miyazaki H, Fujimura K. Production of functional platelets by differentiated embryonic stem (ES) cells *in vitro*. Blood 2003; 102(12): 4044-51.
[160] Eto K, Murphy R, Kerrigan SW, *et al.* Megakaryocytes derived from embryonic stem cells implicate CalDAG-GEFI in integrin signalling. Proc Natl Acad Sci USA 2002; 99(20): 12819-24.
[161] Lieber JG, Keller GM, Worthen GS. The *in vitro* differentiation of mouse embryonic stem cells into neutrophils. Methods Enzymol 2003; 365: 129-42.
[162] Lieber JG, Webb S, Suratt BT, *et al.* The *in vitro* production and characterization of neutrophils from embryonic stem cells. Blood. 2004 Feb 1; 103(3): 852-9.
[163] Hamaguchi-Tsuru E, Nobumoto A, Hirose N, *et al.* Development and functional analysis of eosinophils from murine embryonic stem cells. Br J Haematol 2004; 124(6): 819-27.
[164] de Pooter RF, Cho SK, Zuniga-Pflucker JC. *In vitro* generation of lymphocytes from embryonic stem cells. Methods Mol Biol 2005; 290: 135-47.
[165] Schmitt TM, de Pooter RF, Gronski MA, Cho SK, Ohashi PS, Zuniga-Pflucker JC. Induction of T cell development and establishment of T cell competence from embryonic stem cells differentiated *in vitro*. Nat Immunol 2004; 5(4): 410-7.
[166] de Pooter RF, Cho SK, Carlyle JR, Zuniga-Pflucker JC. *In vitro* generation of T lymphocytes from embryonic stem cell-derived prehematopoietic progenitors. Blood 2003; 102(5): 1649-53.

[167] Moore KJ, Fabunmi RP, Andersson LP, Freeman MW. *In vitro*-differentiated embryonic stem cell macrophages: a model system for studying atherosclerosis-associated macrophage functions. Arterioscler Thromb Vasc Biol 1998; 18(10): 1647-54.

[168] Senju S, Hirata S, Matsuyoshi H, *et al.* Generation and genetic modification of dendritic cells derived from mouse embryonic stem cells. Blood 2003; 101(9): 3501-8.

[169] Fairchild PJ, Brook FA, Gardner RL, *et al.* Directed differentiation of dendritic cells from mouse embryonic stem cells. Curr Biol 2000; 10(23): 1515-8.

[170] Lian RH, Maeda M, Lohwasser S, *et al.* Orderly and nonstochastic acquisition of CD94/NKG2 receptors by developing NK cells derived from embryonic stem cells *in vitro*. J Immunol 2002; 168(10): 4980-7.

[171] Chang KH, Nelson AM, Cao H, *et al.* Definitive-like erythroid cells derived from human embryonic stem cells coexpress high levels of embryonic and fetal globins with little or no adult globin. Blood 2006; 108(5): 1515-23.

[172] Olivier EN, Qiu C, Velho M, Hirsch RE, Bouhassira EE. Large-scale production of embryonic red blood cells from human embryonic stem cells. Exp Hematol 2006; 34(12): 1635-42.

[173] Qiu C, Olivier EN, Velho M, Bouhassira EE. Globin switches in yolk sac-like primitive and fetal-like definitive red blood cells produced from human embryonic stem cells. Blood 2008; 111(4): 2400-8.

[174] Lu SJ, Feng Q, Park JS, *et al.* Biologic properties and enucleation of red blood cells from human embryonic stem cells. Blood 2008; 112(12): 4475-84.

[175] Ma F, Ebihara Y, Umeda K, *et al.* Generation of functional erythrocytes from human embryonic stem cell-derived definitive hematopoiesis. Proc Natl Acad Sci USA 2008; 105(35): 13087-92.

[176] Lapillonne H, Kobari L, Mazurier C, *et al.* Red blood cells generation from human induced pluripotent stem cells: perspectives for transfusion medicine. Haematologica. 2010; 95(10): 1651-9.

[177] Dias J, Gumenyuk M, Kang H, *et al.* Generation of red blood cells from human induced pluripotent stem cells. Stem Cells Dev. 2011 [Epub ahead of print]

[178] Choi ES, Nichol JL, Hokom MM, Hornkohl AC, Hunt P. Platelets generated *in vitro* from proplatelet-displaying human megakaryocytes are functional. Blood 1995; 85(2): 402-13.

[179] Bruno S, Gunetti M, Gammaitoni L, *et al. In vitro* and *in vivo* megakaryocyte differentiation of fresh and ex-vivo expanded cord blood cells: rapid and transient megakaryocyte reconstitution. Haematologica 2003; 88(4): 379-87.

[180] Sullenbarger B, Bahng JH, Gruner R, Kotov N, Lasky LC. Prolonged continuous *in vitro* human platelet production using three-dimensional scaffolds. Exp Hematol 2009; 37(1): 101-10.

[181] Gaur M, Kamata T, Wang S, Moran B, Shattil SJ, Leavitt AD. Megakaryocytes derived from human embryonic stem cells: a genetically tractable system to study megakaryocytopoiesis and integrin function. J Thromb Haemost 2006; 4(2): 436-42.

[182] Takayama N, Nishikii H, Usui J, *et al.* Generation of functional platelets from human embryonic stem cells *in vitro via* ES-sacs, VEGF-promoted structures that concentrate hematopoietic progenitors. Blood 2008; 111(11): 5298-306.

[183] Takayama N, Nishimura S, Nakamura S, *et al.* Transient activation of c-MYC expression is critical for efficient platelet generation from human induced pluripotent stem cells. J Exp Med 2010; 207(13): 2817-30.

[184] Lu SJ, Li F, Yin H, *et al.* Platelets generated from human embryonic stem cells are functional *in vitro* and in the microcirculation of living mice. Cell Res 2011; 21(3): 530-45.

[185] Slukvin, II, Vodyanik MA, Thomson JA, Gumenyuk ME, Choi KD. Directed differentiation of human embryonic stem cells into functional dendritic cells through the myeloid pathway. J Immunol 2006; 176(5): 2924-32.

[186] Su Z, Frye C, Bae KM, Kelley V, Vieweg J. Differentiation of human embryonic stem cells into immunostimulatory dendritic cells under feeder-free culture conditions. Clin Cancer Res 2008; 14(19): 6207-17.

[187] Tseng SY, Nishimoto KP, Silk KM, *et al.* Generation of immunogenic dendritic cells from human embryonic stem cells without serum and feeder cells. Regen Med 2009; 4(4): 513-26.

[188] Bandi S, Akkina R. Human embryonic stem cell (hES) derived dendritic cells are functionally normal and are susceptible to HIV-1 infection. AIDS Res Ther 2008; 5: 1.

[189] Hallett WH, Murphy WJ. Natural killer cells: biology and clinical use in cancer therapy. Cell Mol Immunol 2004; 1(1): 12-21.

[190] Caligiuri MA. Human natural killer cells. Blood 2008; 112(3): 461-9.

[191] Woll PS, Martin CH, Miller JS, Kaufman DS. Human embryonic stem cell-derived NK cells acquire functional receptors and cytolytic activity. J Immunol 2005; 175(8): 5095-103.

[192] Woll PS, Grzywacz B, Tian X, *et al.* Human embryonic stem cells differentiate into a homogeneous population of natural killer cells with potent *in vivo* antitumor activity. Blood 2009; 113(24): 6094-101.

[193] Ni Z, Knorr DA, Clouser C, *et al.* Human pluripotent stem cells produce natural killer cells that mediate anti-HIV-1 activity by utilizing diverse cellular mechanisms. J Virol 2011; 85(1): 43-50.

[194] Seder RA, Ahmed R. Similarities and differences in CD4+ and CD8+ effector and memory T cell generation. Nat Immunol 2003; 4(9): 835-42.

[195] Godfrey DI, Stankovic S, Baxter AG. Raising the NKT cell family. Nat Immunol 2010; 11(3): 197-206.

[196] Galic Z, Kitchen SG, Kacena A, *et al.* T lineage differentiation from human embryonic stem cells. Proc Natl Acad Sci USA 2006; 103(31): 11742-7.

[197] Galic Z, Kitchen SG, Subramanian A, *et al.* Generation of T lineage cells from human embryonic stem cells in a feeder free system. Stem Cells 2009; 27(1): 100-7.

[198] Awong G, Herer E, Surh CD, Dick JE, La Motte-Mohs RN, Zuniga-Pflucker JC. Characterization *in vitro* and engraftment potential *in vivo* of human progenitor T cells generated from hematopoietic stem cells. Blood 2009; 114(5): 972-82.

[199] Timmermans F, Velghe I, Vanwalleghem L, De Smedt M, Van Coppernolle S, Taghon T, *et al.* Generation of T cells from human embryonic stem cell-derived hematopoietic zones. J Immunol 2009; 182(11): 6879-88.

[200] Hume DA. The mononuclear phagocyte system. Curr Opin Immunol 2006; 18(1): 49-53.

[201] Gordon S, Taylor PR. Monocyte and macrophage heterogeneity. Nat Rev Immunol 2005; 5(12): 953-64.

[202] Lieschke GJ, Dunn AR. Development of functional macrophages from embryonal stem cells *in vitro*. Exp Hematol 1995; 23(4): 328-34.

[203] Odegaard JI, Vats D, Zhang L, *et al.* Quantitative expansion of ES cell-derived myeloid progenitors capable of differentiating into macrophages. J Leukoc Biol 2007; 81(3): 711-9.

[204] Anderson JS, Bandi S, Kaufman DS, Akkina R. Derivation of normal macrophages from human embryonic stem (hES) cells for applications in HIV gene therapy. Retrovirology. 2006; 3: 24.

[205] Saeki K, Nakahara M, Matsuyama S, *et al.* A feeder-free and efficient production of functional neutrophils from human embryonic stem cells. Stem Cells 2009; 27(1): 59-67.

[206] Yokoyama Y, Suzuki T, Sakata-Yanagimoto M, *et al.* Derivation of functional mature neutrophils from human embryonic stem cells. Blood 2009; 113(26): 6584-92.

[207] Morishima T, Watanabe K, Niwa A, *et al.* Neutrophil differentiation from human-induced pluripotent stem cells. J Cell Physiol 2011; 226(5): 1283-91.

[208] Laflamme MA, Chen KY, Naumova AV, *et al.* Cardiomyocytes derived from human embryonic stem cells in pro-survival factors enhance function of infarcted rat hearts. Nat Biotechnol 2007; 25(9): 1015-24.

[209] Roy NS, Cleren C, Singh SK, Yang L, Beal MF, Goldman SA. Functional engraftment of human ES cell-derived dopaminergic neurons enriched by coculture with telomerase-immortalized midbrain astrocytes. Nat Med 2006; 12(11): 1259-68.

[210] Hagell P, Piccini P, Bjorklund A, *et al.* Dyskinesias following neural transplantation in Parkinson's disease. Nat Neurosci 2002; 5(7): 627-8.

[211] Swijnenburg RJ, Tanaka M, Vogel H, Baker J, Kofidis T, Gunawan F, *et al.* Embryonic stem cell immunogenicity increases upon differentiation after transplantation into ischemic myocardium. Circulation 2005; 112(9 Suppl): I166-72.

[212] Nussbaum J, Minami E, Laflamme MA, *et al.* Transplantation of undifferentiated murine embryonic stem cells in the heart: teratoma formation and immune response. FASEB J 2007; 21(7): 1345-57.

[213] Drukker M, Katz G, Urbach A, *et al.* Characterization of the expression of MHC proteins in human embryonic stem cells. Proc Natl Acad Sci USA 2002; 99(15): 9864-9.

[214] Hussein SM, Batada NN, Vuoristo S, *et al.* Copy number variation and selection during reprogramming to pluripotency. Nature 2011; 471(7336): 58-62.

[215] Gore A, Li Z, Fung HL, *et al.* Somatic coding mutations in human induced pluripotent stem cells. Nature 2011; 471(7336): 63-7.

[216] Lister R, Pelizzola M, Kida YS, *et al.* Hotspots of aberrant epigenomic reprogramming in human induced pluripotent stem cells. Nature 2011; 471(7336): 68-73.

CHAPTER 10

Pluripotent Stem Cells and Diabetes: Basic Science to Clinical Applications

Khun H. Lie[1*], Methichit Chayosumrit[2#], Anand Hardikar[3] and Kuldip S. Sidhu[1#]

[1]Stem Cell Laboratory, Faculty of Medicine, University of New South Wales, Australia; [2]Siriraj Center of Excellence for Stem Cell Research, Faculty of Medicine Siriraj Hospital, Mahidol University, Thailand and [3]Diabetes and Pancreas Biology Section, The O'Brien Institute, Department of Surgery - St Vincent's Hospital, The University of Melbourne

Abstract: Type 1 diabetes is caused by the loss of insulin-producing (β) cells in pancreas. A significant advance in cell therapy for diabetes has been the development of a protocol for islet transplantation from Dr. James Shapiro and colleagues at the University of Alberta in Edmonton, Canada (the Edmonton protocol). However, lack of suitable organ donors for transplantation is a critical factor that limits this therapy from majority of individuals suffering from diabetes. Research in the last decade has therefore been largely focused on generating insulin-producing cells that can be easily obtained (derived) and used in transplantation setting for replacement therapy in diabetes. Although insulin-producing cells have been obtained from various sources including human ES cells, bone marrow-derived mesenchymal cells, umbilical cord-blood derived mesenchymal cells, transdifferentiation of liver / gallbladder cells, pancreatic duct cells, exocrine cells as well as islet-derived mesenchymal cells, the amount of insulin produced by most of these cell types is significantly less as compared to the amount of insulin produced by a normal adult pancreas. Present research is therefore focused on understanding signalling molecules and processes that would enhance differentiation of these cells. Although embryonic pluripotent/stem cells may be limited due to ethical issues, adult tissue-derived progenitor cells are believed to possess inherent traits that result in "commitment" to a particular phenotype, demonstrated by their relatively restricted differentiation capacity. In this chapter, we discuss the cell types that have been studied for replacement therapy in diabetes with specific reference to their possibility for use in a clinical setting.

Keywords: Diabetes, embryonic stem cells, pancreas, beta cells, differentiation, pluripotency, induced pluripotent stem cells.

1. PANCREAS DEVELOPMENT (REGULATION OF PANCREAS DEVELOPMENT, TRANSCRIPTION FACTORS AND SIGNALLING MOLECULES)

Diabetes mellitus has become one of the most common non-communicable diseases reaching epidemic levels. In 2007, it was estimated that 246 million people worldwide have diabetes and is predicted to rise to over 333 million people by the year 2025 [1]. Type 1 diabetes mellitus (T1DM) is an autoimmune disease caused by the destruction of the insulin-producing (β) cells which are located in the islets of Langerhans of the pancreas [2]. An ideal therapy for T1DM is to replace the missing β-cells. This can be achieved by whole-organ pancreas or isolated islet of Langerhans transplantation from a brain-dead donor. Despite the progress in transplantation protocols, both whole-pancreas and islet of Langerhans transplants are limited by scarcity of cadaveric pancreata [3]and the need for lifelong immunosuppression. One way forward is to generate insulin-producing cells in the laboratory.

The generation of insulin-producing cells from alternate sources including embryonic stem cells [4-6], induced pluripotent stem cells [7, 8], liver cells [9, 10], pancreatic ducts [11, 12], gallbladder [13, 14] and human pancreatic islets themselves [15-17] is presently being pursued throughout several laboratories.

***Address correspondence to Khun H. Lie:** Stem Cell Laboratory Faculty of Medicine, The University of New South Wales Level 2, Room G65, Clinical Sciences Building Prince of Wales Hospital Randwick, NSW 2031Australia; Tel: +61 (2) 9382 2617; E-mail: khun.lie@unsw.edu.au

[#] both the authors contributed equally

Kuldip S. Sidhu (Ed)

Successful therapy requires comprehensive β-cell phenotypes, which are the ability to synthesize, store and secrete insulin in response to blood glucose levels. The following sections explore embryonic signalling pathways and transcription factors involved in pancreas development and recent differentiation protocols to derive insulin-producing cells using extracellular signals, co-culture system and genetic manipulation from different cell types.

1.1 Embryonic Signalling Pathways and Transcription Factors Involved in Pancreas Development

Several transcription factors have been identified as critical regulators that drive pancreatic development and enable cells to function properly. These transcription factors function by binding to DNA sequences in the regulatory elements upstream of promoter region of endocrine genes and subsequently control transcription. Throughout pancreas development, there are a number of transcription factors forming regulatory networks or a hierarchy of factors that specify and maintain each differentiated cell phenotype [18, 19]. Therefore, it is important to understand how these transcription factors interact in order to understand the mechanism by which pancreatic β-cells developed.

1.1.1 Gastrulation and Definitive Endoderm Formation

Early pancreatic development has been well described in mice and chickens. During embryogenesis in mice, the epiblast is derived from the inner cell mass (ICM) of the blastocyst. It gives rise to the three principal germ layers (endoderm, mesoderm and ectoderm) through the process called gastrulation at embryonic day (E) 6.5-7.5 (Fig. **1**). Initially, the primitive streak (PS) is formed at the posterior part of the epiblast [20]. Several genes, including *Brachyury* (*T*) and *Mixl1*, are expressed throughout the PS. Mesendodermal cells that are concentrated towards the anterior node of PS, regulate the expression of several genes that control differentiation of the definitive endoderm (DE) and mesoderm progenitors. The pancreas arises from DE, which starts as a flat sheet of cells from the anterior PS that is specified during gastrulation [21-24].

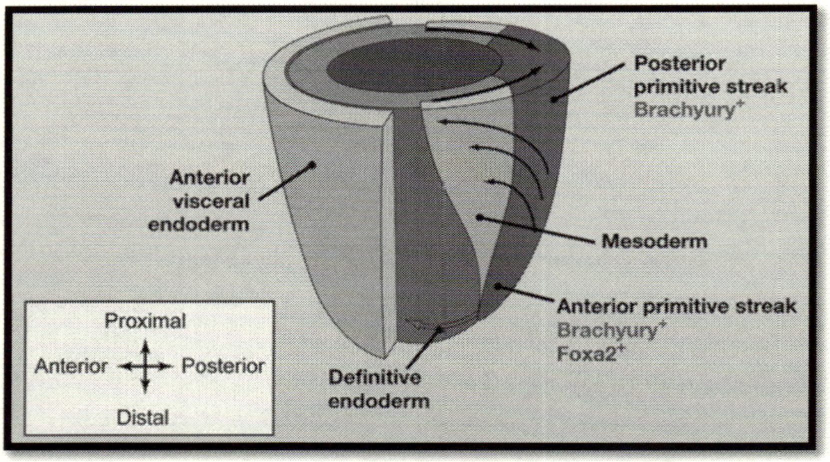

Fig. 1: Gastrulation in the mouse embryo. The posterior region of the primitive streak (PS) expressed *Brachyury*, whereas the anterior PS expressed both *Brachyury* and *Foxa2*. Epiblast cells enter the PS as indicated by the thick black arrows at the top of the embryo. Mesoderm is formed and migrated from the PS as indicated by the thin black arrow. DE is formed from anterior PS. The red arrow shows the movement of DE cells. Adapted from Murry *et al.* [24].

Several families of growth factors regulate gastrulation in mouse. Nodal, the member of the transforming growth factor-beta (TGF-β) super family, is required for establishing the anterior-posterior axis during gastrulation as well as initiating endoderm and mesoderm development by regulating the Wnt, fibroblast growth factors (FGF) and bone morphogenetic protein-4 (BMP-4) pathways [24, 25]. There are a number of transcription factors downstream of the Nodal and Wnt signalling pathway to direct endoderm formation, including Mix-like homeodomain proteins, Gata zinc finger factors (*Gata 4/5/6*), forkhead domain factors (*Foxa2*), and SRY (sex determining region Y)-box containing gene 17 (*Sox17*) [26].

1.1.2 Foregut Endoderm Formation

Embryological and molecular genetic studies in mice revealed that the digestive gut tube is formed at E8.5 (Fig. **2**). The pancreas arises from 3 distinct primodia (1 dorsal (DP) and 2 ventral (VP) buds) of the foregut epithelium (gut tube). Pancreas formation requires extracellular factors that modulate signalling pathways. The early specification of the pancreatic domain (pancreatic buds) requires signals from the notochord (activin-β and FGF-2) and endothelial cells of the aorta as well as the inhibition of Hedgehog (Hh) signalling at E9 of mouse gestation [21, 27, 28].

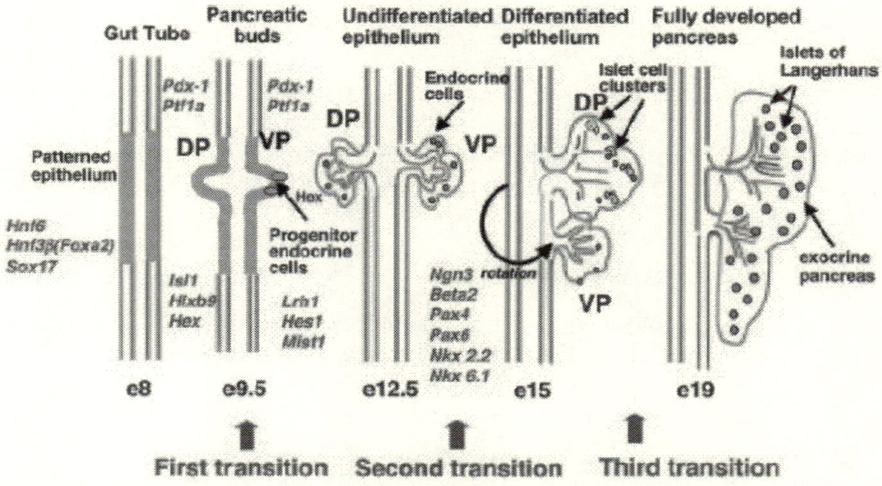

Fig. 2: Schematic diagram of pancreas development in mice. The transcription factors are indicated in each stage of development. DP = dorsal bud and VP = ventral bud. Adapted from Habener *et al.* [29].

Activin-βB, which is a member of TGF-β family and FGF2 have been shown to inhibit the expression of Sonic hedgehog (Shh), an activator of the Hh signalling pathway, in the dorsal pancreatic bud and regulate branching morphogenesis as well as induce expression of *Pdx1* and *Insulin* in dorsal prepancreatic epithelium [30]. The *Pdx1* gene expression at the foregut-midgut junction specifies the territory of the forming pancreas [31]. Furthermore, studies in Xenopus [32-34] and zebrafish [35, 36] demonstrated that retinoic acid (RA) also acts as an instructive signal to inhibit Shh activity and induce both pancreatic gene expression in dorsal endoderm and endocrine differentiation. RA signalling has been demonstrated to regulate *Pdx1* expression [37-39]. There are several other transcription factors involved in foregut patterning, including *Foxa2* (also known as *Hnf3β*), *Hnf6*, *Hnf4α*, *Hnf1β*, *Hlxb9*, *Hex*, *Isl1* and *Ptf1a/p48* [21, 40].

1.1.3 Pancreatic, Endocrine and Exocrine Progenitor Cells

At around E9.5 during the primary transition, the pancreatic epithelium proliferates, invades the surrounding mesenchyme and undergoes a series of differentiation and subsequent branching morphogenesis. These events are regulated by the most characterized mesenchymal factor, FGF10, which is expressed by the dorsal and ventral mesenchyme from E9.5-12.5. FGF10 has been demonstrated to promote proliferation of *Pdx1*+ epithelial pancreatic progenitors and maintain *Ptf1a* expression in the dorsal pancreatic bud [41]. In the absence of FGF10, these cells failed to proliferate [42]. Apart from *Pdx1* and *Ptf1a*, other transcription factors including *Sox9, Pbx1* and *Hlxb9* are also involved in specification of the pancreas [5, 43, 44].

Although glucagon+ cells (α-cells) are evident at E9.5 during the primary transition, most of the hormone-expressing cells, which comprise the endocrine islets (α, β, δ, ε and PP), appear to emerge from the pancreatic duct epithelium at around E13.5-14.5 during the secondary transition. The endocrine progenitor cells continue to migrate into the surrounding mesenchyme and aggregate into cell clusters, while the exocrine cells become amylase-expressing acinar cells [21, 27]. At this stage (E13.5), expression of *Ptf1a* is restricted to cells that will form the exocrine pancreas [44]. On the other hand, expression of *Ngn3* peaking around E15.5 in mice is required for endocrine development. There are several endocrine-specific transcription factors downstream of *Ngn3*, including *NeuroD1/Beta2, Pax4, Pax6, Nkx2.2, Nkx6.1* and *Isl1* [27, 43].

A study in transgenic mice showed that prolonged expression of FGF10 in the embryonic pancreas inhibits pancreatic differentiation by maintaining Notch activation, which results in impaired *Ngn3* expression within the pancreatic epithelium and expansion of undifferentiated progenitor cells [45]. Ectopic activation of the Notch signalling pathway inhibits both endocrine and exocrine development while maintaining pancreatic progenitor cell characteristics [27]. Therefore, Notch signalling is an important regulator to maintain the proper balance of endocrine and exocrine cells from pancreatic progenitor cells, and the loss of Notch signalling allows the endocrine lineage specification by permitting the expression of *Ngn3* [45].

Activinβ signalling has also mediated lineage specification at later stages. Study in rat has shown that activinβ may be permissive for endocrine differentiation since the inhibition by follistatin, an antagonist of activinβ, has been demonstrated to inhibit endocrine differentiation and promote an exocrine lineage [46]. In addition, expression of stabilized Wnt/β-catenin at later stages of pancreas development results in enhanced proliferation of exocrine pancreas, maintenance of *Ptf1a/p48+* exocrine progenitor cells and increase in pancreas size [47-49].

1.1.4 Endocrine Subtype Specification/Maturation

The endocrine cells and their acquisition of full nutrient responsiveness continue for 2-3 weeks after birth [29, 31, 50, 51]. Within the endocrine lineage, there are a number of transcription factors that are restricted to each endocrine subtype. The formation of insulin-producing β-cells is driven by *Nkx2.2, Nkx6.1, Pax4, Pax6, Hlxb9, MafA, Isl1* and *Pdx1* (in a second wave of expression). The formation of glucagon-producing α-cells is driven by *Pax6, Arx, Nkx2.2, Foxa2, MafB* and *Brn4*. Less is known about other cell types but some studies reported that *Pax4* and *Pax6* are expressed in somatostatin-producing δ-cells and *Nkx2.2* is expressed in PP cells. At present, there has been limited information about the transcription factors that regulate the formation of ε-cells. [19, 43, 52, 53].

There is evidence that hepatocyte growth factor (HGF) induces β-cell formation. HGF and its receptor c-met are expressed at high levels in human pancreatic fetal fibroblasts, which are derived from early embryonic mesenchyme. Conditioned medium from human pancreatic fetal fibroblasts has been shown to induce β -cell proliferation and islet-like cell cluster formation [54, 55]. Epidermal growth factor (EGF) is another mesenchymal factor that has been shown to regulate proliferation of the developing pancreas. EGF receptor (EGF-R) is expressed throughout the developing pancreas. Study by Miettinen *et al.* (2000) demonstrated that, β-cell differentiation is delayed and their proliferation is impaired in EGF-R deficient mice. The EGF-R signalling is also required for adequate basement membrane degrading activity and subsequent ductal branching and migration [56].

Studies have shown that glucagon signalling is involved in early differentiation of insulin-producing cells. Members of the glucagon family of peptide hormones are glucagon-like peptide-1 (GLP-1), GLP-2, glucose dependent insulinotrophic peptide (GIP), secretin, vasoactive intestinal peptide (VIP) and others [57, 58]. Among these members, GIP and GLP-1 are the major physiological incretins, which are the intestinal hormones secreting in response to the ingested food that stimulate glucose-dependent insulin secretion [59]. GLP-1 signalling involves the formation of new β-cells through enhanced proliferation of existing β-cells, induction of islet neogenesis [60, 61] as well as an inhibition of β-cell apoptosis [62, 63]. GIP signalling altered expression of pancreatic endocrine transcription factors including *Gata4, Isl1* and *Pdx1* [64]. In addition, GIP has been demonstrated to promote growth and survival of pancreatic β-cells by increasing the anti-apoptotic gene *Bcl-2* and decreasing the pro-apoptotic gene *Bax*, which results in reduced β-cell death [65].

2. STEM/PROGENITOR CELLS AS Β-CELLS SURROGATES

Several attempts have been made to find other alternative means for cell replacement therapy to treat T1DM. The ideal β-cell substitute should be readily available, easily maintained and expanded, and capable of generating therapeutically relevant amount of insulin. Importantly, these cells would need to encompass the glucose sensing mechanism, which regulates insulin secretion and storage at a normal physiological level. It is possible that genetic modifications would be required to alter their immune characteristics in order to protect them from autoimmune destruction post-transplantation, unless being protected by immune-suppression therapy. Transplantation of derived surrogate β-cells from these sources may resolve the problem with the limited supply of islets and offer several cell sources that are suitable for the treatment of T1DM. A detail about various stem cells and their therapeutic potential is described in chapter 1.

2.1 Adult Pancreatic Stem Cells

It has been demonstrated that pancreatic endocrine progenitor cells are present during embryonic and fetal stages of development [66], however the nature and existence of adult pancreatic stem cells (APSC) has been a subject of controversy. The ability of adult cells to differentiate along other cell lineages suggests for the presence of multipotent stem cells residing in various tissues. It is anticipated that tissues within or adjacent to the epithelial compartment of the adult human pancreas such as ductal, acinar and islet cells themselves may retain a small populations of APSC which are capable of regenerating and differentiating into an endocrine phenotype. It has been demonstrated that highly purified non-endocrine pancreatic epithelial cells which consist of mostly acinar and ductal cells are capable of undergoing differentiation to a beta cell phenotype under certain conditions [67]. The application of APSC does not raise ethical concerns and promotes the potential use of autologous cells which would solve the problem of allogeneic rejection. Moreover, non-endocrine pancreatic cells are also more readily accessible and could therefore be regarded as an alternative source of progenitor cells for β-cell replacement therapy.

2.2 β-Cells Replication

It is believed that adult stem cells contribute to β-cell maintenance and repair *in vivo.* The extent to which APSC's actually involved in β-cell neogenesis remains debatable due to the unclear origin and location of these cells. A study on the dynamics of the β-cell mass in the rodent has found that increment in β-cell mass is due to robust β-cell replication which is maintained throughout adult life [68]. Further study in mice development has reported the corresponding expression of cyclin D2 in parallel to endocrine cell replication and islet mass increment [69]. Based on this study, it is suggested that cyclin D2 plays an essential role in the regulation of β-cell regeneration.

A genetic lineage tracing study in adult mice demonstrated that a majority of new β-cells are formed by the self replication of pre-existing β–cells and suggest little to no role for stem cells in adult β-cells regeneration [70]. Similar findings indicate that microRNAs, a class of non-coding RNAs negatively regulate this process of stem cell differentiation by targeting the transcriptional activity of Neurogenin3 (Ngn3), thereby allowing proliferation of existing beta cells [71, 72]. Fluorescently-labelled β-cells were observed for the changes in clonal size and loss of expression throughout cell division. The uniform loss of fluorescence and similar clonal sizes conclude that all β-cells promote similar proliferative potential and expand from a homogenous pool of mature β-cells [73]. In contrast to the studies in adult mouse and rat, much less information is known about human's β-cell replication. A direct comparison of cell cycle kinetics through the exposure to alloxan has shown a significantly higher uptake of tritiated thymidine labeling in the rodent beta cells when compared to human β-cells [74]. This outcome suggests a lower rate of β-cell replication in the human compared to rodents. A case study of an 89-years old T1DM patient showed an increase in the frequency of Ki67 stained β-cells which is evidence of attempted regeneration through β-cell replication [75]. Two recent studies have applied lineage tracing approaches to demonstrate the ability of human β-cells to proliferate *in vitro* [76, 77]. These reports have indicated the ability of pancreatic β-cells to retain their proliferative capacity both *in vitro* and *in vivo*.

It was demonstrated that although human beta cells proliferate *in vitro*, they are not able to retain their differentiated state during proliferation. During *in vitro* expansion, human pancreatic β-cells undergo epithelial-to-mesenchymal transition (EMT) to produce proliferating population of islet-derived precursor cells (hIPCs) [78]. The phenomenon of EMT is observed during normal embryonic development as well as during metastasis. Here, epithelial cells transition to acquire all the phenotypic and molecular signatures of mesenchymal cells [79-81]. The concept of EMT in pancreatic islet cultures was first introduced through our previous work [78]. Generation of mesenchymal-like progenitor cells from β-cells was also reported shortly after this initial presentation [82]. In this study, single cells obtained from human islets were cultured *in vitro* and Pdx1 producing human β-cells were seen to co-express mesenchymal proteins during expansion, thus suggesting β-cell origin of mesenchymal progenitors [82]. The entire concept of EMT of beta cells was strongly opposed by a series of articles [83-86]. In these studies Cre/LoxP based genetic lineage tracing system were used to label β-cells and their progeny *in vitro*. These reports [83-86] convincingly demonstrate that labelled cells do not exist in proliferating cultures *in vitro* and thus suggested that mouse β-cells neither proliferate nor undergo EMT *in vitro*. These studies also propose that the mesenchymal cells arising in the culture originate from pre-existing/contaminating mesenchymal cell in the islet preparation. Two of these articles [85, 86] noted that labelled β-cells de-differentiate to a mesenchymal-like phenotype *in vitro*, but fail to proliferate. Since previous studies in human islets lacked lineage-tracing analysis [78, 82], findings in mouse [83-86] were extrapolated to human islets and it became a consensus that terminally differentiated islet β-cells do not proliferate *in vitro*. Last year, the group of Shimon Efrat, used a dual lentiviral system to irreversibly label human β-cells *in vitro* [76]. They observed that labelled cells, indicating the progeny of β-cells, are seen in the proliferating population at the same frequency for up to 16 population doublings. This group further demonstrated that adult human islet β-cells undergo EMT [87] and express mesenchymal stem cell surface antigens. We confirmed these findings in human foetal islets using multiple approaches, including immunostaining and FISH, single cell PCR, clonal expansion analysis, assessment of heritable marks of insulin-promoter region and thymidine-analog based lineage tracing analysis [77]. We also demonstrated that members of the miR-30 family of microRNAs, that are strategically located in the intron of nf-yc gene, are involved in regulation of EMT in human islets by targeting several mesenchymal genes including vimentin, snail1 and snail2 [88]. All these studies confirm that human pancreatic insulin-producing cells proliferate and undergo EMT *in vitro*. However, when the same dual lentiviral system was used to label mouse β-cells, it was observed that mouse β-cells do not proliferate *in vitro* [76]. We believe that species-specific differences in mouse and human may explain the different behaviour of pancreatic beta cells in cultures. Although the contribution of human β-cells in generation of islet-derived progenitor cells *via* EMT has been demonstrated through these 2 publications, contribution of non-β cells and pre-existing mesenchymal cells cannot be refuted, until lineage tracing data for these cells are available.

The discrepancy in ability of β-cell to proliferate *in vitro* may be due to intricate differences in the species. Nonetheless, there are several studies that confirm the ability of mouse β-cells to proliferate *in vivo* following exposure to varied physio [89, 90] chemical stimuli [70, 73, 91, 92]. Whether such *in vivo* proliferation involves a mesenchymal intermediate, remains yet unclear [93]. Although studies discussed above [70, 73, 91, 92] confirm proliferation of mouse β-cells *in vivo*, attempts to mimic these *in vitro* have failed [83-86]. Human pancreatic islet-derived progenitor cells are known to be committed to differentiate into endocrine pancreatic lineage as they inherit the active marks that define an open chromatin conformation at insulin promoter region. Such islet-derived mesenchymal-like cells therefore offer great potential for cell replacement therapy in diabetes.

2.3 Pancreatic Duct Cells

It has been proposed that APSC reside within the adult pancreatic epithelium due to their origin and shared similarities with embryonic primitive ducts, thus allowing these cells to retain their ability to generate endocrine phenotype [94-96]. Studies in rodent pancreatic regeneration also support the concept that islet neogenesis might normally occur in the ductal epithelium during normal embryonic development [97, 98] as well as in diabetic mice [99, 100].

The concept of islet neogenesis from ductal cells was challenged by a lineage tracing study that demonstrated an early separation between the adult duct progenitors and the endocrine cell lineages [101]. These authors have observed that many of the cells within 'duct-like structures' in early embryonic pancreata are not duct progenitors and do not develop into adult ducts, instead they may possibly be endocrine progenitors which could cluster and form islets. It is still difficult to depict the early stages of ductal cell differentiation due to lack of duct cell-specific markers and uncertainty of its origin during embryonic development. More extensive studies however have supported the idea of pancreatic ductal cells being the main source of APSC and further demonstrated the generation of glucose-responsive pancreatic islet-like structures *in vitro* through the expansion and differentiation of both human adult ductal tissue [94, 102] and human fetal pancreatic duct cells [103].

Arguments were raised suggesting that insulin production could have originated from a mixed population that contained contaminating islet cells when isolating the duct cells. To confirm that pancreatic duct cells are *in vitro* progenitors, a study has shown that the addition of stromal cells to a highly purified ductal cell population increases the cells expansion rate, formation of cell aggregates and improves insulin production when transplanted under the kidney capsules of normoglycemic NOD/SCID mice [104].

Adenovirus-mediated expression of transcription factors PDX1 [105] and Neurogenin3 (NGN3) [106] in adult human pancreatic duct cells were also found to induce β-cell neogenesis and promote their differentiation into endocrine phenotype cells that express insulin. It still remains unclear as to how these pancreatic transcription factors are activated naturally in the adult ductal cells and further promote their differentiation towards endocrine phenotype. Convincingly, all these studies suggest that pancreatic duct cells are a possible alternative β-cells surrogate for the generation of insulin producing cells.

2.4 Acinar Cells

Pancreatic exocrine cells which comprise the bulk of adult pancreatic tissue provide an attractive source for the generation of transplantable surrogate β-cells by trans-differentiation. There have been many studies reporting the phenotypic plasticity of pancreatic acinar cells. A study focusing on the morphological and physiological changes of rodent β-cells in prolonged hyperglycemia has shown that β-cells may arise from acinar cells directly and not from ductal progenitor cells [107]. This study also observed a reduction in the duct cell proliferation rate and no increment in the duct-associated β-cell frequency, which indicate that the increase in β-cell neogenesis does not originate from duct-associated stem cells.

It is likely that pancreatic acinar cells are indirectly involved in β-cell regeneration as studies using rodent and human pancreas have shown evidence that pancreatic acinar cells can be trans-differentiated to ductal cells [108, 109]. A lineage tracing study involving cre/lox mediated cell labelling strongly support the acinar-to-ductal trans-differentiation and demonstrated the generation of nestin-expressing progenitor cells, which are generally observed in early pancreatic development. [110]. Trans-differentiation of pancreatic acinar cells into ductal cells and insulin-secreting cells has been demonstrated both *in vivo* and *in vitro* under different conditions [111-114]. This phenomenon is commonly followed by changes in their phenotypic state into a transitional ductal-like cell which is characterized by the increasing proliferation rate of duct-like cells, reduction of amylase expression and increasing expression of cytokeratin and insulin. Nevertheless, a contradicting *in vivo* lineage tracing study has failed to observe any acinar-beta cells trans-differentiation after a partial pancreatectomy in the adult mice [115]. Instead they detected a preference in the pre-existing acinar cells to replicate and generate new acinar cells. There are many studies reporting the optimization of culture conditions known to promote β-cell regeneration through the acinar-to-ductal conversion, yet there are fewer studies identifying the molecular and cellular mechanisms involved.

A recent genetic study has made progress and identifies three transcription factors (NGN3, PDX1 and Mafa) which could play a major role in β-cell neogenesis [116]. This combination of three factors known to be involved in pancreas development was found to reprogram adult mice exocrine cells into cells with similar morphological structures, genetic expressions and functions as islet β-cells. However, as mentioned previously, it remains uncertain as to how activation of these genetic factors can be initiated naturally in the adult acinar cells.

2.5 Extra-Pancreatic cells

Trans-differentiation is the conversion of cells from one differentiated cell type to another. Gastrulation and patterning studies of endoderm development in vertebrates have pointed out that liver and the ventral part of the pancreas are derived from the ventral foregut which is adjacent to the notochord and the cardiac mesoderm [117]. Analysis in mouse embryo indicates that the adjacent cardiac mesoderm secrete signalling molecules that control a common precursor cell to develop into either a pancreas or liver cell fate [118, 119]. This led to the possibility of a common endodermal stem cell existence within the adult liver that is capable of giving rise to either hepatic or pancreatic lineages. In addition, both liver and pancreas are found to share several similar characteristics such as responsiveness to glucose and the expression of similar genetic transcription factors [120]. The shared characteristics, origin and adjacent location between liver and pancreas suggest that liver cells represent one of the most promising extra-pancreatic cells that can trans-differentiate into pancreatic β-cells and be used to generate transplantable insulin-producing cells.

Several groups have demonstrated the trans-differentiation of both adult and fetal liver cells into β-like cells through virally-mediated PDX1 expression *in vitro* [121] and *in vivo* [122-124]. These results have confirmed that PDX1 expressing liver cells could produce glucose-responsive insulin-producing cells that express transcription factors required for β-cell development and function. Furthermore, insulin induced through the activated PDX1 expression was processed to a mature functional state and could partly or fully ameliorate hyperglycemia in the streptozotocin-induced diabetic mice [122-124]. In contrast, it was reported that PDX1 induction in human hepatoma cell line, HepG2, was not sufficient to promote insulin production either *in vitro* or after the transplantation under renal capsule of streptozotocin-induced diabetic nude mice [125]. They reasoned that the absence of transcription factor NeuroD1 in HepG2 cell line was due to the diversity within the cells population that may reduce the differentiation potential and development toward pancreatic cells. This reason was later confirmed by a viral-mediated combined expression of PDX1/VP16 together with NeuroD or NGN3 in HepG2 cell line which has markedly induces the insulin production and ameliorates glucose tolerance in streptozotocin-induced diabetic mice [126].

Alternatively, the expression of human insulin in human hepatoma cell line HUH7 resulted in HUH7-ins cells which could synthesize, store and secrete insulin in response to glucose [127]. The endogenous expression of NeuroD in HUH7 cells however has indicated that these cells already possessed pancreatic attributes, which may have enhanced the ability of HUH7-ins to form pancreatic secretory granules-like structure that regulate insulin. In another study, it was demonstrated that hyperglycaemia could be reversed in diabetic mice transplanted with human fetal liver progenitor cells which had been trans-differentiated into insulin-producing cells [128].The unique capability of liver cells to trans-differentiate may provide a valuable approach to the use of autologous liver cells in individuals with diabetes to generate 'self' surrogate β-cells.

Insulin-producing cells have also been derived from human umbilical cord blood stem cells [129-131], human neural progenitor cells [132], human bone marrow stem cells [133-135], human adipose tissue-derived mesenchymal stem cells [136-138] and placenta-derived multipotent stem cells [139] using different induction protocols. These data demonstrated that adult stem cells could provide important sources of stem cells because there are little ethical concerns, no risk to the donors and low risk of graft-versus-host disease (GVHD).

It has been proposed that autologous nonmyeloablative hematopoietic stem cell (AHSCT) transplantation in newly diagnosed T1DM patients may improve β-cell function, and a recent study has demonstrated that this treatment induced prolonged insulin independence in the majority of the patients [140]. After a two-year follow up, it was found that C-peptide levels remained significantly increased and the majority of patients were insulin independent with good glycaemic control [141]. Despite these successful results obtained, the biological mechanisms of AHSCT remain unclear whether it is from partial reprogramming of the patients' immune system or stem cell-mediated β-cell regeneration [142]. A better understanding of these mechanisms is required before these cells can be applied for the treatment of T1DM.

2.6 Embryonic Stem Cells (ESCs)

2.6.1 Differentiation of ESCs to Endoderm Lineage

ESCs represent a suitable model for studying the lineage specification mechanisms involved in human embryo development. One of the earliest cell fate decisions made in vertebrate embryogenesis is gastrulation, which resulted in the formation of three primary germ layers: mesoderm, endoderm and ectoderm. Embryonic stem cells (ESCs) are unique due to their abilities to undergo both long-term proliferation and differentiation into derivatives of all three embryonic germ layers. Furthermore, cells derived from hESCs could potentially provide an unlimited source of cells for cell replacement therapies. In the absence of feeder cells and anti-differentiating agents, ESCs will spontaneously develop into a complex cell-structure embryoid bodies (EBs) which is comprised of derivatives of all the germ layers and recapitulates certain aspects of early embryogenesis [143]. Nonetheless, EB spontaneous differentiation is generally unsystematic and frequently results in the formation of a variety of lineages [144]. Differentiation through EBs therefore, proved to be unsuitable for the generation of a uniform lineage-specific population. In contrast, directed differentiation of ESCs through germ layer specification may provide cells which are suitable for cell regenerative studies.

Several groups have effectively recapitulated the endoderm lineage-induction event in early embryo and demonstrated the capability of ESCs to differentiate *in vitro* towards endoderm lineage [145-147]. The endoderm germ layer gives rise to the digestive system and its associated organs, including the pancreas. There are two types of endoderm observed in early development: visceral endoderm, which diverges directly from the ICM and gives rise to extra-embryonic endoderm [148]; and definitive endoderm which develops from the anterior primitive streak corresponding to early and mid gastrula organizer [117, 149, 150]. These studies have suggested that the specification of extraembryonic endoderm occurs earlier in the blastocyst stage, whereas the specification of definitive endoderm occurs later during gastrulation. Both visceral and definitive endoderm share many similar molecular properties and are therefore believed to respectively represent the outer and inner layers of the ICM. Interestingly, a lineage-tracing study has demonstrated that the ICM consist of a heterogeneous cell population that could differentiate into either extraembryonic endoderm or pluripotent epiblast cells [151].

Since visceral and definitive endoderms are derived from different cell sources, it is possible to trace back and define molecular markers expressed in each tissue. A comparative gene-expression profiling was performed to analyze the transcriptional similarities and differences between the definitive endoderm and visceral endoderm tissues [152]. This profiling study has suggested a specific gene-expression signature for visceral and definitive endoderm that would assist in the detection and the commitment of the ESC-derived endoderm. ESCs derived from the pre-gastrulation ICM are capable of recapitulating the early endoderm germ layer development. Furthermore, the selective induction of definitive endoderm from ESCs may require the suppression of visceral endoderm formation. Through ontogeny-based studies, hESCs were effectively differentiated to a homogenous definitive endoderm population and found to be essential for the generation of pancreatic cells [147, 153].

Studies in vertebrate endoderm development and gene expression analysis have revealed the presence of a bi-potent mesendoderm population which is capable of generating both mesoderm and definitive endoderm cells [154, 155]. At present, the characterization and differentiation process of mesendoderm cells in human embryo development remain uncertain. A number of studies have demonstrated the effectiveness of using genetically-labeled ESCs for monitoring mesendoderm development in differentiation cultures and identifying the molecular factors which affect its formation [146, 156, 157]. These findings have effectively characterized the signalling pathways and genetic factors that are required for the generation of definitive endoderm in ESC differentiation cultures.

2.6.2 Differentiation of ESCs to Pancreatic β-Cells

Due to their limitless proliferation and differentiation potential, ESCs are regarded as a possible cell source for therapeutic application in regenerative medicine particularly for T1DM. It has been reported that the *in vitro* spontaneous differentiation of mESCs and hESCs could result in the generation of cells with characteristics of insulin-producing β-cells [158-163]. Nevertheless, the level of insulin content and the ratio of the insulin producing cells generated under those conditions were very low, hence making them unsuitable for further investigation as a possible cell source for therapy. There are other concerns associated with the spontaneous differentiation of ESCs into insulin-producing cells (IPCs) such as the induction of non-specific cells from multiple germ lineages. Based on these results, several strategies have been proposed to enhance the differentiation process of ESCs towards IPCs. Genetic engineering has proven to be effective for the identification of molecular growth factors that enhance the lineage commitment of ESCs into IPCs. The use of a transgenic system provides a reliable tool for understanding the differentiation mechanism of pancreatic β-cells, however it cannot be used to derive functional β-cells that are suitable for cell transplantation or drug discovery assays. Thus, the application of defined differentiation factors and selection of lineage-specific progenitor cells appear to be essential for directing the differentiation of ESCs into β-cell surrogate phenotype (Table **1**).

A better approach for generating β-cells from ESCs is to recapitulate the signalling factors that are known to be involved in the *in vivo* development of pancreatic β–cells. Full understanding of the mechanisms that control each stage of β–cell development would allow progress towards the derivation of functional beta cells from hESCs. Critical to the early pancreas developmental processes is the specification of endoderm during gastrulation. The endoderm subsequently differentiates to form the pancreatic epithelium, which then branches into the pancreatic mesenchyme to form islet clusters of endocrine cells. These endocrine precursor cells mature into β–cells, which are capable of producing insulin and other pancreatic hormones.

Studies on vertebrates have uncovered the central role of nodal signalling for the development of endoderm progenitors [149]. Both activin and nodal belong to the TGF-β superfamily which trigger Smad intracellular signalling. It was first demonstrated in mESCs, that definitive endoderm cells could be induced by stimulation with activin A under serum-free conditions [145]. Another characterization study has also shown that either nodal or activin could selectively induce the differentiation of mESC towards definitive endoderm [156]. It is therefore believed that the differentiation of ESCs to mature pancreatic cells requires the sequential application of activin followed by other endoderm/pancreas-specialized growth factors. A simple three-step protocol utilizing activin, retinoic acid (RA), bFGF and nicotinamide was used to stimulate the differentiation of mESC EBs into glucose-responsive IPCs [164]. The transplantation of these IPCs was sufficient to normalize the BGLs of the streptozotocin-induced diabetic mice. Tumorigenesis was however frequently observed in the transplanted site, which suggested the immaturity of the differentiated cells.

Another well-defined multi-step protocol (Fig. **3**) was recently established based on the differentiation of hESCs into definitive endoderm and pancreatic endocrine cells by factors that are involved in *in vivo* pancreatic development. [147]. Directed differentiation of hESCs towards definitive endoderm (SOX17+/CXCR4+) was obtained through the supplementation of activin and Wnt3a in a serum-reduced medium. The reduction of serum from the media was shown to be critical for the reduction of PI3K signalling, which further increased the ability of hESCs to differentiate in response to activin. Alternatively, the addition of activin and PI3K signalling inhibitors (LY294002) in serum-containing medium has also resulted in the formation of definitive endoderm cells [165]. Both of these studies have demonstrated that activin signalling and low PI3K activity were substantially required for the endoderm formation from hESCs.

In the next stage of differentiation, the combination of fibroblast growth factor 10 (FGF10) and retinoic acid (RA) as well as inhibition of sonic hedgehog (SHH) signalling prove to be vital for the patterning of foregut endoderm ($HNF1\beta^+$/$HNF4\alpha^+$) and pancreatic epithelium ($PDX1^+$/$NKX6.1^+$) [147]. Studies in mouse and human pancreas have indicated the expression of FGF10 within the mesenchyme adjacent to the pancreatic epithelial progenitor cells [167, 168]. These studies have demonstrated that FGF10 signalling is required for the proliferation of $PDX1^+$ pancreatic precursor cells and the specification of pancreas development. On the contrary, the activation of SHH signalling has been reported to interfere with the early pancreas organogenesis [169]. The application of cyclopamine, a steroid alkaloid that inhibits SH signalling causes ectopic pancreatic buds to form in cultured gut tube explants [170]. These reports further suggested that pancreas development was initiated in a restricted region of posterior foregut in which SHH expression was repressed. Study in mouse embryonic pancreatic cultures has shown that the application of RA could stimulate the pancreatic epithelium to differentiate into primarily endocrine cells [171]. Furthermore, it has been demonstrated that RA treatment induces the expression of the PDX1 gene in differentiating mESCs [39, 164]. Importantly, the application of RA also induced the coordinate expression of other transcription factors (HNF6, HLXB9, SOX9) which closely mimicked the timed expression of these genes during pancreas development [147].

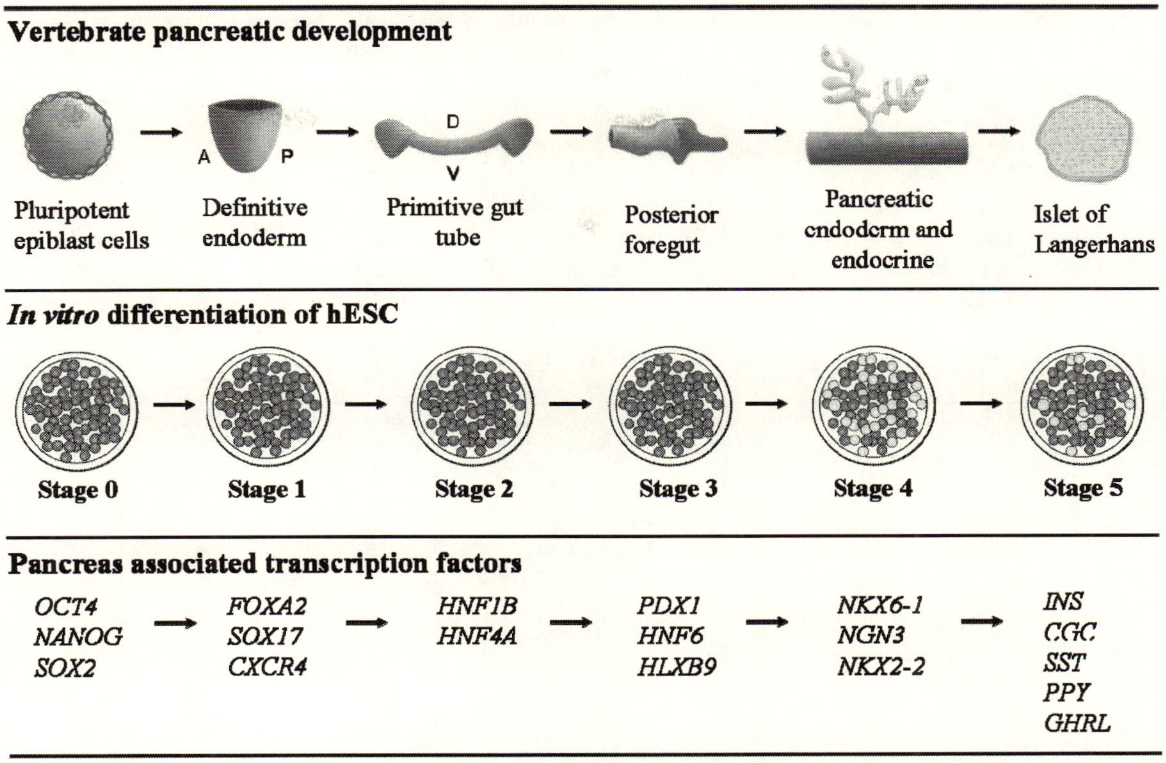

Fig. 3: Schematic representation of the strategy for differentiation of hESCs to insulin-producing cells. The figure orchestrates the differentiation of hESCs to hormone-expressing endocrine cells by mimicking pancreas development in vertebrates. Upper panel, fundamental stages of pancreatic endocrine development. Middle panel, stepwise differentiation of hESCs. Lower panel, several markers characteristics of each cell population. The colors of the tissues correspond to the cell types generated in the differentiation protocol. Modified from Madsen and Serup [166].

Following the removal of RA, there are continuing signalling processes that regulate the differentiation of $PDX1^+$ pancreatic progenitor cells toward the endocrine precursor cells. Application of the Glucagon-Like Peptide-1 (GLP1) receptor agonist exendin-4, and inhibition of notch signalling by N-[N-(3,5-difluorophenacetyl)-L-alanyl]-(S)-phenylglycine t-butyl ester (DAPT) resulted in the up-regulation of endocrine specific markers [147]. Moreover, the temporary-overlapping expression of pancreatic progenitor (PDX1) and endocrine markers (NGN3, NKX2.2, NKX6.1 and PAX4) was proven to be consistent with the expression of these genes *in vivo*. The transient expression of endocrine markers was further reduced in extended culture and followed by the formation of early hormone-expressing cells.

In the final stage of the ontogeny-based differentiation protocol, it has been demonstrated that the endocrine cells derived from hESCs were capable of producing pancreatic hormones. Addition of exendin-4, hepatocyte growth factor and insulin growth factor to the endocrine-like cells resulted in the formation of multiple hormone expressing cells (insulin, glucagon, somatostatin, pancreatic polypeptide and ghrelin) [147]. It has been further demonstrated that the hESC-derived endocrine cells were capable of producing high amounts of insulin that was comparable to that in adult islet. These cells were also shown to secrete C-peptide in response to multiple secretory stimuli. In spite of this, they were not able to secrete insulin or C-peptide upon glucose stimulation and were therefore considered to resemble immature fetal β-cells [147]. In their later studies, the same group has reported an improved hESCs differentiation strategy which produced cells that were functional in a xenotransplant diabetic mouse model [172]. This strategy was established based on the differentiation studies of human fetal pancreatic tissues and their abilities to reverse hyperglycemia in diabetic mice model [173-175]. In the new approach, hESCs were differentiated *in vitro* into endocrine precursors and implanted in SCID mice to mature *in vivo*. The transplanted hESC-derived endocrine cells resulted in the specification of glucose-responsive insulin secreting cells and prevented the destruction of mouse β-cells upon streptozotocin treatment. Interestingly, the study observed an immature teratoma formation in one out of forty-six grafts which suggested for the presence of undifferentiated hESCs [172].

Similarly, there were other hESC-derived ontogeny-based approaches that have resulted in the significant production of multi-hormonal cells. These studies demonstrated the production of mature cell types that were either capable of secreting C-peptide in response to glucose [153], or capable of lowering hyperglycaemia in diabetic animals [176]. Based on these ontogeny-based differentiation results, it is believed that hESCs represent the prospective renewable source of pancreatic β-cells for diabetes cell-replacement therapies.

Table 1. Summary of studies reporting the step-wise differentiation of hESCs into insulin-producing cells.

Cell Lines	Culture Conditions	Culture Period	Expression of Pancreatic Proteins	Gsis (c-Peptide)	Amelioration of Hyperglycaemia	References
CyT203, CyT25, CyT49, BG01, BG02 and BG03	Adherent on MEFs in ActA, Wnt3A, FGF10, CYC, RA, DAPT, Ex-4, B27, IGF1, HGF	15-18 days	PDX1, NKX6.1, NKX2.2, NGN3, INS, PAX6, ISL1, GHRL, C-PEP, SST, PPY, GCG, IAPP	No	ND	[177]
HUES 2, HUES 4, HUES 8 and HUES 9	Adherent on MEFs in ActA, Wnt3A, FGF10, CYC, RA, ILV, N2, B27, bFGF, albumin fraction V	21 days	PDX1, AMY, C-PEP, GCG, INS	*In vitro*	ND	[178]
H1, H7 and H9	Adherent on Matrigel in ActA, B27, Na-butyrate, EGF, bFGF, Noggin, IGF2, Nic	36 days	PDX1, C-PEP, GCG, SST	*In vitro*	ND	[179]
H1	Adherent on Matrigel in ActA, B27, Na-butyrate, EGF, bFGF, Noggin, IGF2, Nic	36 days	PDX1, INS, GCG, C-PEP, PC1/3, PC2, CK19	*In vivo*	No	[180]
Miz-hES4 and Miz-hES6	EB formation in ActA, RA followed by plating EB in ITS medium containing fibronectin	20 days	PDX1, INS, C-PEP, GCG	ND	Yes	[181]
HES1, HES2, HES3, HES4, HUES7, and HUES9	EB formation in ActA, BMP4, B27, FGF18, heparin, EGF, TGF-α, IGF-1, IGF-2, VEGF, HGF, PPY, forskolin	34 days	PDX1, INS, NKX6.1, ISL1, C-PEP, SST, GCG, GLUT2, CgA	*In vitro* (marginal)	No	[182]
H1 and H9	Adherent on Matrigel in ActA, RA, bFGF, Nic followed by suspension culture	21 days	PDX1, AMY, NGN3, GCG, C-PEP, SST	*In vitro* and *in vivo*	Yes (30% rescue efficiency)	[183]
CyT49 and CyT203	Adherent on MEFs in ActA, Wnt3A, KGF, Noggin, CYC, B27	12 days *in vitro* and >100 days *in vivo*	PDX1, NKX6.1, MAFA, GCG, INS, SST, C-PEP, GHRL, PP,	*In vivo*	Yes	[184]

Abbreviations: ActA: activin A; AMY: Amylase; bFGF: basic fibroblast growth factor; BMP4: bone morphogenetic protein 4; CgA: Chromogranin A; CK19: Cytokeratin 19; C-PEP: C-peptide; CYC: cyclopamine; DAPT: γ-secretase inhibitor; EGF: epidermal growth factor; Ex-4: Exendin-4; FGF10: Fibroblast growth factor 10; FGF18: Fibroblast growth factor 18; GCG: Glucagon; GHRL: Ghrelin; GLUT2: Glucose transporter 2; GSIS: Glucose-stimulated insulin secretion; HGF: Hepatocyte growth factor; IAPP: Islet amyloid polypeptide; IGF1: Insulin-like growth factor 1; IGF2: Insulin-like growth factor 2; ILV: (-)-indolactam V; INS: Insulin; ISL1: Islet1; KGF: Keratinocyte growth factor; MEFs: Mouse embryonic fibroblasts; ND: Not determined; NGN3: Neurogenin3; Nic: Nicotinamide; PC1/3: Pro-hormone convertase 1/3; PC/2: Pro-hormone convertase 2; PDX1: Pancreatic and duodenal homeobox 1; PP: Pancreatic polypeptide; PPY: Pancreatic polypeptide Y; RA: all-trans retinoic acid; SST: Somatostatin; TGF-α: Transforming growth factor α; VEGF: Vascular endothelial growth factor.

2.7 Patient-Specific Embryonic Stem Cells

Although hESCs may provide an unlimited renewable source of surrogate cells for human therapy, the need for life-long immunosuppression remains a significant drawback. The generation of patient-specific ESCs would allow better understanding of human disease mechanisms and drug discovery, and be the potential renewable sources for various diseases and injuries without rejection [185].

Currently, patient-specific ESCs can be achieved by reprogramming of adult somatic (non-pluripotent) cells. Two main approaches are being investigated: somatic cell nuclear transfer (SCNT) [186-188] and induced pluripotent stem cells (iPSCs) [189, 190]. These approaches would provide ESCs that are genetically identical to the patients. However, generation of nuclear transfer human embryo by SCNT remains in its infancy since it is technically difficult and involves the use of human oocytes, which requires legal regulation [185]. So far, there are a few reports of the derivation of nuclear transfer human blastocysts but a nuclear transfer embryonic stem cell (ntESC) line is yet to be achieved [186-188]. A study by Jiang *et al.* (2008) demonstrated that mouse ntESCs can be differentiated into functional pancreatic β-cells. The mouse ntESC-derived insulin-producing cells were transplanted into recipient diabetic mice and normalised the blood glucose level for at least 2 months [191]. The results show the potential of SCNT for cell therapy of T1DM in a mouse model.

In order to avoid the ethical issues regarding the use of human oocytes and embryos, researchers have created pluripotent human stem cells by genetic reprogramming of human adult somatic cells. The reprogrammed cells are called iPSCs [189, 190]. There are reports of the generation of insulin-producing islet-like clusters from iPSCs derived from human skin fibroblasts. The differentiated cells secreted C-peptide upon glucose stimulation [7, 8]. More recently, skin fibroblasts from patients with T1DM were used to generate iPSCs. These T1DM-iPSCs have the ability to differentiate into insulin-producing cells. These cells could become renewable source of β-cells for autologous transplantation and for disease modelling [192]. Although there are some technical challenges, such as oncogenic and viral integration, genome instability and clonal variability to be overcome, this iPSC technology offers great promise for drug development, disease models and transplantation in the future [193].

3. PROSPECTS OF STEM CELL THERAPY FOR DIABETES

As already described earlier in this chapter, Type 1 diabetes mellitus (T1DM) is an autoimmune disease in which the insulin-producing β-cells of the pancreatic islets are specifically destroyed by the immune system. People with T1DM are unable to produce insulin, and require exogenous insulin to regulate their blood glucose. By contrast, Type 2 diabetes mellitus (T2DM) does not have an autoimmune component: in its early stages, normal insulin secretion is met with insulin resistance, which leads, over time, to a loss of functional β-cell mass. A proportion of people with severe T2DM require exogenous insulin injections. Stem cells hold a great promise for providing an abundant source of cells for β-cell replacement. Potential target patient populations for treatment with these β-cells might be patients with T1DM, pancreatitis, severe T2DM, brittle (*i.e.*, difficult to control) diabetes or more controlled T1DM. It is the consensus that the juvenile T1DM population will probably be the last group to receive cell replacement treatment, owing to safety and ethical considerations. As in the development of islet transplantation [194], patients without autoimmune concerns, such as those with pancreatitis or severe T2DM, represent the most straightforward potential patient population. While people with brittle diabetes would have the most to gain from a cell therapy, it has been argued that for initial trials there may be a greater chance of success for the transplanted stem cell-derived β-cells if they were transplanted into a patient with less severe diabetes.

The generation of human embryonic stem cell (hESC) lines, more than a decade ago, raised hopes for an unlimited supply of all cell types needed in regenerative medicine. However, a number of issues such as safety, graft rejection and generation of enough such cells in homogeneous population that are functional and that respond to elevated glucose levels must be addressed before realizing the full potential of pluripotent stem cells. Nevertheless, harnessing the potential of embryonic stem cells for diabetes cell therapy remains a very attractive prospect. Recently, induced pluripotent stem cells (iPSCs), derived from various somatic cells such as skin by reprogramming including trans differentiation have joined the ranks of autologous source of pluripotent stem cells. As summarised in Table **2** their relative advantages, and disadvantages, in the generation of various differentiated cell types is a heated topic of debate [116, 195]. As also described earlier in the chapter, other sources of stem cells, including fetal and adult mesenchymal stem cells (MSCs), including umbilical cord blood stem cells are also being intensely investigated for their potential to form insulin-producing cells, or for their indirect use to support tissue regeneration and transplantation. The trans differentiation potential of several pancreatic cell types, including duct, acinar, and islet α-cells, as well as cells from developmentally-related organs (*e.g.* liver and intestine), is being explored [116]. The existence of stem/progenitor cells in adult pancreas remains a controversial issue. However, the plasticity of mature pancreatic cells appears to be higher than originally expected. This may be exploited for both *in vivo* β-cell neogenesis, and *ex vivo* nuclear reprogramming. Finally, the renewal potential of the β-cells themselves is being studied for *in vivo* regeneration from the residual β-cell mass, and *ex vivo* expansion of cells from human islet donors [91].

Table 2. Summary of advantages and disadvantages of different types of stem cells for clinical application of Type 1 diabetes.

Stem Cells	Advantages	Disadvantages
Adult Stem Cells	Found in many tissues and therefore, patient-specific, no risk to the donors and low risk of graft-versus-host disease. Lineage committed nature to such cells has also been looked upon as an advantage to generate cells of specific interest Little ethical concerns and do not require oocytes or the destruction of an embryo	Lineage-restricted (multipotent) Limited proliferation and therefore, it is difficult to generate sufficient numbers of specialized cells for transplantation purposes.
Embryonic Stem Cells	Pluripotent Unlimited proliferation	Controversial, requires destruction of an embryo Tumorigenic potential after transplantation
Nuclear Transfer Embryonic Stem Cells	Pluripotent Unlimited proliferation Patient-specific and therefore, no immune rejection after transplantation.	Controversial, requires oocytes and involves destruction of an embryo Technically difficult
Induced Pluripotent Stem Cells	Pluripotent and unlimited proliferation Patient-specific and therefore, no immune rejection after transplantation. Procedure is quick and becoming increasingly efficient Do not require oocytes or the destruction of an embryo	Oncogenic and viral integration Genome instability and clonal variability Tumorigenic potential after transplantation

A hit on clinical trial webpage (http://www.clinicaltrials.gov) lists 77 clinical trials involving stem cells for diabetes, of which 22 have been already completed. Majority of the completed trials involve studies relate to immunomodulatory properties of cells to achieve their desired therapeutic goal by using various autologous sources of stem cells such as mesenchymal and umbilical cord blood stem cells. Some of these relevant studies (completed/recruiting) are listed in Table **3**. The remaining clinical trials in progress are primarily focussed on ameliorating the peripheral neuropathy involved in diabetes by using various types of adult stem cells. Currently there are no clinical trials with insulin producing cells derived from pluripotent stem cells for diabetes, however, there is a significant progress in a number of laboratories with future potentials.

Cell replacement therapy for diabetes focuses on the replacement of functional β-cell mass. Proof of principle for the efficacy of this therapy is provided by the experience of islet transplantation from cadaveric donors [194]. At present, islet transplantation is an experimental procedure, and is only used to treat a limited number of people with T1DM and life-threatening hypoglycemia. Nonetheless, data are emerging that suggest that islet transplantation can prevent the progression of the complications of diabetes [196, 197], and that transplant recipients have a much improved quality of life [198]. These improvements have spurred efforts to develop replacement therapies despite the high costs and the requirement for immunosuppression. Different types of pluripotent stem cells have been proposed as the starting material for deriving β-cells [199]. The advantage of starting with pluripotent stem cells is their ability to differentiate into all cells of the body, including β-cells. However, the risk of teratoma formation from undifferentiated stem cells is well known. In theory, iPSCs are also able to develop into β-cells, with the additional potential for individualized therapy, but questions remain to be answered about the pluripotency, phenotypic stability and ultimate safety of these cells [200]. Although proof of concept for a stem cell-derived pancreatic progenitor cell therapy for diabetes has been achieved in small animal models, a fully functional β-cell *in vitro* has not yet been achieved. Understanding the developmental biology of the β-cell is required for a functional β-cell to be derived, whatever the starting material [201]. Much of the early work on β-cell development was performed by academic groups with government funding using mouse models. Many of these groups were constrained by government restrictions on research using human pluripotent stem cells. This provided an opportunity for commercial entities to undertake the development of β-cells from hESCs [147, 172, 202]. A current protocol from Viacyte (*i.e.*, Novocell, Cythera) starts with pluripotent cells and differentiates them into definitive endoderm, through a primitive gut, then posterior foregut, to pancreatic endoderm-like cells. These pancreatic progenitor cells are implanted into mice and allowed to mature for 3 months *in situ*, resulting in insulin-expressing cells that exhibit many properties of functional β-cells [172]. This protocol is optimized with Viacyte's own Cythera49 hESC line. Independent verification of these results, especially with different starting cell lines, has not yet been published. It is likely that a commercial entity would either want an exclusive license for the use of a cell line for a particular product or would develop its own verified cGMP line. Although substantial progress has been made, the later stages of development and the functional maturation of the human β-cell remain active areas of study by both academic and commercial groups.

Table 3. Some selective clinical trials (completed/recruiting) with stem cells for diabetes (Source: www.clinicaltrials.gov)

Conditions	Intervention	Sponsor/Study Start Date/Completion Date	Phase	Outcome/Publication
Peripheral Vascular Diseases Diabetic Foot	Autologous Bone Marrow Mononuclear Cells	University Hospital Virgen Macarena, Seville, Spain, 41007/December 2007/May 2009	Phase I Phase II	Not available/completed
Diabetic Patients With Critical Limb Ischemia	Induced Wound Healing by Application of Expanded Bone Marrow Stem Cells	Herz- und Diabeteszentrum Nordrhein Westfalen, Bad Oeynhausen, Germany, 32545/August 2005/February 2009	Phase II	Not available/completed
Treatment of Limb Ischemia and Diabetic Neuropathy	Hematopoietic Stem Cell Transplantation	Hospital Universitario "José E. González", Monterrey, Nuevo León, Mexico, 64460/March 2006/December 2007	Phase I	Diabetes Care. 2005 Sep;28(9):2155-60/completed
Risk Factors for Diabetes After Stem Cell Transplantation	Allogenic Stem Cell Transplantation	Vanderbilt University Medical Center/November 2005/March 2010	Phase I Phase II	Not available/completed
Efficacy Of Autologous Bone Marrow Derived Stem Cell Transplantation In Patients T2DM	To improve blood glucose control in T2DM patients.	Post Graduate Institute of Medical Education and Research Chandigarh, Chandigarh UT, India/November 2009/July 2011	Phase II Phase III	Recruiting

Table 3: cont....

Hematopoietic Stem Cell Transplantation in T1DM	To determine if autologous HSC transplantation prolonge and significant increases C-peptide levels with absence of or reduction of daily insulin.	Hospital Universitario Dr. José Eleuterio González, Monterrey, Nuevo Leon, Mexico, 64460/May 2010/March 2011	Phase I Phase II	Recruiting
Safety and Efficacy of Autologous Adipose-Derived Stem Cell Transplantation in Patients With T1DM	To determine whether intravenous administration of autologous adipose stem cells is safe and beneficial in patients with T1DM.	Beverly Hills Medical Group, Makati City, Manila, Philippines/November 2007	Phase I Phase II	Recruiting
Cotransplantation of Islet and Mesenchymal Stem Cell in T1DM Patients	Stem Cells infusion for protection of islet from inflammatory damage, immunological modulation, engraftment promotion, improve β-cell function.	Fuzhou General Hospital, Fuzhou, Fujian, China, 350025/January 2008/January 2011	Phase I Phase II	Recruiting
Autologous Bone Marrow Mononuclear Cells Transplantation in Treating Diabetes Patients	β-cells regeneration by directly differentiated from the transplanted autologous BMMCs or stimulated local stem cells regeneration and thus decrease or eliminate the need of exogenous insulin and improve β-cells function.	Qilu Hospital of Shandong University, Jinan, Shandong, China, 250012	Phase I Phase II	Recruiting
Cord Blood Infusion for T1DM	Goal is to transfuse autologous umbilical cord blood into 23 children (Germany 10 and 20 controls) with T1DM in an attempt to regenerate pancreatic islet insulin producing beta cells and improve blood glucose control.	Forschergruppe Diabetes der Technischen Universität, München, Bavaria, Germany, 80804/October 2009	Phase I	Recruiting

The work leading to the development of a reproducible and efficient β-cell product is far from finished, although many now feel that is a reachable goal. The time required for *in vitro* differentiation of the final β-cell product will have a large impact on scale-up costs. Estimates of the costs for production are unknown. For small-scale experiments, both two dimensional (2D) and 3D cultures have been used; 3D or suspension cultures may be necessary for scale-up, although this has not yet been shown. Characteristics for QC/release criteria should ideally correlate with outcomes in patients, but will have to be agreed with regulatory authorities [203].

More than 20 years of experience in improving cadaveric islet transplantation [204] has provided insights into the number of islet equivalents required, the site of transplantation, delivery and immunosuppression regimens. FDA guidance criteria exist for islet transplantation. This knowledge base will be utilized in the design of clinical trials for stem cell-derived β-cells. The requirement for immune protection will affect replacement therapy based on stem cell-derived β-cells [200]. In addition to cell source, the platform technology to protect the β-cells from transplant rejection will be essential to the success of any commercialization strategy. If used to treat T1DM, it should be kept in mind that there is evidence of recurrent autoimmunity in whole-organ transplants, even in the face of significant immunosuppression [205]. Clinical translation of stem cell therapies for diabetes requires unravelling the underlying biology of pancreatic development and islet function, and then recapitulating development in a defined, scalable and commercially viable manner. Encapsulation for β-cells/islets has been investigated for some years, both at the level of microencapsulation coatings of individual islets and macro devices encompassing a larger cell population [206, 207]. These encapsulation strategies are designed to protect the β-cells from the immune system, allowing engraftment, and nutrient and oxygen exchange. The device should ideally be biologically inert and retrievable. In the context of β-cells derived from stem cells, encapsulation may also prevent the escape of cells that are not fully differentiated, which may produce teratomas. However, the presence of a capsule means that regulatory agencies will be considering a combination cell and device product. As both the scale-up of β-cells and the encapsulation device are still being developed, the possible regulatory complexity increases.

The use of xeno islets from pigs might be considered as an alternative or even interim model between the use of human cadaveric donor islets and β-cells derived from human pluripotent stem cells. Indeed, a health-economic analysis of porcine islet xenotransplantation found that porcine islet cell xenotransplantation is cost effective when compared with the standard treatment with insulin and the development of complications [208]. In this case, the possible eligible patient profile would be similar (*i.e.*, those with severe hypoglycemia unawareness) [209].

CONCLUSIONS AND PROSPECTS

The potential of stem cell therapies for the treatment of diabetes has emerged as an important agenda for academic research programs of both public and private sectors. Clinical translation of stem cell therapies for diabetes impinges up on understanding first the underlying biology of pancreatic development and then recapitulating development in a defined, scalable and commercially viable manner. The overall health benefits in terms of, quality of life and cost savings from improved glucose homeostasis make a compelling case for cell replacement therapy for the treatment of diabetes. Although proof of concept for a stem cell-derived pancreatic progenitor cell therapy for diabetes has been achieved in small animal models, a fully functional β-cell *in vitro* has not yet been achieved. Even though independent groups have shown repeatedly that insulin-producing cells can be derived from hESCs *in vitro*, none of these end-stage products responded adequately to exogenous glucose stimuli to meet the requirements for surrogate β-cells. Based on the observed expression of multiple endocrine hormones per cell, it has been argued that these cells may represent immature endocrine cells that might develop eventually into fully functional glucose-responsive cells. One successful approach to the treatment of experimental diabetes with hESC-derived cells did not use fully differentiated insulin-producing cells, but a heterogeneous population of cells containing pancreatic endoderm-like tissue [172]. The additional weeks of development *in vivo* that were required before endocrine cells fully differentiated and normoglycemia was achieved in these grafts of pancreatic endoderm suggest that the final differentiation process may take longer than allowed for the *in vitro* protocols. Thus final maturation may simply depend on time.

Given the strong foundation in basic biology and continued interest in developing cell therapies for diabetes from academia, biotech and industry, a phenomenal expansion of technology is thus anticipated sooner than latter. There are significant regulatory hurdles to address, including overcoming teratoma risk, the need for immunosuppression, and designing and obtaining approval for a possible combination product. However, two recent FDA approvals for clinical trial with hESC-derived cells for spinal cord injury and Macula degenerating patients has strengthened the case for more such clinical trials in future. Stem cell-derived β-cells as treatment are evolving rapidly albeit in competition with islet transplantation from cadaveric donors and with xenotransplantation using pig islets.

REFERENCES

[1] IDF. Diabetes Atlas. 3rd ed. Brussels: Int Diabetes Fed; 2007.

[2] Daneman D. Type 1 diabetes. The Lancet [doi: DOI: 10.1016/S0140-6736(06)68341-4] 2006; 367(9513): 847-58.

[3] Croon AC, Karlsson R, Bergström C, *et al.* Lack of donors limits the use of islet transplantation as treatment for diabetes. Transplant Proc 2003; 35(2): 764.

[4] Baetge EE. Production of b-cells from human embryonic stem cells. Diabetes Obes Metab 2008; 10(SUPPL. 4): 186-94.

[5] Best M, Carroll M, Hanley NA, Piper Hanley K. Embryonic stem cells to beta-cells by understanding pancreas development. Mol Cell Endocrinol 2008; 288(1-2): 86-94.

[6] Stanley EG, Elefanty AG. Building Better Beta Cells. Cell Stem Cell 2008; 2(4): 300-1.

[7] Tateishi K, He J, Taranova O, Liang G, D'Alessio AC, Zhang Y. Generation of insulin-secreting islet-like clusters from human skin fibroblasts. J Biol Chem 2008; 283(46): 31601-7.

[8] Zhang D, Jiang W, Liu M, *et al.* Highly efficient differentiation of human ES cells and iPS cells into mature pancreatic insulin-producing cells. Cell Res 2009; 19(4): 429-38.

[9] Motoyama H, Ogawa S, Kubo A, *et al. In vitro* reprogramming of adult hepatocytes into insulin-producing cells without viral vectors. Biochem Biophys Res Commun 2009; 385(1): 123-8.

[10] Thowfeequ S, Li WC, Slack JMW, Tosh D. Reprogramming of liver to pancreas. 2009. p. 407-18.

[11] Bonner-Weir S, Inada A, Yatoh S, *et al.* Transdifferentiation of pancreatic ductal cells to endocrine β-cells. Biochem Soc Trans 2008; 36(3): 353-6.

[12] Yatoh S, Dodge R, Akashi T, *et al.* Differentiation of affinity-purified human pancreatic duct cells to β-cells. Diabetes 2007; 56(7): 1802-9.

[13] Sahu S, Joglekar MV, Dumbre R, Phadnis SM, Tosh D, Hardikar AA. Islet-like cell clusters occur naturally in human gall bladder and are retained in diabetic conditions. J Cell Mol Med 2009; 13(5): 999-1000.

[14] Eberhard D, Tosh D, Slack JMW. Origin of pancreatic endocrine cells from biliary duct epithelium. Cell Mol Life Sci 2008; 65(21): 3467-80.

[15] Joglekar MV, Hardikar AA. Epithelial-to-mesenchymal transition in pancreatic islet beta cells. Cell Cycle 2010; 9(20): 4077-9.

[16] Davani B, Ariely S, Ikonomou L, Oron Y, Gershengorn MC. Human islet-derived precursor cells can cycle between epithelial clusters and mesenchymal phenotypes. J Cell Mol Med 2009; 13(8 B): 2570-81.

[17] Russ HA, Ravassard P, Kerr-Conte J, Pattou F, Efrat S. Epithelial-mesenchymal transition in cells expanded *in vitro* from lineage-traced adult human pancreatic beta cells. PLoS One 2009; 4(7).

[18] Chakrabarti SK, Mirmira RG. Transcription factors direct the development and function of pancreatic b cells. Trends Endocrinol Metab 2003; 14(2): 78-84.

[19] Wilson ME, Scheel D, German MS. Gene expression cascades in pancreatic development. Mech Dev 2003; 120(1): 65-80.

[20] Tam PPL, Behringer RR. Mouse gastrulation: The formation of a mammalian body plan. Mech Dev 1997; 68(1-2): 3-25.

[21] Oliver-Krasinski JM, Stoffers DA. On the origin of the b cell. Genes Dev 2008; 22(15): 1998-2021.

[22] Rivas-Carrillo JD, Okitsu T, Tanaka N, Kobayashi N. Pancreas development and b-cell differentiation of embryonic stem cells. Curr Med Chem 2007; 14(14): 1573-8.

[23] Lawson KA, Pedersen RA, Van De Geer S. Cell fate, morphogenetic movement and population kinetics of embryonic endoderm at the time of germ layer formation in the mouse. Development 1987; 101(3): 627-52.

[24] Murry CE, Keller G. Differentiation of Embryonic Stem Cells to Clinically Relevant Populations: Lessons from Embryonic Development. Cell 2008; 132(4): 661-80.

[25] Zorn AM, Wells JM. Vertebrate endoderm development and organ formation. Annu Rev Cell Dev Biol 2009; 25: 221-51.

[26] Zorn AM, Wells JM. Molecular Basis of Vertebrate Endoderm Development Int Rev Cytol 2007; 259: 49-111.

[27] Cano DA, Hebrok M, Zenker M. Pancreatic Development and Disease. Gastroenterology 2007; 132(2): 745-62.

[28] Lammert E, Cleaver O, Melton D. Induction of pancreatic differentiation by signals from blood vessels. Science 2001; 294(5542): 564-7.

[29] Habener JF, Kemp DM, Thomas MK. Minireview: Transcriptional regulation in pancreatic development. Endocrinology 2005; 146(3): 1025-34.

[30] Hebrok M, Kim SK, Melton DA. Notochord repression of endodermal sonic hedgehog permits pancreas development. Genes Dev 1998; 12(11): 1705-13.

[31] Schwitzgebel VM. Programming of the pancreas. Mol Cell Endocrinol 2001; 185(1-2): 99-108.

[32] Chen Y, Pan FC, Brandes N, Afelik S, Sölter M, Pieler T. Retinoic acid signalling is essential for pancreas development and promotes endocrine at the expense of exocrine cell differentiation in Xenopus. Dev Biol 2004; 271(1): 144-60.

[33] Pan FC, Chen Y, Bayha E, Pieler T. Retinoic acid-mediated patterning of the pre-pancreatic endoderm in Xenopus operates *via* direct and indirect mechanisms. Mech Dev 2007; 124(7-8): 518-31.

[34] Stafford D, Hornbruch A, Mueller PR, Prince VE. A conserved role for retinoid signalling in vertebrate pancreas development. Dev Genes Evol 2004; 214(9): 432-41.

[35] Stafford D, Prince VE. Retinoic acid signalling is required for a critical early step in zebrafish pancreatic development. Curr Biol 2002; 12(14): 1215-20.

[36] Stafford D, White RJ, Kinkel MD, Linville A, Schilling TF, Prince VE. Retinoids signal directly to zebrafish endoderm to specify insulin-expressing β-cells. Development 2006; 133(5): 949-56.

[37] Shen CN, Marguerie A, Chien CY, Dickson C, Slack JMW, Tosh D. All-trans retinoic acid suppresses exocrine differentiation and branching morphogenesis in the embryonic pancreas. Differentiation 2007; 75(1): 62-74.

[38] Micallef SJ, Li X, Janes ME, *et al.* Endocrine cells develop within pancreatic bud-like structures derived from mouse ES cells differentiated in response to BMP4 and retinoic acid. Stem Cell Res 2007; 1(1): 25-36.

[39] Micallef SJ, Janes ME, Knezevic K, Davis RP, Elefanty AG, Stanley EG. Retinoic acid induces Pdx1-positive endoderm in differentiating mouse embryonic stem cells. Diabetes 2005; 54(2): 301-5.

[40] Jensen J. Gene Regulatory Factors in Pancreatic Development. Dev Dyn 2004; 229(1): 176-200.

[41] Jacquemin P, Yoshitomi H, Kashima Y, Rousseau GG, Lemaigre FP, Zaret KS. An endothelial-mesenchymal relay pathway regulates early phases of pancreas development. Dev Biol 2006; 290(1): 189-99.

[42] Norgaard GA, Jensen JN, Jensen J. FGF10 signalling maintains the pancreatic progenitor cell state revealing a novel role of Notch in organ development. Dev Biol 2003; 264(2): 323-38.

[43] Docherty K, Bernardo AS, Vallier L. Embryonic stem cell therapy for diabetes mellitus. Semin Cell Dev Biol 2007; 18(6): 827-38.

[44] Krapp A, Knöfler M, Ledermann B, *et al.* The bHLH protein PTF1-p48 is essential for the formation of the exocrine and the correct spatial organization of the endocrine pancreas. Genes Dev 1998; 12(23): 3752-63.

[45] Hart A, Papadopoulou S, Edlund H. Fgf10 maintains notch activation, stimulates proliferation, and blocks differentiation of pancreatic epithelial cells. Dev Dyn 2003; 228(2): 185-93.

[46] Miralles F, Czernichow P, Scharfmann R. Follistatin regulates the relative proportions of endocrine versus exocrine tissue during pancreatic development. Development 1998; 125(6): 1017-24.

[47] Heiser PW, Lau J, Taketo MM, Herrera PL, Hebrok M. Stabilization of β-catenin impacts pancreas growth. Development 2006; 133(10): 2023-32.

[48] Murtaugh LC, Law AC, Dor Y, Melton DA. β-Catenin is essential for pancreatic acinar but not islet development. Development 2005; 132(21): 4663-74.

[49] Wells JM, Esni F, Boivin GP, *et al.* Wnt/β-catenin signalling is required for development of the exocrine pancreas. BMC Dev Biol 2007; 7.

[50] Slack JMW. Developmental biology of the pancreas. Development 1995; 121(6): 1569-80.

[51] Sander M, German MS. The β cell transcription factors and development of the pancreas. J Mol Med 1997; 75(5): 327-40.

[52] Servitja JM, Ferrer J. Transcriptional networks controlling pancreatic development and beta cell function. Diabetologia 2004; 47(4): 597-613.

[53] Spence JR, Wells JM. Translational embryology: Using embryonic principles to generate pancreatic endocrine cells from embryonic stem cells. Dev Dyn 2007; 236(12): 3218-27.

[54] Otonkoski T, Cirulli V, Beattie GM, *et al.* A Role for Hepatocyte Growth Factor/Scatter Factor in Fetal Mesenchyme-Induced Pancreatic β-Cell Growth. Endocrinology 1996; 137(7): 3131-9.

[55] Otonkoski T, Beattie GM, Lopez AD, Hayek A. Use of hepatocyte growth factor/scatter factor to increase transplantable human fetal islet cell mass. Transplant Proc 1994; 26(6): 3334.

[56] Miettinen PJ, Huotari MA, Koivisto T, *et al.* Impaired migration and delayed differentiation of pancreatic islet cells in mice lacking EGF-receptors. Development 2000; 127(12): 2617-27.

[57] Gittes GK. Developmental biology of the pancreas: A comprehensive review. Dev Biol 2009; 326(1): 4-35.

[58] Drucker DJ. Does exenatide provide a meaningful new treatment option for patients with type 2 diabetes? Nat Clin Pract End Met [10.1038/ncpendmet0026] 2005; 1(2): 70-1.

[59] Brubaker PL, Drucker DJ. Minireview: Glucagon-like peptides regulate cell proliferation and apoptosis in the pancreas, gut, and central nervous system. Endocrinology 2004; 145(6): 2653-9.

[60] Xu G, Kaneto H, Lopez-Avalos MD, Weir GC, Bonner-Weir S. GLP-1/exendin-4 facilitates b-cell neogenesis in rat and human pancreatic ducts. Diabetes Res Clin Pract 2006; 73(1): 107-10.

[61] Tourrel C, Bailbe D, Meile M-J, Kergoat M, Portha B. Glucagon-Like Peptide-1 and Exendin-4 Stimulate b-Cell Neogenesis in Streptozotocin-Treated Newborn Rats Resulting in Persistently Improved Glucose Homeostasis at Adult Age. Diabetes 2001; 50(7): 1562-70.

[62] Farilla L, Bulotta A, Hirshberg B, *et al.* Glucagon-like peptide 1 inhibits cell apoptosis and improves glucose responsiveness of freshly isolated human islets. Endocrinology 2003; 144(12): 5149-58. 63.

[63] Li Y, Hansotia T, Yusta B, Ris F, Halban PA, Drueker DJ. Glucagon-like peptide-1 receptor signalling modulates b cell apoptosis. J Biol Chem 2003; 278(1): 471-8.

[64] Jepeal LI, Fujitani Y, Boylan MO, Wilson CN, Wright CV, Wolfe MM. Cell-specific expression of glucose-dependent-insulinotropic polypeptide is regulated by the transcription factor PDX-1. Endocrinology 2005; 146(1): 383-91.

[65] McIntosh CHS, Widenmaier S, Kim S. Glucose-Dependent Insulinotropic Polypeptide (Gastric Inhibitory Polypeptide; GIP). Vitamins and Hormones Elsevier Inc.; 2009. p. 409-71.

[66] Lee CS, De Leon DD, Kaestner KH, Stoffers DA. Regeneration of pancreatic islets after partial pancreatectomy in mice does not involve the reactivation of neurogenin-3. Diabetes 2006; 55(2): 269-72. 67.

[67] Hao E, Tyrberg B, Itkin-Ansari P, *et al.* Beta-cell differentiation from nonendocrine epithelial cells of the adult human pancreas. Nat Med 2006; 12(3): 310-6.

[68] Montanya E, Nacher V, Biarnes M, Soler J. Linear correlation between beta-cell mass and body weight throughout the lifespan in Lewis rats: role of beta-cell hyperplasia and hypertrophy. Diabetes 2000; 49(8): 1341-6.

[69] Georgia S, Bhushan A. Beta cell replication is the primary mechanism for maintaining postnatal beta cell mass. J Clin Invest 2004; 114(7): 963-8.

[70] Dor Y, Brown J, Martinez OI, Melton DA. Adult pancreatic beta-cells are formed by self-duplication rather than stem-cell differentiation. Nature 2004; 429(6987): 41-6.

[71] Joglekar MV, Parekh VS, Hardikar AA. New pancreas from old: microregulators of pancreas regeneration. Trends Endocrinol Metab 2007; 18(10): 393-400.

[72] Joglekar MV, Parekh VS, Mehta S, Bhonde RR, Hardikar AA. MicroRNA profiling of developing and regenerating pancreas reveal post-transcriptional regulation of neurogenin3. Dev Biol 2007; 311(2): 603-12.

[73] Brennand K, Huangfu D, Melton D. All beta cells contribute equally to islet growth and maintenance. PLoS Biol 2007; 5(7): e163.

[74] Tyrberg B, Andersson A, Borg LA. Species differences in susceptibility of transplanted and cultured pancreatic islets to the beta-cell toxin alloxan. Gen Comp Endocrinol 2001; 122(3): 238-51.

[75] Meier JJ, Lin JC, Butler AE, Galasso R, Martinez DS, Butler PC. Direct evidence of attempted beta cell regeneration in an 89-year-old patient with recent-onset type 1 diabetes. Diabetologia 2006; 49(8): 1838-44.

[76] Russ HA, Bar Y, Ravassard P, Efrat S. *In vitro* proliferation of cells derived from adult human beta-cells revealed by cell-lineage tracing. Diabetes 2008; 57(6): 1575-83.

[77] Joglekar MV, Joglekar VM, Joglekar SV, Hardikar AA. Human fetal pancreatic insulin-producing cells proliferate *in vitro*. J Endocrinol 2009; 201(1): 27-36.

[78] Gershengorn MC, Hardikar AA, Wei C, Geras-Raaka E, Marcus-Samuels B, Raaka BM. Epithelial-to-mesenchymal transition generates proliferative human islet precursor cells. Science 2004; 306(5705): 2261-4.

[79] Kalluri R, Weinberg RA. The basics of epithelial-mesenchymal transition. J Clin Invest 2009; 119(6): 1420-8.

[80] Thiery JP, Acloque H, Huang RY, Nieto MA. Epithelial-mesenchymal transitions in development and disease. Cell 2009; 139(5): 871-90.

[81] Zeisberg M, Neilson EG. Biomarkers for epithelial-mesenchymal transitions. J Clin Invest 2009; 119(6): 1429-37.

[82] Ouziel-Yahalom L, Zalzman M, Anker-Kitai L, *et al.* Expansion and redifferentiation of adult human pancreatic islet cells. Biochem Biophys Res Commun 2006; 341(2): 291-8.

[83] Atouf F, Park CH, Pechhold K, Ta M, Choi Y, Lumelsky NL. No evidence for mouse pancreatic beta-cell epithelial-mesenchymal transition *in vitro*. Diabetes 2007; 56(3): 699-702.

[84] Chase LG, Ulloa-Montoya F, Kidder BL, Verfaillie CM. Islet-derived fibroblast-like cells are not derived *via* epithelial-mesenchymal transition from Pdx-1 or insulin-positive cells. Diabetes 2007; 56(1): 3-7.

[85] Morton RA, Geras-Raaka E, Wilson LM, Raaka BM, Gershengorn MC. Endocrine precursor cells from mouse islets are not generated by epithelial-to-mesenchymal transition of mature beta cells. Mol Cell Endocrinol 2007; 270(1-2): 87-93.

[86] Weinberg N, Ouziel-Yahalom L, Knoller S, Efrat S, Dor Y. Lineage tracing evidence for *in vitro* dedifferentiation but rare proliferation of mouse pancreatic beta-cells. Diabetes 2007; 56(5): 1299-304.

[87] Russ HA, Ravassard P, Kerr-Conte J, Pattou F, Efrat S. Epithelial-mesenchymal transition in cells expanded *in vitro* from lineage-traced adult human pancreatic beta cells. PLoS One 2009; 4(7): e6417.

[88] Joglekar MV, Patil D, Joglekar VM, *et al.* The miR-30 family microRNAs confer epithelial phenotype to human pancreatic cells. Islets 2009; 1(2): 137-47.

[89] Rieck S, Kaestner KH. Expansion of beta-cell mass in response to pregnancy. Trends Endocrinol Metab 2010; 21(3): 151-8.

[90] Mercado AB, Castells S. Pancreatic beta-cell hyperactivity in morbidly obese adolescents. Pediatr Endocrinol Rev 2006; 3 Suppl 4: 560-3.

[91] Nir T, Melton DA, Dor Y. Recovery from diabetes in mice by beta cell regeneration. J Clin Invest 2007; 117(9): 2553-61.

[92] Teta M, Rankin MM, Long SY, Stein GM, Kushner JA. Growth and regeneration of adult beta cells does not involve specialized progenitors. Dev Cell 2007; 12(5): 817-26.

[93] Cole L, Anderson M, Antin PB, Limesand SW. One process for pancreatic beta-cell coalescence into islets involves an epithelial-mesenchymal transition. J Endocrinol 2009; 203(1): 19-31.

[94] Bonner-Weir S, Taneja M, Weir GC, *et al. In vitro* cultivation of human islets from expanded ductal tissue. Proc Natl Acad Sci U S A 2000; 97(14): 7999-8004.

[95] Bonner-Weir S, Sharma A. Pancreatic stem cells. J Pathol 2002; 197(4): 519-26.

[96] Grapin-Botton A. Ductal cells of the pancreas. Int J Biochem Cell Biol 2005; 37(3): 504-10.

[97] Bonner-Weir S, Baxter LA, Schuppin GT, Smith FE. A second pathway for regeneration of adult exocrine and endocrine pancreas. A possible recapitulation of embryonic development. Diabetes 1993; 42(12): 1715-20.

[98] Gu D, Sarvetnick N. Epithelial cell proliferation and islet neogenesis in IFN-g transgenic mice. Development 1993; 118(1): 33-46.

[99] Hardikar AA, Karandikar MS, Bhonde RR. Effect of partial pancreatectomy on diabetic status in BALB/c mice. J Endocrinol 1999; 162(2): 189-95.

[100] Finegood DT, Weir GC, Bonner-Weir S. Prior streptozotocin treatment does not inhibit pancreas regeneration after 90% pancreatectomy in rats. Am J Physiol Endocrinol Metabol 1999; 276(5 39-5).

[101] Gu G, Dubauskaite J, Melton DA. Direct evidence for the pancreatic lineage: NGN3+ cells are islet progenitors and are distinct from duct progenitors. Development 2002; 129(10): 2447-57.

[102] D'Alessandro JS, Lu K, Fung BP, Colman A, Clarke DL. Rapid and efficient *in vitro* generation of pancreatic islet progenitor cells from nonendocrine epithelial cells in the adult human pancreas. Stem Cells Dev 2007; 16(1): 75-89.

[103] Yao ZX, Qin ML, Liu JJ, Chen XS, Zhou DS. *In vitro* cultivation of human fetal pancreatic ductal stem cells and their differentiation into insulin-producing cells. World J Gastroenterol 2004; 10(10): 1452-6.

[104] Yatoh S, Dodge R, Akashi T, *et al.* Differentiation of affinity-purified human pancreatic duct cells to beta-cells. Diabetes 2007; 56(7): 1802-9.

[105] Taniguchi H, Yamato E, Tashiro F, Ikegami H, Ogihara T, Miyazaki J. beta-cell neogenesis induced by adenovirus-mediated gene delivery of transcription factor pdx-1 into mouse pancreas. Gene Ther 2003; 10(1): 15-23.

[106] Heremans Y, Van De Casteele M, in't Veld P, *et al.* Recapitulation of embryonic neuroendocrine differentiation in adult human pancreatic duct cells expressing neurogenin 3. J Cell Biol 2002; 159(2): 303-12.

[107] Lipsett M, Finegood DT. beta-cell neogenesis during prolonged hyperglycemia in rats. Diabetes 2002; 51(6): 1834-41.

[108] Wang RN, Kloppel G, Bouwens L. Duct- to islet-cell differentiation and islet growth in the pancreas of duct-ligated adult rats. Diabetologia 1995; 38(12): 1405-11.

[109] Gmyr V, Kerr-Conte J, Belaich S, *et al.* Adult human cytokeratin 19-positive cells reexpress insulin promoter factor 1 *in vitro*: further evidence for pluripotent pancreatic stem cells in humans. Diabetes 2000; 49(10): 1671-80.

[110] Means AL, Meszoely IM, Suzuki K, *et al.* Pancreatic epithelial plasticity mediated by acinar cell transdifferentiation and generation of nestin-positive intermediates. Development 2005; 132(16): 3767-76.

[111] Lardon J, Huyens N, Rooman I, Bouwens L. Exocrine cell transdifferentiation in dexamethasone-treated rat pancreas. Virchows Arch 2004; 444(1): 61-5.

[112] Rooman I, Bouwens L. Combined gastrin and epidermal growth factor treatment induces islet regeneration and restores normoglycaemia in C57Bl6/J mice treated with alloxan. Diabetologia 2004; 47(2): 259-65.

[113] Song KH, Ko SH, Ahn YB, *et al. In vitro* transdifferentiation of adult pancreatic acinar cells into insulin-expressing cells. Biochem Biophys Res Commun 2004; 316(4): 1094-100.

[114] Baeyens L, De Breuck S, Lardon J, Mfopou JK, Rooman I, Bouwens L. *In vitro* generation of insulin-producing beta cells from adult exocrine pancreatic cells. Diabetologia 2005; 48(1): 49-57.

[115] Desai BM, Oliver-Krasinski J, De Leon DD, *et al.* Preexisting pancreatic acinar cells contribute to acinar cell, but not islet beta cell, regeneration. J Clin Invest 2007; 117(4): 971-7.

[116] Zhou Q, Brown J, Kanarek A, Rajagopal J, Melton DA. *In vivo* reprogramming of adult pancreatic exocrine cells to beta-cells. Nature 2008; 455(7213): 627-32.

[117] Wells JM, Melton DA. Vertebrate endoderm development. Annu Rev Cell Dev Biol 1999; 15: 393-410.

[118] Gualdi R, Bossard P, Zheng M, Hamada Y, Coleman JR, Zaret KS. Hepatic specification of the gut endoderm *in vitro*: cell signalling and transcriptional control. Genes Dev 1996; 10(13): 1670-82.

[119] Deutsch G, Jung J, Zheng M, Lora J, Zaret KS. A bipotential precursor population for pancreas and liver within the embryonic endoderm. Development 2001; 128(6): 871-81.

[120] Grompe M. Pancreatic-hepatic switches *in vivo*. Mech Dev 2003; 120(1): 99-106.

[121] Kawasaki H, Mizuguchi T, Kikkawa Y, *et al. In vitro* transformation of adult rat hepatic progenitor cells into pancreatic endocrine hormone-producing cells. J Hepatobiliary Pancreat Surg 2008; 15(3): 310-7.

[122] Ferber S, Halkin A, Cohen H, *et al.* Pancreatic and duodenal homeobox gene 1 induces expression of insulin genes in liver and ameliorates streptozotocin-induced hyperglycemia. Nat Med 2000; 6(5): 568-72.

[123] Zalzman M, Gupta S, Giri RK, *et al.* Reversal of hyperglycemia in mice by using human expandable insulin-producing cells differentiated from fetal liver progenitor cells. Proc Natl Acad Sci USA 2003; 100(12): 7253-8.

[124] Li H, Li X, Lam KS, Tam S, Xiao W, Xu R. Adeno-associated virus-mediated pancreatic and duodenal homeobox gene-1 expression enhanced differentiation of hepatic oval stem cells to insulin-producing cells in diabetic rats. J Biomed Sci 2008; 15(4): 487-97.

[125] Lu S, Wang WP, Wang XF, *et al.* Heterogeneity in predisposition of hepatic cells to be induced into pancreatic endocrine cells by PDX-1. World J Gastroenterol 2005; 11(15): 2277-82.

[126] Kaneto H, Nakatani Y, Miyatsuka T, *et al.* PDX-1/VP16 fusion protein, together with NeuroD or Ngn3, markedly induces insulin gene transcription and ameliorates glucose tolerance. Diabetes 2005; 54(4): 1009-22.

[127] Tuch BE, Szymanska B, Yao M, *et al.* Function of a genetically modified human liver cell line that stores, processes and secretes insulin. Gene Ther 2003; 10(6): 490-503.

[128] Zalzman M, Gupta S, Giri RK, *et al.* Reversal of hyperglycemia in mice by using human expandable insulin-producing cells differentiated from fetal liver progenitor cells. Proc Natl Acad Sci USA 2003; 100(12): 7253-8.

[129] Sun B, Roh KH, Lee SR, Lee YS, Kang KS. Induction of human umbilical cord blood-derived stem cells with embryonic stem cell phenotypes into insulin producing islet-like structure. Biochem Biophys Res Commun 2007; 354(4): 919-23.

[130] Hou P, Zhao DY, Li GF, Li XF, Li JP. Differentiation of human umbilical cord blood stem cells into islet-like cells following *in vitro* induction of various cytokine combination. J Clin Rehabil Tissue Eng Res 2009; 13(6): 1073-6.

[131] Koblas T, Zacharovová K, Berková Z, *et al. In vivo* differentiation of human umbilical cord blood-derived cells into insulin-producing β cells. Folia Biol 2009; 55(6): 224-32.

[132] Hori Y, Gu X, Xie X, Kim SK. Differentiation of insulin-producing cells from human neural progenitor cells. PLoS Med 2005; 2: 0347-56.

[133] Zhao M, Amiel SA, Ajami S, *et al.* Amelioration of streptozotocin-induced diabetes in mice with cells derived from human marrow stromal cells. PLoS ONE 2008; 3(7).

[134] Ianus A, Holz GG, Theise ND, Hussain MA. *In vivo* derivation of glucose-competent pancreatic endocrine cells from bone marrow without evidence of cell fusion. J Clin Invest 2003; 111(6): 843-50.

[135] Li Y, Zhang R, Qiao H, *et al.* Generation of insulin-producing cells from PDX-1 gene-modified human mesenchymal stem cells. J Cell Physiol 2007; 211(1): 36-44.

[136] Timper K, Seboek D, Eberhardt M, *et al.* Human adipose tissue-derived mesenchymal stem cells differentiate into insulin, somatostatin, and glucagon expressing cells. Biochem Biophys Res Commun 2006; 341(4): 1135-40.

[137] Lin G, Wang G, Liu G, *et al.* Treatment of type 1 diabetes with adipose tissue-derived stem cells expressing pancreatic duodenal homeobox 1. Stem Cells Dev 2009; 18(10): 1399-406.

[138] Okura H, Komoda H, Fumimoto Y, *et al.* Transdifferentiation of human adipose tissue-derived stromal cells into insulin-producing clusters. J Artificial Organs 2009; 12(2): 123-30.

[139] Chang CM, Kao CL, Chang YL, *et al.* Placenta-derived multipotent stem cells induced to differentiate into insulin-positive cells. Biochem Biophys Res Commun 2007; 357(2): 414-20.

[140] Voltarelli JC, Couri CEB, Stracieri ABPL, *et al.* Autologous Nonmyeloablative Hematopoietic Stem Cell Transplantation in Newly Diagnosed Type 1 Diabetes Mellitus. JAMA 2007; 297(14): 1568-76.

[141] Couri CEB, Oliveira MCB, Stracieri ABPL, *et al.* C-peptide levels and insulin independence following autologous nonmyeloablative hematopoietic stem cell transplantation in newly diagnosed type 1 diabetes mellitus. JAMA - J Am Med Assoc 2009; 301(15): 1573-9.

[142] Limbert C, Path G, Jakob F, Seufert J. Beta-cell replacement and regeneration: Strategies of cell-based therapy for type 1 diabetes mellitus. Diabetes Res Clin Pract 2008; 79(3): 389-99.

[143] Keller GM. *In vitro* differentiation of embryonic stem cells. Curr Opin Cell Biol 1995; 7(6): 862-9.

[144] Conley BJ, Young JC, Trounson AO, Mollard R. Derivation, propagation and differentiation of human embryonic stem cells. Int J Biochem Cell Biol 2004; 36(4): 555-67.

[145] Kubo A, Shinozaki K, Shannon JM, *et al.* Development of definitive endoderm from embryonic stem cells in culture. Development 2004; 131(7): 1651-62.

[146] Gadue P, Huber TL, Paddison PJ, Keller GM. Wnt and TGF-beta signalling are required for the induction of an *in vitro* model of primitive streak formation using embryonic stem cells. Proc Natl Acad Sci U S A 2006; 103(45): 16806-11.

[147] D'Amour KA, Bang AG, Eliazer S, *et al.* Production of pancreatic hormone-expressing endocrine cells from human embryonic stem cells. Nat Biotechnol 2006; 24(11): 1392-401.

[148] Lu CC, Brennan J, Robertson EJ. From fertilization to gastrulation: axis formation in the mouse embryo. Curr Opin Genet Dev 2001; 11(4): 384-92.

[149] Tam PP, Kanai-Azuma M, Kanai Y. Early endoderm development in vertebrates: lineage differentiation and morphogenetic function. Curr Opin Genet Dev 2003; 13(4): 393-400.

[150] Kinder SJ, Tsang TE, Wakamiya M, *et al.* The organizer of the mouse gastrula is composed of a dynamic population of progenitor cells for the axial mesoderm. Development 2001; 128(18): 3623-34.

[151] Chazaud C, Yamanaka Y, Pawson T, Rossant J. Early lineage segregation between epiblast and primitive endoderm in mouse blastocysts through the Grb2-MAPK pathway. Dev Cell 2006; 10(5): 615-24.

[152] Sherwood RI, Jitianu C, Cleaver O, *et al.* Prospective isolation and global gene expression analysis of definitive and visceral endoderm. Dev Biol 2007; 304(2): 541-55.

[153] Jiang W, Shi Y, Zhao D, *et al. In vitro* derivation of functional insulin-producing cells from human embryonic stem cells. Cell Res 2007; 17(4): 333-44.

[154] Kimelman D, Griffin KJ. Vertebrate mesendoderm induction and patterning. Curr Opin Genet Dev 2000; 10(4): 350-6.

[155] Grapin-Botton A, Melton DA. Endoderm development: from patterning to organogenesis. Trends Genet 2000; 16(3): 124-30.

[156] Tada S, Era T, Furusawa C, *et al.* Characterization of mesendoderm: a diverging point of the definitive endoderm and mesoderm in embryonic stem cell differentiation culture. Development 2005; 132(19): 4363-74.

[157] Yasunaga M, Tada S, Torikai-Nishikawa S, *et al.* Induction and monitoring of definitive and visceral endoderm differentiation of mouse ES cells. Nat Biotechnol 2005; 23(12): 1542-50.

[158] Assady S, Maor G, Amit M, Itskovitz-Eldor J, Skorecki KL, Tzukerman M. Insulin Production by Human Embryonic Stem Cells. Diabetes 2001; 50(8): 1691-7.

[159] Kahan BW, Jacobson LM, Hullett DA, *et al.* Pancreatic precursors and differentiated islet cell types from murine embryonic stem cells: an *in vitro* model to study islet differentiation. Diabetes 2003; 52(8): 2016-24.

[160] Xu X, Kahan B, Forgianni A, *et al.* Endoderm and pancreatic islet lineage differentiation from human embryonic stem cells. Cloning Stem Cells 2006; 8(2): 96-107.

[161] Segev H, Fishman B, Ziskind A, Shulman M, Itskovitz-Eldor J. Differentiation of human embryonic stem cells into insulin-producing clusters. Stem Cells 2004; 22(3): 265-74.

[162] Baharvand H, Jafary H, Massumi M, Ashtiani SK. Generation of insulin-secreting cells from human embryonic stem cells. Dev Growth Differ 2006; 48(5): 323-32.

[163] Brolen GK, Heins N, Edsbagge J, Semb H. Signals from the embryonic mouse pancreas induce differentiation of human embryonic stem cells into insulin-producing beta-cell-like cells. Diabetes 2005; 54(10): 2867-74.

[164] Shi Y, Hou L, Tang F, *et al.* Inducing embryonic stem cells to differentiate into pancreatic beta cells by a novel three-step approach with activin A and all-trans retinoic acid. Stem Cells 2005; 23(5): 656-62.

[165] McLean AB, D'Amour KA, Jones KL, *et al.* Activin A efficiently specifies definitive endoderm from human embryonic stem cells only when phosphatidylinositol 3-kinase signalling is suppressed. Stem Cells 2007; 25(1): 29-38.

[166] Madsen OD, Serup P. Towards cell therapy for diabetes. Nat Biotechnol 2006; 24(12): 1481-3.

[167] Bhushan A, Itoh N, Kato S, *et al.* Fgf10 is essential for maintaining the proliferative capacity of epithelial progenitor cells during early pancreatic organogenesis. Development 2001; 128(24): 5109-17.

[168] Ye F, Duvillie B, Scharfmann R. Fibroblast growth factors 7 and 10 are expressed in the human embryonic pancreatic mesenchyme and promote the proliferation of embryonic pancreatic epithelial cells. Diabetologia 2005; 48(2): 277-81.

[169] Hebrok M. Hedgehog signalling in pancreas development. Mech Dev 2003; 120(1): 45-57.

[170] Kim SK, Melton DA. Pancreas development is promoted by cyclopamine, a hedgehog signalling inhibitor. Proc Natl Acad Sci U S A 1998; 95(22): 13036-41.

[171] Tulachan SS, Doi R, Kawaguchi Y, *et al.* All-trans retinoic acid induces differentiation of ducts and endocrine cells by mesenchymal/epithelial interactions in embryonic pancreas. Diabetes 2003; 52(1): 76-84.

[172] Kroon E, Martinson LA, Kadoya K, *et al.* Pancreatic endoderm derived from human embryonic stem cells generates glucose-responsive insulin-secreting cells *in vivo*. Nat Biotechnol 2008; 26(4): 443-52.

[173] Tuch BE, Osgerby KJ, Turtle JR. Reversal of diabetes in hyperglycaemic nude mice by human fetal pancreas. Transplant Proc 1989; 21(1 Pt 3): 2665-6.

[174] Castaing M, Duvillie B, Quemeneur E, Basmaciogullari A, Scharfmann R. *Ex vivo* analysis of acinar and endocrine cell development in the human embryonic pancreas. Dev Dyn 2005; 234(2): 339-45.

[175] Castaing M, Peault B, Basmaciogullari A, Casal I, Czernichow P, Scharfmann R. Blood glucose normalization upon transplantation of human embryonic pancreas into beta-cell-deficient SCID mice. Diabetologia 2001; 44(11): 2066-76.

[176] Shim JH, Kim SE, Woo DH, *et al.* Directed differentiation of human embryonic stem cells towards a pancreatic cell fate. Diabetologia 2007; 50(6): 1228-38.

[177] D'Amour KA, Bang AG, Eliazer S, *et al.* Production of pancreatic hormone-expressing endocrine cells from human embryonic stem cells. Nat Biotech [10.1038/nbt1259] 2006; 24(11): 1392-401.

[178] Chen S, Borowiak M, Fox JL, *et al.* A small molecule that directs differentiation of human ESCs into the pancreatic lineage. Nat Chem Biol 2009; 5(4): 258-65.

[179] Jiang J, Au M, Lu K, *et al.* Generation of Insulin-Producing Islet-Like Clusters from Human Embryonic Stem Cells. Stem Cells 2007; 25(8): 1940-53.

[180] Eshpeter A, Jiang J, Au M, *et al. In vivo* characterization of transplanted human embryonic stem cell-derived pancreatic endocrine islet cells. Cell Prolif 2008; 41(6): 843-58.

[181] Shim J, Kim S, Woo D, *et al.* Directed differentiation of human embryonic stem cells towards a pancreatic cell fate. Diabetologia 2007; 50(6): 1228-38.

[182] Phillips BW, Hentze H, Rust WL, *et al.* Directed differentiation of human embryonic stem cells into the pancreatic endocrine lineage. Stem Cells Dev 2007; 16(4): 561-78.

[183] Jiang W, Shi Y, Zhao D, *et al. In vitro* derivation of functional insulin-producing cells from human embryonic stem cells. Cell Res 2007; 17(4): 333-44.

[184] Kroon E, Martinson LA, Kadoya K, *et al.* Pancreatic endoderm derived from human embryonic stem cells generates glucose-responsive insulin-secreting cells *in vivo*. Nat Biotech 2008; 26(4): 443-52.

[185] Cervera RP, Stojkovic M. Commentary: Somatic cell nuclear transfer - Progress and promise. Stem Cells 2008; 26(2): 494-5.

[186] Heindryckx B, De Sutter P, Gerris J, Dhont M, Van der Elst J. Embryo development after successful somatic cell nuclear transfer to *in vitro* matured human germinal vesicle oocytes. Hum Reprod 2007; 22(7): 1982-90.

[187] French AJ, Adams CA, Anderson LS, Kitchen JR, Hughes MR, Wood SH. Development of human cloned blastocysts following somatic cell nuclear transfer with adult fibroblasts. Stem Cells 2008; 26(2): 485-93.

[188] Stojkovic M, Stojkovic P, Leary C, *et al.* Derivation of a human blastocyst after heterologous nuclear transfer to donated oocytes. Reprod Biomed Online 2005; 11(2): 226-31.

[189] Takahashi K, Tanabe K, Ohnuki M, *et al.* Induction of pluripotent stem cells from adult human fibroblasts by defined factors. Cell 2007; 131(5): 861-72.

[190] Yu J, Vodyanik MA, Smuga-Otto K, *et al.* Induced pluripotent stem cell lines derived from human somatic cells. Science 2007; 318(5858): 1917-20.

[191] Jiang W, Bai Z, Zhang D, *et al.* Differentiation of mouse nuclear transfer embryonic stem cells into functional pancreatic beta cells. Diabetologia 2008: 1-9.

[192] Maehr R, Chen S, Snitow M, *et al.* Generation of pluripotent stem cells from patients with type 1 diabetes. Proc Natl Acad Sci U S A 2009; 106(37): 15768-73.

[193] Yamanaka S. A Fresh Look at iPS Cells. Cell 2009; 137(1): 13-7.

[194] Robertson RP. Islet transplantation a decade later and strategies for filling a half-full glass. Diabetes 2010; 59(6): 1285-91.

[195] Maehr R, Chen S, Snitow M, *et al.* Generation of pluripotent stem cells from patients with type 1 diabetes. Proc Natl Acad Sci U S A 2009; 106(37): 15768-73.

[196] Tharavanij T, Betancourt A, Messinger S, *et al.* Improved long-term health-related quality of life after islet transplantation. Transplantation 2008; 86(9): 1161-7.

[197] Thompson DM, Begg IS, Harris C, *et al.* Reduced progression of diabetic retinopathy after islet cell transplantation compared with intensive medical therapy. Transplantation 2008; 85(10): 1400-5.

[198] Fiorina P, Shapiro AM, Ricordi C, Secchi A. The clinical impact of islet transplantation. Am J Transplant 2008; 8(10): 1990-7.

[199] Aguayo-Mazzucato C, Bonner-Weir S. Stem cell therapy for type 1 diabetes mellitus. Nat Rev Endocrinol 2010; 6(3): 139-48.

[200] McKernan R, McNeish J, Smith D. Pharma's developing interest in stem cells. Cell Stem Cell 2010; 6(6): 517-20.

[201] Puri S, Hebrok M. Cellular plasticity within the pancreas--lessons learned from development. Dev Cell 2010; 18(3): 342-56.

[202] Jiang J, Au M, Lu K, *et al.* Generation of insulin-producing islet-like clusters from human embryonic stem cells. Stem Cells 2007; 25(8): 1940-53.

[203] Placzek MR, Chung IM, Macedo HM, *et al.* Stem cell bioprocessing: fundamentals and principles. J R Soc Interface 2009; 6(32): 209-32.

[204] McCall MD, Toso C, Baetge EE, Shapiro AM. Are stem cells a cure for diabetes? Clin Sci (Lond) 2010; 118(2): 87-97.

[205] Vendrame F, Pileggi A, Laughlin E, *et al.* Recurrence of type 1 diabetes after simultaneous pancreas-kidney transplantation, despite immunosuppression, is associated with autoantibodies and pathogenic autoreactive CD4 T-cells. Diabetes 2010; 59(4): 947-57.

[206] Korsgren O, Nilsson B. Improving islet transplantation: a road map for a widespread application for the cure of persons with type I diabetes. Curr Opin Organ Transplant 2009; 14(6): 683-7.

[207] Chayosumrit M, Tuch B, Sidhu K. Alginate microcapsule for propagation and directed differentiation of hESCs to definitive endoderm. Biomaterials 2010; 31(3): 505-14.

[208] Beckwith J, Nyman JA, Flanagan B, Schrover R, Schuurman HJ. A health-economic analysis of porcine islet xenotransplantation. Xenotransplantation 2010; 17(3): 233-42.

[209] O'Connell PJ. The International Xenotransplantation Association consensus statement on conditions for undertaking clinical trials of porcine islet products in type 1 diabetes--chapter 6: Patient selection for pilot clinical trials of islet xenotransplantation. Xenotransplantation 2009; 16(4): 249-54.

CHAPTER 11

Stem Cells and Ocular Disorders: Basic Science to Clinical Applications

Indumathi Mariappan[1], Subhash Gaddipati[1], Taraprasad Das[2], Geeta K. Vemuganti[1] and Virender S. Sangwan[3*]

[1]Sudhakar and Sreekanth Ravi Stem Cell Biology Laboratory, L.V. Prasad Eye Institute, Hyderabad, India; [2]Retina and Vitreous Service, L V Prasad Eye Institute, Bhubaneswar, Orissa, India and [3]Cornea and Anterior Segment Services, L.V. Prasad Eye Institute, Hyderabad, India

Abstract: The visual function of an eye gets affected due to defects involving some of its major parts, such as the cornea, lens and retina. Simple refractive error corrections and advanced surgical procedures like cataract surgeries, corneal transplantations and others have revolutionized the treatment of ocular diseases in the past. However, the past decade has witnessed the promise of adult limbal stem cell transplantation (LSCT) for the reconstruction of ocular surface that has been damaged by chemical and thermal burns. LSCT has now become a widely accepted treatment modality in several clinics around the world. While this approach works well for the corneal epithelium, reconstruction of other damaged ocular tissues pose a greater challenge. The current interest is therefore to look for various sources of stem cells that can generate corneal endothelial cells for corneal endothelial dystrophies, glandular cells of the lacrimal epithelium for various dry eye conditions, retinal neurons and the retinal pigmented epithelium (RPE) for treating glaucoma-induced vision loss and other age-related and hereditary retinal dystrophies. Apart from some of the adult ocular stem cell sources, the embryonic stem cells (ESCs) and the recently introduced induced pluripotent stem cells (iPSCs) have come to the forefront and have kindled a lot of hope for ocular regenerative medicine in the future.

Keywords: Ocular disorders, induced pluripotent stem cells, retinal pigmented epithelium, embryonic stem cell, limbal stem cells.

INTRODUCTION

The eye is an important organ which enables us to see the outside world. The eye balls sit in sockets on either side of the skull, protected by the surrounding bone, fat and the muscle tissues. The eye lids and eyelashes protect the eye in the front and also control the entry of light. Along with these, tear secretions of the lacrimal gland located just above the outside corner of each eye help lubricate and protect the ocular surface. The eye ball comprises three major layers. The outermost layer comprises the cornea and the sclera, the middle layer comprises the lens and the uveal tract (iris, ciliary body and choroid) and the innermost layer comprises the retina (Fig. **1**).

The cornea is the central transparent, avascular and dome-shaped structure on the ocular surface that allows light to enter the eye and together with the lens helps to focus the light onto the retina at the back of the eye. Cornea contributes to about 60% of the total refractive power of an eye. It consists of a stratified squamous epithelial layer of 5-6 cells in thickness on the outer surface, an outer basement membrane called the Bowman's layer, a highly ordered collagenous stroma with keratocytes in the middle, an inner basement membrane called the Descemet's membrane and a monolayer of endothelium on the inner surface. The highly ordered collagen fibrils and the avascularity of the cornea enable smooth light penetration. The endothelial layer acts as a two-way fluid pump to enable corneal water balance which is important for corneal nourishment and for maintenance of its light refractive properties. The sclera is the fibrous white portion of the ocular surface and is covered by the bulbar conjunctiva. The conjunctival epithelium also extends and covers the inner lining of the eye lids. Apart from the epithelial cells, the goblet cells of the conjunctiva secrete mucins that help in anchoring the tear film to the epithelial surface. A narrow zone of about 2 mm thickness between the corneal epithelium and the bulbar conjunctiva is known as the limbus and is widely accepted as the niche for the corneal epithelial stem cells [1-3]. The limbal epithelium is 10 to

***Address correspondence to Virender S. Sangwan:** Virender Singh Sangwan, MD, Associate Director, L V Prasad Eye Institute, Head, Cornea and Anterior Segment, Ocular Immunology & Uveitis Services, Road No.2, Banjara Hills, Hyderabad - 500 034, India; Tel: +91-40- 30612611, Fax: +91-40-2354 8271; E-mail: vsangwan@lvpei.org

Kuldip S. Sidhu (Ed)

12 cell layers thick, and contains epithelial crypts and rete ridges called the 'palisades of Vogt' [4] harboring melanocytes, Langerhans cells and an underlying network of blood vessels in the stroma to nourish the limbal epithelial stem cells. The palisades are present 360° all around the corneo-scleral rim but are most well defined inferiorly and superiorly. The pigmented limbus demarcates the avascular corneal epithelium from the vascular conjunctiva. The corneal surface epithelium together with the limbal and conjunctival epithelium, protects the ocular surface.

However the central corneal surface appears black due to the presence of the pigmented iris tissue beneath. The iris sits between the anterior chamber and the posterior chamber and controls the size of the pupil thereby controlling the amount of light that enters the eye. The anterior and the posterior chambers are filled with aqueous humor, which is continuously secreted by the epithelial cells of the ciliary body and are drained away through the trabecular meshwork at the junction of the iris and the ciliary body. The aqueous humor nourishes the avascular lens and the corneal cells and the critical balance between the fluid inflow and outflow rates dictates the ocular pressure. The lens is a clear, flexible structure that changes its shape to enable focusing of objects seen at varying distances (Fig. **1**). It lies below the posterior chamber and is connected to the ciliary body by suspensory ligaments. The contractions of the ciliary muscles and the suspensory ligaments help in adjusting the focus by changing the shape of the lens.

The retina lines the back side of the eye and acts as the light screen. The retina is a multilayered tissue that comprises the neuroretina and the RPE. The neuroretina contains six different light-sensitive neurons and a glial cell type. The photoreceptors are the primary cell types that perceive light signals and convert them to nerve impulses through multiple steps of signal amplification *via* the other retinal neurons. The nerve fibers from the retinal neurons converge to form the optic nerve at the optic disc or the blind spot and the nerve impulses are then sent to the cortex region of the brain for image perception. The RPE cell layer at the back establishes the blood-retinal barrier and helps in nourishing the photoreceptor cells and also takes part in the photo-chemical visual cycle. The choroid is the thick membrane found between the sclera and the retina below the RPE layer and contains blood vessels that supply oxygen and nutrients to the retina and is highly pigmented with melanocytes (Fig. **1**). The pigmented choroidal layer together with the retinal RPE cell layer helps to absorb and prevent scattering of the excess light. The chamber between the lens and the retina is filled with a highly viscous jelly called the vitreous humor, which helps in keeping the retinal screen in place by pressing it against the choroid. It also serves to nourish the retinal cells. The Fig. **1** shows the cross section of an eye ball highlighting the different parts of an eye and hematoxylin and eosin (H&E) dye stained sections of some of the important regions harboring stem cells.

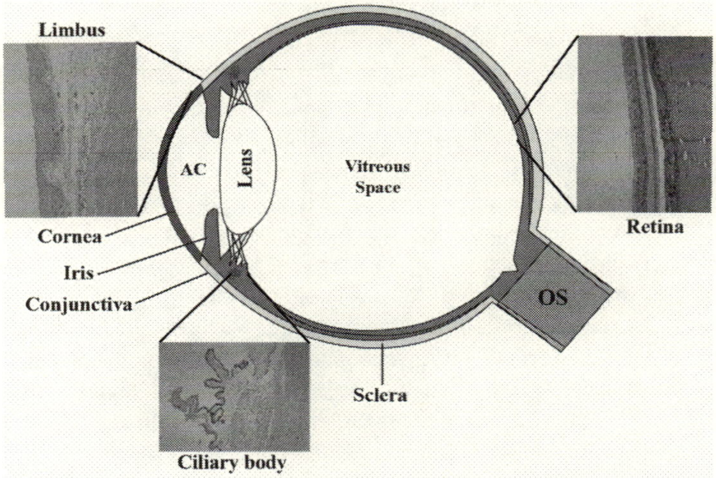

Fig. 1: Parts of an adult human eye.

The cornea, limbus and the conjunctiva forms the ocular surface and the corneal stem cells reside in the limbal region and migrate centripetally towards the central cornea upon activation and help in the corneal epithelial regeneration and wound healing. The pigmented epithelial cells of the ciliary body harbor retinal progenitors which are capable of generating retinal neurons. The Muller glial cells of the neuroretina and the retinal pigmented epithelium from the fetal sources are being explored for their potential in regenerative medicine. The anterior chamber (AC) (between the cornea and the iris) and the posterior chamber (between the iris and the lens) are filled with the aqueous humor and the space between the lens and the retina is filled with a highly viscous vitreous humor. The nerve fibers from the retinal neurons converge to form the optic stalk (OS) which connects the eye with the brain.

Ocular Disorders and the Promise of Stem Cell Therapy

The visual function of an eye gets affected if the defects involve the cornea, the lens or the retina. Defects in the cornea will compromise the ocular surface integrity and invite infections and influence its light refractive properties as well. Corneal defects may involve the surface epithelium, the underlying stroma and/or the endothelium. Simple refractive error corrections are generally done using spectacles and contact lenses or by surgical corrections through laser-assisted *in situ* keratomileusis (LASIK). Full thickness corneal defects can also be easily treated by penetrating keratoplasty, generally termed as corneal transplantation. Cornea being an avascular tissue, donor corneal transplantation has become a successful organ transplantation modality which, in most cases does not face any immune rejection issues. However in the case of limbal stem cell deficiencies (LSCD), corneal transplantations are not successful and primarily require limbal stem cell transplantations. In case of endothelial dysfunction, surgical procedures like Descemet's Stripping Endothelial Keratoplasty (DSEK) involving the use of donor corneas remain an option. Several groups are also looking at options of using donor derived or cultured endothelial cell transplantations for treating this condition [5, 6]. The existence of transiently amplifying (TA) cells and slow cycling cells located close to the Schwalbe line at the corneal periphery and in the regions of the trabecular meshwork has been reported. Research is also in progress to use the stem cell approach to treat xerophthalmia or dry eye conditions using the cultured lacrimal gland and conjunctival goblet cells to reconstitute the tear fluid secretion and function.

Cataract surgeries and the use of intraocular lenses solve most of the lens related problems. Glaucoma involves the retina, uveal and the aqueous flow tract and some of which are successfully managed using drugs that regulate the aqueous flow rate and by surgical interventions. However, glaucoma induced damage to the retinal ganglion cells (RGCs) and the loss of retinal neurons and RPE cells due to age-related and hereditary retinal dystrophies remain a major challenge. Currently, the quest is on to look out for various sources of stem cells that can generate retinal neurons and RPE cells for their use in regenerative medicine.

Cultured Limbal Epithelial Transplantation (CLET) for the Treatment of (LSCD) - The Success Story

One of the major advances made in translational research is in the field of ocular surface reconstruction using cell therapy. It owes its success to innovations in surgical techniques as well as progress in the field of stem cell biology. It was discovered over twenty years ago that the limbus contains stem cells that regenerates the corneal epithelium. These cells play an important role in the maintenance of corneal epithelium and tissue homeostasis [1-3]. The corneal epithelium is a highly dynamic structure and is constantly renewed all through the adulthood of an eye. Activated stem cells from the limbus migrate centripetally to the central cornea and helps in tissue homeostasis. As proposed by the XYZ theory of corneal maintenance [7], the combined rate of cell proliferation and centripetal migration of the activated limbal stem cells (XY component) is equal to the rate of cell loss by desquamation (Z component). The centripetal migration is the characteristic of the limbal basal epithelial cells. Upon activation, the basal cells proliferate and migrate centripetally towards the central cornea and also move upward towards the corneal surface. Once the epithelial cells leave the limbal basal layer or the limbal niche, they get differentiated into TA cells with lesser proliferative capacity. Prior to or during the course of their upward migration, these TA cells become terminally differentiated and contribute to the process of corneal wound healing.

Several factors *e.g.*, chemical (alkali/acid) or thermal injury, UV and other ionizing radiations, repeated surgical interventions, extensive microbial infection and immunological disorders like Steven Johnson's Syndrome (SJS), and even long term usage of contact lenses in some cases may lead to LSCD. The complete loss of limbal epithelial stem cells in LSCD leads to the conjunctivalization of the cornea, neovascularization, sub epithelial scarring and the adhesion of the eyelid(s) to the eyeball. These conditions lead to corneal opacity and visual impairment apart from varying degrees of discomfort including redness, irritation and watering in the affected eye. In unilateral LSCD, the treatment option includes transplantation of (i) the healthy limbal tissue [8] or (ii) *in vitro* expanded autologous limbal epithelial cells cultured from a small biopsy taken from the unaffected fellow eye. Limbal epithelial cells are cultured *in vitro* using various substrates such as, mitotically inactivated feeder layers of mouse NIH 3T3 cells [9] or human limbal stromal fibroblasts, human amniotic membrane (hAM) [10] and also synthetic, bio-compatible and bio-degradable matrices made of fibrin gel [11], myogel [12], plasma polymer–coated surfaces [13], human recombinant collagen [14], and temperature-responsive culture inserts [15]. These substrates act as scaffolds on which a clean, transparent layer of the corneal epithelium if formed ready for transplantation. CLET enables stem cell expansion and thereby minimizes the size of the donor tissue required for successfully generating an epithelial sheet for reconstructing the diseased ocular surface and thus helps in replenishing the depleted stem cell pool, restoring the ocular surface and improving the long term visual outcome [9, 16-20]. A schematic representation of the various steps involved in CLET is shown in Fig. **2**.

In case of unilateral LSCD, a small 2x2 mm fresh autologous limbal biopsy is taken from the unaffected eye of the patient in a sterile operating room by a trained corneal surgeon, with informed consent from the patient and the approval of the Institutional Review Board (IRB). All procedures using human biological material are done in accord with the tenets of the Declaration of Helsinki on ethical principles of medical research involving human subjects, as approved in the World Medical Association (WMA) meeting held in 2008. The biopsied tissue is then taken to a sterile tissue culture facility where it is further processed and explanted on to a de-epithelialized and infection-free (hAM) of about 3x2.5 cm. The cultures are then grown using human corneal epithelial growth medium containing the autologous serum and maintained at 37ºC in a CO_2 incubator with 5% CO_2 supply. The growth is monitored regularly and the culture medium is changed every alternate day till a confluent monolayer is generated in about 10-15 days [21]. The hAM, along with the generated corneal epithelial cell sheet is then transplanted on to the corneal surface of the affected eye after removing the pannus and the scar tissue. Once the ocular surface becomes stable and gets successfully reconstructed, the patient may undergo a penetrating keratoplasty (corneal transplantation) for improving the visual outcome further.

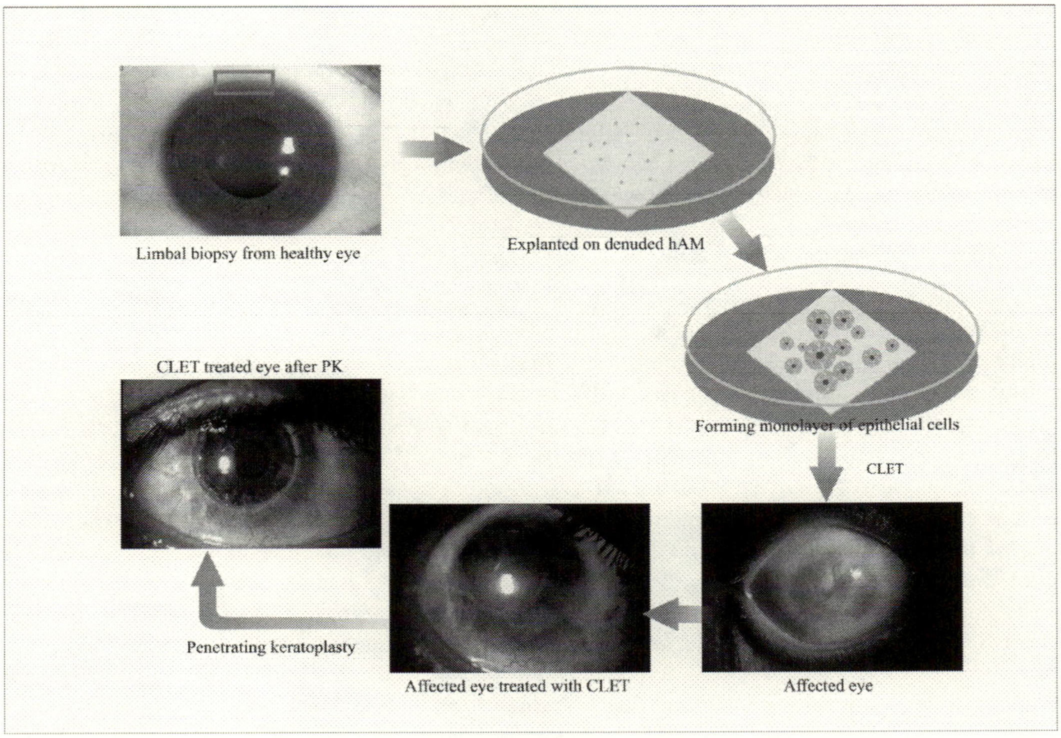

Fig. 2: Schematic representation of various steps involved in CLET.

However in cases of bilateral LSCD, allografting is done using limbal tissues from a live related donor [22, 23] or from cadaveric sources, but this might require the continuous usage of immune suppressive drugs. Therefore, to obviate the need for immune suppression regime in allogenic transplantations, a few groups have attempted to use other autologous sources, such as conjunctival [24, 25] or oral mucosal epithelial cells [26, 27] as an alternative to limbal epithelium for corneal surface reconstruction in humans. Patients with severe bilateral limbal stem cell deficiency have been successfully treated with autologous cultured oral mucosal epithelial transplantation, with good clinical outcomes in terms of ocular surface stabilization and marginal improvement in terms of visual acuity [28, 29].

Retinal Dystrophies - the Major Challenge Ahead

Considerable progress has been made in other blinding conditions involving cornea and the lens. However the most untreatable form of blindness as yet, is the retinal dystrophies, notably caused by the loss of rod and cone photoreceptor cells and the adjacent retinal pigment epithelial cells. Age-related macular degeneration (ARMD), retinitis pigmentosa (RP), Leber's congenital amaurosis (LCA) and Stargardt's disease are some of the major hereditary forms of retinal dystrophies. Retinal dystrophy can be congenital as in the case of LCA, where the photoreceptors are completely absent at birth or can be a progressive and age related disorder as in the case of RP and ARMD. Millions of people world over suffer visual disability as a result of retinal dystrophy. ARMD causes irreversible vision loss in patients over sixty years of age worldwide [30, 31]. RP is the most common heritable form of retinal degeneration and it affects about one in 5000 individuals worldwide [32]. Patients with RP lose vision initially due to the loss of rod photoreceptors with subsequent degeneration of the cone photoreceptors [33]. It is a progressive disorder wherein the photoreceptors made during embryonic development and present at birth undergo gradual degeneration as the individual ages. It is typically characterized by initial symptoms of night blindness during early adulthood, which progresses to severe visual impairment during later stages. Clinically the retinal changes include the presence of attenuated vasculature, intraretinal pigmentary deposits and a gradual diminution of electroretinographic (ERG) responses within a few years of life. Mutations in several genes involved in phototransduction pathway, vitamin A metabolism, photoreceptor-specific structural and transport proteins, retina-specific transcription factors and splicing factors are linked to retinal dystrophies which results in the loss of cellular function, leading to apoptotic cell death. These are terminally differentiated cells like neurons and therefore possess only a limited capacity to self-repair. This results in the accumulation of cell loss over a period, which ultimately renders the patient legally blind. Most of the existing treatment modalities including dietary supplements and gene therapy aim only to slow down the disease progression by preventing or delaying further cell death. But in congenital forms of retinal dystrophy and in cases wherein majority of the photoreceptor cells were lost, cell replacement therapy holds the promise. However unlike the corneal surface, an adult retina does not harbor any stem cells and the major challenge lies in identifying a suitable cell source that can generate retinal cell types suitable for clinical applications.

Adult Neural Stem Cells and Fetal RPE Cells for Retinal Disorders

Many studies have shown that retina is amenable for cell replacement therapy. Experiments using animal models have shown that visual function can be preserved or improved after transplantation of retinal progenitors [34] precursors [35] or ES cell derived retinal pigment epithelium [36]. However, in each of these cases, the clinical application is beset with the lack of a reliable, renewable, safe and uncontroversial source of donor cells. Though human ES cells are a preferred stem cell source, the associated ethical, immunological and neoplastic concerns pose a significant obstacle to their clinical use.

Initial studies using adult hippocampal stem cell progenitors revealed that when transplanted in the vitreous of neonatal or adult eyes, they survive and incorporate into the laminar structure of the host retina [37, 38]. However, despite successful integration of the transplanted cells within the host retina and their morphological similarities to various retinal cell types, they failed to express any retina-specific markers. Therefore, the attention was directed towards ocular tissues for the search of adult stem cells with retinal potential.

In amphibians, RPE cells play an important role in retinal regeneration. Upon retinal damage, the RPE cells dedifferentiate to generate retinal neurons by the downregulation of RPE markers, lose their pigmentation and concomitantly upregulate of the expression of neuroretinal markers by a process termed as transdifferentiation [39, 40]. However, the RPE cells do not possess this retinal regenerative capacity in fishes, birds and mammals. But fetal retinas, with or without the intact RPE cell layer, have been transplanted as a sheet or as isolated progenitors and shown to offer some promise for treating retinal degenerative problems [41]. Pilot studies involving human subjects have established the clinical safety of fetal derived cells [42, 43]. However the fetal-derived tissue sources are not abundant and also their usage face some ethical concerns. Such barriers have initiated a search for renewable sources of adult stem cells from which the retinal progenitors can be derived for clinical use.

Adult Ocular Stem Cells for Retinal Disorders

i) Ciliary Epithelial Cells and the Muller Glial Cells

During early eye development, at the optic vesicle stage, the cells at the leading edge of the vesicle are in close proximity to the cells of the ocular surface ectoderm (OSE) which expresses FGFs. Exposure to FGF signalling leads to a retinal commitment of the leading edge cells. During the optic cup stage the leading edge cells that are exposed to the FGF signals invaginate to form the inner layer of the optic cup which later develops into the future retina. The cells at the distal edge of the optic vesicle receive inputs from the surrounding mesenchyme and get committed to the RPE fate. It was therefore hypothesized that a population of cells at the peripheral retina or the boundary region called the ciliary marginal zone (CMZ) remain in a progenitor state, having incompletely influenced by the ocular surface ectoderm and the surrounding mesenchyme. In lower vertebrates like amphibians (Xenopus sp.) and fishes (zebra fish), the CMZ harbors retinal progenitors which participate in retinal regeneration throughout their adulthood and helps in the repair process upon injuries [44].

However in birds like chicks and quails, the stem cells in the CMZ have limited potency and can only differentiate into bipolar and amacrine interneurons but not into ganglion cells or photoreceptors. They do not participate in the regeneration process following injury [45]. Adjacent to the ciliary margin zone of the retina, the non-pigmented epithelium of the ciliary body, which is a continuation of the retina, also harbors some quiescent stem cells. These cells do not normally proliferate or participate in the regeneration process but they get activated when stimulated with growth factors and differentiate into early retinal neurons like the ganglion cells [46]. This differential potency of the ciliary epithelial cells and the CMZ cells suggests that a gradient of multipotency might exist in the peripheral chick retina. The most peripheral and less committed cells of the ciliary body might retain multipotency and are capable of generating the early cell types, while the more proximal and committed precursors in the CMZ might have a restricted potency capable of generating only the later cell types [47].

Stem cells have been reported to occur in the pigmented ciliary epithelium of adult mammals with retinal properties [48-51]. *In vivo* BrdU labeling analysis revealed that the pigmented portion of the ciliary body of adult rats contains cells with proliferative potential [52]. When cultured in the presence of FGF2 and/or EGF, these cells proliferate and self renew giving rise to clonally derived neurospheres containing nestin-positive cells. These cells are multipotent too, and can differentiate along the neuronal and glial lineages. They also express the retinal progenitor markers like Chx10, Rx, and Pax6, which suggests that they possess retina-specific properties and can differentiate into retinal cells when cultured under suitable conditions [53]. Another source of retinal progenitors is the Muller glial cells which are shown to participate in the regenerative processes in fish and chicks. When induced chemically with neurotoxins or growth factors, they re-enter the cell cycle and migrate to the outer nuclear layer (ONL), lose the glial specific markers and begin to express the retinal progenitor markers [54, 55]. However, the accessibility of CPE cells and Muller glial cells remains an issue for clinical applications.

ii) Limbal Stem Cells – Retinal Regeneration by Culture Induced Trasdifferentiation

A proven renewable and accessible source of stem cells in adult ocular tissue is the limbal epithelium that participates in cornea regeneration [1, 2, 56]. Limbal epithelium, like neural epithelium, is of ectodermal origin. During embryonic development, the cells of the embryonic ectoderm differentiate into the neural lineage by default pathway unless they are exposed to signalling by the molecule, bone morphogenetic protein (BMP), which induces their differentiation towards the epidermal lineage [57]. It was hypothesized that the removal of these stem cells from the limbal niche would expose their neural properties. Therefore when rat LSCs were subjected to neurosphere culture in the presence of bFGF2 and Noggin (an inhibitor of BMP signalling), they acquired neural progenitor properties and differentiated along the neuronal and glial lineages, both *in vitro* and *in vivo* [58]. These reprogrammed cells first expressed Notch1; subsequently a subset of these Notch1-positive cells expressed the neuroectodermal stem cell marker, Nestin. Based on their self-renewal capacity, it is proposed that these nestin-positive progenitors are derived from limbal stem cells rather than TA cells which have limited proliferating capacity. Subsequently, the neural properties of LSCs was also demonstrated in humans suggesting that the neural potential of LSCs is evolutionarily conserved [59]. More recently, it was demonstrated that the rat LSC-derived neural progenitors could give rise to neurons with functional properties when co-cultured with embryonic day 18 (E18) hippocampal cells. In addition, it was shown that these neural progenitors could differentiate along photoreceptor lineage when co-cultured with postnatal day 1 (PN1) retinal cells [60]. These observations suggest that adult LSCs might be a renewable source of retinal progenitors for an autologous stem cell approach to treat blinding degenerative diseases such as the RP and ARMD.

Pluripotent Stem Cells- A Versatile and Promising Alternative

However, in case of genetic disorders, genetic manipulations would be required to address some of the basic questions related to the disease and also to use these adult stem cells in regenerative medicine. This proves to be difficult as most of the adult stem cells cannot be maintained in culture for too long to enable genetic manipulations. Human ES cells are a valuable pluripotent stem cell source and can differentiate into most cell types of the body. They have unlimited self renewal capacity and are amenable for all kinds of *in vitro* genetic manipulations. *In vitro* protocols to differentiate hES cells into retinal cell types in a step-wise manner using defined growth factors and/or ectopically expressed genes have been reported by several groups [61-67]. Embryonic stem cell derived retinal cell types have also been shown to rescue disease phenotype to some extent when transplanted in animal models [35, 68]. However, the problems of host immune response and rejection following transplantation of cells types derived from ESCs, the lack of enough approved ESC lines for HLA matching of diverse ethnic groups and the associated ethical concerns pose a significant obstacle to their clinical use. Recently, it was shown that adult somatic cells such as the dermal fibroblasts can be reprogrammed directly to a pluripotent state very similar to the ES cells by ectopic expression of certain transcription factors like the Oct4, Sox2 Klf4 and cMyc to generate *"induced pluripotent stem cells"* (iPSCs) [69]. Adult somatic cells from patients have been used to derive disease-specific iPS cells [70, 71] and these cells can serve as valuable *in vitro* disease models for screening new drugs, to conduct pre-clinical trials and to study the physiology of genetic disorders for which there exists no adequate human *in vitro* or animal model systems. In a sickle cell anemia mouse model, it was shown that the genetic defect could be corrected *in vitro* by site-directed recombination methods and the genetically corrected iPS cells are differentiated to hematopoietic progenitors and successfully used for autologous cell replacement therapy bypassing any immune rejection problems [72]. Similar attempts to correct some genetic defects have been reported by two other groups using human iPS cells [73, 74]. It was also shown that human iPS cells are capable of giving rise to most of the retinal cell types (RPE and retinal neuronal cell types) when differentiated under defined culture conditions [65, 75-77]. Pluripotent stem cells capable of differentiating to functional neurons have also been derived using limbal stem cells by culturing them with growth factors and the conditioned medium of embryonic stem cells without using any genetic factors [78, 79]. In another recent advancement and a step ahead, two groups have shown that somatic cells such as the dermal fibroblasts can be directly reprogrammed to functional neurons and cardiomyocytes using defined factors that are cell type-specific master regulators [80, 81]. Thus the adult somatic cell derived iPS cells have kindled a lot of hope for ocular regenerative medicine in the future.

Bottlenecks and Future Directions

Though cell-based therapies offer a great hope for treating some of the blinding conditions, it is important to establish an in depth understanding of the behavior and stability of the *in vitro* derived cell types before going to the clinics. When pluripotent stem cell sources are used, robust protocols to differentiate them to the desired cell type has to be established and the sorting efficiencies should be good enough to eliminate the unwanted cell types and the undifferentiated cells in order to reduce the risk of tumor or teratoma formation. The *in vitro* derived cells should be comparable to the adult or fetal cell types in terms of their morphology, cell polarity, marker gene expression and other physiological functions. In this context, it was reported that both the ESC and iPSC derived cell types were fetal-like and were not comparable to the adult mature cells in terms of marker gene expression profiles [82]. This requires the establishment of standard markers and functional assays for comparative evaluation and for effective cell sorting. Though several groups have reported the derivation of RPE and retinal neurons *in vitro* from different sources, they remain immature and lack phenotypic differentiation in spite of expressing the committed cell markers. This might require the use of different culture substrates, specific growth factors and protocols that mimic the native signalling cues that dictate step-wise differentiation and maturation during the normal eye development. Also in the case of RPE cells, establishment of cell polarity is very important for its physiological function. The transplanted cells tend to form aggregates and do not seem to establish a uniform monolayer of honeycomb-shaped cells resembling the native RPE cells. Nevertheless, in preclinical studies on rat and mice models, the transplanted cells seem to have some protective effects on the integrity of retinal neurons and were shown to provide marginal improvements in visual outcomes [66, 76, 83]. This problem can be overcome by culturing RPE cells on suitable substrates, *e.g.*, synthetic bio-degradable PLGA membranes (poly lactic glycolic acid polymers) that are surface-modified to promote cell growth and the intact cell sheet can then be transplanted along with the substrate to enable better engraftment. Another concern is with regard to the stage at which the *in vitro* derived cells can be used for transplantation. Transplanting completely differentiated cells will avoid the risk of tumor formation, but their survival, homing, migration and integration will remain an issue apart from the requirement for large number of cells to account for cell death in order to have at least some minimal effect. However, transplanting the committed precursor cells that can proliferate will enable *in vivo* cell expansion and the native niche might direct the terminal differentiation process. This might require less cell numbers and could enable better homing, integration, phenotypic maturation and functional recovery. But how do we define a committed precursor? Suitable markers that define lineage commitment have to be established and stringent sorting methods coupling both negative and positive sorting can help in enriching the desired cell population. And all this becomes possible only if the microenvironment or the niche is permissive for the transplanted cells to fill in the empty space. If the supporting niche gets completely damaged, even the best of stem cells may not survive to do the job. Therefore, it becomes important to understand the changes in the retina of patients with retinal degeneration in order to measure the extent of damage and to identify the target patient groups. Preservation of at least the inner retinal layers may be important to attempt any cell-based therapy. This can be assessed using certain parameters like the measurement of visual acuity, ERG, the thickness of the intact retinal layers (inner and outer layers) by Optical Coherence Tomography (OCT) examinations and others. Most studies in animal models have shown that the transplanted cells die within 10-15 weeks, and the survival of cells for up to 20-30 weeks remains an exception [66, 76, 83]. Immunological issues may be of importance while considering the case of ARMD-patients where the retina is already flooded with inflammatory cytokines and could either promote cell proliferation or may lead to immune-mediated rejection and apoptosis of the transplanted cells. A damaged tissue or an injury is also likely to provide the cues that can mobilize the transplanted cells to migrate and integrate into host. This was reported in some of the animal studies where widespread migration and incorporation of adult hippocampal neural stem cells progenitors was observed only in the adult retina that was either dystrophic [38] or injured [84]. Therefore a degenerating retina may offer a more conducive environment for differentiation as well as efficient integration of the transplanted retinal progenitor cells and it remains to be checked in humans.

CONCLUSIONS AND PROSPECTS

As discussed in detail above, limbal stem cell therapy for the treatment of LSCD had been a major success. Multiple sources of stem cells are being explored for possible application in the treatment of corneal endothelial and retinal dystrophies. Though human embryonic stem cells and induced pluripotent stem cell sources offer a great promise, particularly for genetic disorders, both in terms of culture expansion and *in vitro* manipulations, they still need to evolve and refine further to reach the clinical footsteps. Though the use of iPSCs and the concept of individualized cell therapy may not be cost effective, it is possible to establish a global cell banking system using well characterized and HLA-typed pluripotent cell lines of varied ethnic groups to cater to the needs of the entire human race. Moreover, the establishment of disease-specific induced pluripotent stem cells may serve as wonderful disease models to understand the basic biology of various genetic diseases that may eventually help in better drug development, disease prevention or therapy. Some of the major issues with the use of pluripotent stem cells are already being addressed and a FDA approved Phase I clinical safety trial has been initiated in November this year for the treatment of Stargardt's disease and macular degeneration using human ESC-derived RPE cells. This has been a major breakthrough and has become a beacon of hope for millions of people suffering from retinal dystrophies around the world.

REFERENCES

[1] Schermer A, Galvin S, Sun TT. Differentiation-related expression of a major 64K corneal keratin *in vivo* and in culture suggests limbal location of corneal epithelial stem cells. J Cell Biol 1986; 103(1): 49-62.

[2] Cotsarelis G, Cheng SZ, Dong G, Sun TT, Lavker RM. Existence of slow-cycling limbal epithelial basal cells that can be preferentially stimulated to proliferate: implications on epithelial stem cells. Cell 1989; 57(2): 201-9.

[3] Huang AJ, Tseng SC. Corneal epithelial wound healing in the absence of limbal epithelium. Invest Ophthalmol Vis Sci 1991; 32(1): 96-105.

[4] Van Buskirk EM. The anatomy of the limbus. Eye (Lond) 1989; 3 (Pt 2): 101-8.

[5] Ishino Y, Sano Y, Nakamura T, *et al.* Amniotic membrane as a carrier for cultivated human corneal endothelial cell transplantation. Invest Ophthalmol Vis Sci 2004; 45(3): 800-6.

[6] Yokoo S, Yamagami S, Yanagi Y, *et al.* Human corneal endothelial cell precursors isolated by sphere-forming assay. Invest Ophthalmol Vis Sci 2005; 46(5): 1626-31.

[7] Thoft RA, Friend J. The X, Y, Z hypothesis of corneal epithelial maintenance. Invest Ophthalmol Vis Sci 1983; 24(10): 1442-3.

[8] Kenyon KR. Limbal autograft transplantation for chemical and thermal burns. Dev Ophthalmol 1989; 18: 53-8.

[9] Pellegrini G, Traverso CE, Franzi AT, *et al.* Long-term restoration of damaged corneal surfaces with autologous cultivated corneal epithelium. Lancet 1997; 349(9057): 990-3.

[10] Tsai RJ, Li LM, Chen JK. Reconstruction of damaged corneas by transplantation of autologous limbal epithelial cells. N Engl J Med 2000; 343(2): 86-93.

[11] Rama P, Bonini S, Lambiase A, *et al.* Autologous fibrin-cultured limbal stem cells permanently restore the corneal surface of patients with total limbal stem cell deficiency. Transplantation 2001; 72(9): 1478-85.

[12] Francis D, Abberton K, Thompson E, Daniell M. Myogel supports the ex-vivo amplification of corneal epithelial cells. Exp Eye Res 2009; 88(3): 339-46.

[13] Deshpande P, Notara M, Bullett N, *et al.* Development of a surface-modified contact lens for the transfer of cultured limbal epithelial cells to the cornea for ocular surface diseases. Tissue Eng Part A 2009; 15(10): 2889-902.

[14] Merrett K, Fagerholm P, McLaughlin CR, *et al.* Tissue-engineered recombinant human collagen-based corneal substitutes for implantation: performance of type I versus type III collagen. Invest Ophthalmol Vis Sci 2008; 49(9): 3887-94.

[15] Hayashida Y, Nishida K, Yamato M, *et al.* Transplantation of tissue-engineered epithelial cell sheets after excimer laser photoablation reduces postoperative corneal haze. Invest Ophthalmol Vis Sci 2006; 47(2): 552-7.

[16] Solomon A, Ellies P, Anderson DF, *et al.* Long-term outcome of keratolimbal allograft with or without penetrating keratoplasty for total limbal stem cell deficiency. Ophthalmology 2002; 109(6): 1159-66.

[17] Sangwan VS, Matalia HP, Vemuganti GK, *et al.* Clinical outcome of autologous cultivated limbal epithelium transplantation. Indian J Ophthalmol 2006; 54(1): 29-34.

[18] Fatima A., Vemuganti GK, Iftekhar G, Rao GN, Sangwan VS. *In vivo* survival and stratification of cultured limbal epithelium. Clin Experiment Ophthalmol 2007; 35, 96-8.

[19] Miri A, Al-Deiri B, Dua HS. Long-term outcomes of autolimbal and allolimbal transplants. Ophthalmology 2010; 117(6): 1207-13.

[20] Rama P, Matuska S, Paganoni G, *et al.* Limbal stem-cell therapy and long-term corneal regeneration. N Engl J Med 2010; 363(2): 147-55.

[21] Mariappan I, Maddileti S, Savy S, *et al.* *In vitro* culture and expansion of human limbal epithelial cells. Nat Protoc 2010; 5(8): 1470-9.

[22] Tsai RJ, Tseng SC. Human allograft limbal transplantation for corneal surface reconstruction. Cornea 1994; 13(5): 389-400.

[23] Rao SK, Rajagopal R, Sitalakshmi G, Padmanabhan P. Limbal autografting: comparison of results in the acute and chronic phases of ocular surface burns. Cornea 1999; 18(2): 164-71.
[24] Daya SM, Ilari FA. Living related conjunctival limbal allograft for the treatment of stem cell deficiency. Ophthalmology 2001; 108(1): 126-33; discussion 33-4.
[25] Tanioka H, Kawasaki S, Yamasaki K, *et al.* Establishment of a cultivated human conjunctival epithelium as an alternative tissue source for autologous corneal epithelial transplantation. Invest Ophthalmol Vis Sci 2006; 47(9): 3820-7.
[26] Nishida K, Yamato M, Hayashida Y, *et al.* Corneal reconstruction with tissue-engineered cell sheets composed of autologous oral mucosal epithelium. N Engl J Med 2004; 351(12): 1187-96.
[27] Nakamura T, Inatomi T, Sotozono C, *et al.* Transplantation of cultivated autologous oral mucosal epithelial cells in patients with severe ocular surface disorders. Br J Ophthalmol 2004; 88(10): 1280-4.
[28] Inatomi T, Nakamura T, Koizumi N, *et al.* Midterm results on ocular surface reconstruction using cultivated autologous oral mucosal epithelial transplantation. Am J Ophthalmol 2006; 141(2): 267-75.
[29] Inatomi T, Nakamura T, Kojyo M, *et al.* Ocular surface reconstruction with combination of cultivated autologous oral mucosal epithelial transplantation and penetrating keratoplasty. Am J Ophthalmol 2006; 142(5): 757-64.
[30] Klein R, Peto T, Bird A, Vannewkirk MR. The epidemiology of age-related macular degeneration. Am J Ophthalmol 2004; 137(3): 486-95.
[31] Woo JH, Sanjay S, Au Eong KG. The epidemiology of age-related macular degeneration in the Indian subcontinent. Acta Ophthalmol 2009; 87(3): 262-9.
[32] Weleber R, Gregory Evans, K. Retinitis pigmentosa and allied disorders. Mosby: St. Louis; 2000.
[33] Verhoeff FH. Microscopic Observations in a Case of Retinitis Pigmentosa. Trans Am Ophthalmol Soc 1930; 28: 176-95.
[34] Klassen HJ, Ng TF, Kurimoto Y, *et al.* Multipotent retinal progenitors express developmental markers, differentiate into retinal neurons, and preserve light-mediated behavior. Invest Ophthalmol Vis Sci 2004; 45(11): 4167-73.
[35] MacLaren RE, Pearson RA, MacNeil A, *et al.* Retinal repair by transplantation of photoreceptor precursors. Nature 2006; 444(7116): 203-7.
[36] Lund RD, Wang S, Klimanskaya I, *et al.* Human embryonic stem cell-derived cells rescue visual function in dystrophic RCS rats. Cloning Stem Cells 2006; 8(3): 189-99.
[37] Takahashi M, Palmer TD, Takahashi J, Gage FH. Widespread integration and survival of adult-derived neural progenitor cells in the developing optic retina. Mol Cell Neurosci 1998; 12(6): 340-8.
[38] Young MJ, Ray J, Whiteley SJ, Klassen H, Gage FH. Neuronal differentiation and morphological integration of hippocampal progenitor cells transplanted to the retina of immature and mature dystrophic rats. Mol Cell Neurosci 2000; 16(3): 197-205.
[39] Sakaguchi DS, Janick LM, Reh TA. Basic fibroblast growth factor (FGF-2) induced transdifferentiation of retinal pigment epithelium: generation of retinal neurons and glia. Dev Dyn 1997; 209, 387-98.
[40] Yoshii C, Ueda Y, Okamoto M, Araki M. Neural retinal regeneration in the anuran amphibian Xenopus laevis post-metamorphosis: transdifferentiation of retinal pigmented epithelium regenerates the neural retina. Dev Biol 2007; 303, 45-56.
[41] Radtke ND, Seiler MJ, Aramant RB, Petry HM, Pidwell DJ. Transplantation of intact sheets of fetal neural retina with its retinal pigment epithelium in retinitis pigmentosa patients. Am J Ophthalmol 2002; 133(4): 544-50.
[42] Das T, Del Cerro M, Jalali S, *et al.* The transplantation of human fetal neuroretinal cells in advanced retinitis pigmentosa patients: results of a long-term safety study. Exp Neurol 1999; 157, 58-68.
[43] Radtke ND, Aramant RB, Petry HM, *et al.* Vision improvement in retinal degeneration patients by implantation of retina together with retinal pigment epithelium. Am J Ophthalmol 2008; 146(2): 172-82.
[44] Reh TA, Nagy T. Characterization of Rana germinal neuroepithelial cells in normal and regenerating retina. Neurosci Res Suppl 1989; 10: S151-61.
[45] Fischer AJ, Reh TA. Identification of a proliferating marginal zone of retinal progenitors in postnatal chickens. Dev Biol 2000; 220(2): 197-210.
[46] Fischer AJ, Reh TA. Growth factors induce neurogenesis in the ciliary body. Dev Biol 2003; 259(2): 225-40.
[47] Fischer AJ, Reh TA. Identification of a proliferating marginal zone of retinal progenitors in postnatal chickens. Dev Biol 2000; 220, 197-210.

[48] Ahmad I, Das AV, James J, Bhattacharya S, Zhao X. Neural stem cells in the mammalian eye: types and regulation. Semin Cell Dev Biol 2004; 15(1): 53-62.

[49] Tropepe V, Coles BL, Chiasson BJ, *et al.* Retinal stem cells in the adult mammalian eye. Science 2000; 287(5460): 2032-6.

[50] Coles BL, Angenieux B, Inoue T, *et al.* Facile isolation and the characterization of human retinal stem cells. Proc Natl Acad Sci U S A 2004; 101(44): 15772-7.

[51] Vossmerbaeumer U, Kuehl S, Kern S, *et al.* Induction of retinal pigment epithelium properties in ciliary margin progenitor cells. Clin Experiment Ophthalmol 2008; 36(4): 358-66.

[52] Zhao X, Das AV, Soto-Leon F, Ahmad I. Growth factor-responsive progenitors in the postnatal mammalian retina. Dev Dyn 2005; 232(2): 349-58.

[53] Das AV, James J, Rahnenfuhrer J, *et al.* Retinal properties and potential of the adult mammalian ciliary epithelium stem cells. Vision Res 2005; 45(13): 1653-66.

[54] Fischer AJ, Reh TA. Muller glia are a potential source of neural regeneration in the postnatal chicken retina. Nat Neurosci 2001; 4(3): 247-52.

[55] Del Debbio CB, Balasubramanian S, Parameswaran S, *et al.* Notch and Wnt signalling mediated rod photoreceptor regeneration by Muller cells in adult mammalian retina. PLoS One 2010; 5(8): e12425.

[56] Pellegrini G, Golisano O, Paterna P, *et al.* Location and clonal analysis of stem cells and their differentiated progeny in the human ocular surface. J Cell Biol 1999; 145(4): 769-82.

[57] Hemmati-Brivanlou A, Melton D. Vertebrate embryonic cells will become nerve cells unless told otherwise. Cell 1997; 88(1): 13-7.

[58] Zhao X, Das AV, Thoreson WB, *et al.* Adult corneal limbal epithelium: a model for studying neural potential of non-neural stem cells/progenitors. Dev Biol 2002; 250(2): 317-31.

[59] Seigel GM, Sun W, Salvi R, *et al.* Human corneal stem cells display functional neuronal properties. Mol Vis 2003; 9: 159-63.

[60] Zhao X, Das AV, Bhattacharya S, *et al.* Derivation of neurons with functional properties from adult limbal epithelium: implications in autologous cell therapy for photoreceptor degeneration. Stem Cells 2008; 26(4): 939-49.

[61] Ikeda H, Osakada F, Watanabe K, *et al.* Generation of Rx+/Pax6+ neural retinal precursors from embryonic stem cells. Proc Natl Acad Sci U S A 2005; 102(32): 11331-6.

[62] Lamba DA, Karl MO, Ware CB, Reh TA. Efficient generation of retinal progenitor cells from human embryonic stem cells. Proc Natl Acad Sci U S A 2006; 103(34): 12769-74.

[63] Osakada F, Ikeda H, Mandai M, *et al.* Toward the generation of rod and cone photoreceptors from mouse, monkey and human embryonic stem cells. Nat Biotechnol 2008; 26(2): 215-24.

[64] Osakada F, Jin ZB, Hirami Y, *et al. In vitro* differentiation of retinal cells from human pluripotent stem cells by small-molecule induction. J Cell Sci 2009; 122(Pt 17): 3169-79.

[65] Osakada F, Ikeda H, Sasai Y, Takahashi M. Stepwise differentiation of pluripotent stem cells into retinal cells. Nat Protoc 2009; 4(6): 811-24.

[66] Idelson M, Alper R, Obolensky A, *et al.* Directed differentiation of human embryonic stem cells into functional retinal pigment epithelium cells. Cell Stem Cell 2009; 5(4): 396-408.

[67] Meyer JS, Shearer RL, Capowski EE, *et al.* Modelling early retinal development with human embryonic and induced pluripotent stem cells. Proc Natl Acad Sci U S A 2009; 106(39): 16698-703.

[68] Lamba DA, Gust J, Reh TA. Transplantation of human embryonic stem cell-derived photoreceptors restores some visual function in Crx-deficient mice. Cell Stem Cell 2009; 4(1): 73-9.

[69] Takahashi K, Tanabe K, Ohnuki M, *et al.* Induction of pluripotent stem cells from adult human fibroblasts by defined factors. Cell 2007; 131(5): 861-72.

[70] Dimos JT, Rodolfa KT, Niakan KK, *et al.* Induced pluripotent stem cells generated f rom patients with ALS can be differentiated into motor neurons. Science 2008; 321(5893): 1218-21.

[71] Park IH, Arora N, Huo H, *et al.* Disease-specific induced pluripotent stem cells. Cell 2008; 134(5): 877-86.

[72] Hanna J, Wernig M, Markoulaki S, *et al.* Treatment of sickle cell anemia mouse model with iPS cells generated from autologous skin. Science 2007; 318(5858): 1920-3.

[73] Raya A, Rodriguez-Piza I, Guenechea G, *et al.* Disease-corrected haematopoietic progenitors from Fanconi anaemia induced pluripotent stem cells. Nature 2009; 460(7251): 53-9.

[74] Zou J, Maeder ML, Mali P, *et al.* Gene targeting of a disease-related gene in human induced pluripotent stem and embryonic stem cells. Cell Stem Cell 2009; 5(1): 97-110.

[75] Buchholz DE, Hikita ST, Rowland TJ, *et al.* Derivation of functional retinal pigmented epithelium from induced pluripotent stem cells. Stem Cells 2009; 27(10): 2427-34.

[76] Carr AJ, Vugler AA, Hikita ST, *et al.* Protective effects of human iPS-derived retinal pigment epithelium cell transplantation in the retinal dystrophic rat. PLoS One 2009; 4(12): e8152.

[77] Hirami Y, Osakada F, Takahashi K, *et al.* Generation of retinal cells from mouse and human induced pluripotent stem cells. Neurosci Lett 2009; 458(3): 126-31.

[78] Dravida S, Pal R, Khanna A, *et al.* The transdifferentiation potential of limbal fibroblast-like cells. Brain Res Dev Brain Res 2005; 160(2): 239-51.

[79] Balasubramanian S, Babai N, Chaudhuri A, *et al.* Non cell-autonomous reprogramming of adult ocular progenitors: generation of pluripotent stem cells without exogenous transcription factors. Stem Cells 2009; 27(12): 3053-62.

[80] Vierbuchen T, Ostermeier A, Pang ZP, *et al.* Direct conversion of fibroblasts to functional neurons by defined factors. Nature 2010; 463(7284): 1035-41.

[81] Ieda M, Fu JD, Delgado-Olguin P, *et al.* Direct reprogramming of fibroblasts into functional cardiomyocytes by defined factors. Cell 2010; 142(3): 375-86.

[82] Liao JL, Yu J, Huang K, *et al.* Molecular signature of primary retinal pigment epithelium and stem-cell-derived RPE cells. Hum Mol Genet 2010; 19(21): 4229-38.

[83] Lu B, Malcuit C, Wang S, *et al.* Long-term safety and function of RPE from human embryonic stem cells in preclinical models of macular degeneration. Stem Cells 2009; 27(9): 2126-35.

[84] Nishida A, Takahashi M, Tanihara H, *et al.* Incorporation and differentiation of hippocampus-derived neural stem cells transplanted in injured adult rat retina. Invest Ophthalmol Vis Sci 2000; 41(13): 4268-74.

CHAPTER 12

Micro-RNAs, Pluripotent Stem Cells and Germ Cells: Basic Science to Clinical Applications

Jinlian Hua[1*], Shun Zhang[1], L. Wang[1], H. Cao[1], H. Zhu[1], J. Sun[1], Y. Hu[1] and Kuldip S.Sidhu[2]

[1]*College of Veterinary Medicine, Shaanxi Centre of Stem Cells Engineering & Technology, Key Lab for Animal Biotechnology of Agriculture Ministry of China, Northwest A&F University and* [2]*Stem Cell Laboratory, Faculty of Medicine, University of New South Wales, Australia*

Abstract: Micro RNAs (miRNAs), recently discovered specific small molecules, have emerged as crucial regulators of diverse developmental processes, such as cell maintenance, proliferation, differentiation, apoptosis, organ formation, tumor formation and so on. They exert widespread functions as regulatory molecules mainly in post-transcriptional levels (also in translation levels). Germ cells, as a kind of unique cells, carry the responsibility for passing genetic information to the next generation. miRNAs play a crucial role in germline development including the formation of primordial germ cells (PGCs), germline stem cells (GSCs) and gametes. We have reviewed here mainly the effects of miRNAs on germ cell development and its related progress.

Keywords: Micro RNAs (miRNAs), differentiation, germ cells, primordial germ cells (PGCs), germline stem cells (GSCs), gametes.

INTRODUCTION

Pluripotent stem cells, which include embryonic stem cells (ESCs) derived from the inner cell mass of an embryo, Epiblast Stem Cells (EpiSCs), Embryonic Germ Cells (EGCs) and adult Germline Stem Cells (GSCs), have unrestricted developmental potential and thus can give rise to all cell types, containing the potential to differentiate into germ cells [1]. Recently, scientists have found a type of small RNAs (miRNAs), which play a significant role in developmental processes, have been considered to regulate the fate of pluripotent stem cells, germ cells and other stem cells. In this chapter, we mainly discuss the progress in miRNAs and the development of pluripotent stem cells and germ cells.

Differentiation of Pluripotent Stem Cells (ESCs/iPSCs) to Germ Cells

In recent years, scientists have made great progress in researching on the establishment of human ES cell lines and their differentiation [2, 3]. The primordial germ cells (PGCs) and the embryonic germ cells (EGCs) derived from them share many of their properties with ESCs [3]. ESC lines have now been derived from several stages of germ cell development and they could differentiate into functional gametes which have the ability to participate in embryonic development, even including the production of live pups after transplantation in mice [4]. Conversely, germ cells can also be derived from ESCs. It has been demonstrated that mouse and human embryonic stem cells can differentiate into PGCs and subsequently to early gametes (oocytes and sperms) and blastocysts [5-8]. Immature sperm cells derived from mESCs in culture have been reported to produce live offspring [9]. In addition, preliminary research has indicated that human embryonic stem cells (hESCs) and induced pluripotent stem cells (iPSCs) probably have the potential to differentiate into germ cells by some cytokines, co-culture with specific cells or transfecting germ cell genes [10-16]. Adult stem cells have also been reported to differentiate into mature germ cells *in vitro* [17]. Thus, stem cells may offer a valuable *in vitro* model for the investigation of germ cell development and early stages of human gametogenesis including epigenetic modifications of the germline. The following Table **1** is the progress of germ cells derivation from stem cells.

***Address correspondence to Jinlian Hua:** College of Veterinary Medicine, Shaanxi Centre of Stem Cells Engineering & Technology, Key Lab for Animal Biotechnology of Agriculture Ministry of China, Northwest A&F University, Yangling, Shaanxi 712100, China; E-mail:jinlianhua@nwsuaf.edu.cn

Kuldip S. Sidhu (Ed)

Table 1. Progress in production of male/female gametes *in vitro* from stem cells

Authors	Source	Method	End product
Male Gamete Production *In Vitro* from Stem Cells			
Yu *et al.* [18]	129/SV mESC—XY	Monolayer induction, combined with ectopic expression of Dazl for more than 5 days	motile tailed-sperms, Oocyte-like cells and ovarian follicle-like structures
Kee *et al.* [14]	hESC-XX,XY(VASA-reporter)	adherent differentiation combined with over-expression of DAZL, DAZ and BOULE for 14 days	haploid gametes
Aflatoonian *et al.* [19]	hESC	EBs formation, culture medium with BMP4, RA and neonatal mouse testes conditioned media (14d)	Round spermatid (1–5%), Follicle-like structures, but failed to detect zona pellucida in these structures
Yamauchi *et al.* [20]	Monkey ESC	EB induction, combined with RA, gonadal cell-conditioned medium and growth factors	haploid gametes
Zohreh *et al.* [21]	129/SV mESC—XY	Dissociated-EB induction, combined with MEF feeder and BMP4 (5ng/ml) for 4 days	Late-stage germ cells
West *et al.* [13]	hESC-XY	MEF feeder and basic fibroblast growth factor (bFGF) for 10, 16 and 30 days	haploid germ-like cells
Kerkis *et al.*[22]	129/SV mESC—XY	EB induction, combined with Neurobasal (NB) medium、B27、0.1 μM RA	FE-J1$^+$ cells, Follicular-like structures, presumptive oocytes and blastocysts
Lee *et al.* [23]	male germ stem cell-like cells (GSC-LC) from the testicular tissue of patients with non-obstructive azoospermia (NOA).	RA、RE、vitamin C、vitamin E、FSH、testosterone addition for up to 6 weeks	haploid male germ cells
Nayernia *et al.* [9]	SSC7,SSC12 cell lines derived from ESC	RA induction (10^{-5}), FACS selection of Stra8-EGFP$^+$ cells; transfected with the Prm1-DsRed fusion vector, RA induction for 72hr	Prm1$^+$ haploid gametes were derived and live offsprings were got by ICSI
Geijsen *et al.* [8]	mESC-XX,XY	EB formation, combined with RA induction. On the day 20, FE-J1$^+$ cells isolated by FACS and injected into oocytes by ICSI	blastocysts were derived after transplantation (0.01%)
Nayernia *et al.* [24]	mSSC1(stra8-reporter)	RA(10^{-6}) induction for 12hr, with two FACS selections by expression of Stra8-EGFP, then were transplanted into host testis for 8 weeks.	functional sperms were found in the inner cell layer of the tubules
Toyooka *et al.* [7]	129/SV mESC-XY	EB induction combined with BMP4-producing cells for 8~10 days, selected MVH-positive cells and transplanted into host testis capsule for 6~8 weeks.	functional sperms
Female Gamete Production *In Vitro* from Stem Cells			
Hua and Sidhu [24]	hESCs	Monolayer induction plus testicular extract, hormones	Oocyte-like structures
Hübner *et al.* [25]	mESC-XX, XY (Pou5f1-reporter)	Monolayer, no feeder cells, spontaneous differentiation (43d)	Oocyte-like cells (20% produced oocytes larger than 40μm), blastocyst-like structure after parthenogenic activation
Clark *et al.* [10]	hESC-XX, XY	EBs formation，spontaneous differentiation (14d)	Oocyte-like cells
Johnson *et al.* [26]	GSC in OSE	Ovarian fragments graft	Formation of ovarian follicles after graft
Johnson *et al.* [27]	GSC in female bone marrow and blood	Bone marrow and blood transplantations	Formation of ovarian follicles after graft
Bukovsky *et al.* [28]	GSC in OSE	Monolayer, culture medium with estrogenic stimuli	Oocyte-like cells and granulosa-like Cells
Dyce *et al.* [29]	fetal porcine skin stem cell	Aggregate formation, medium with follicular fluid, gonadotrophins (40d)	Oocyte-like cells (0.014%) and blastocyst-like structures
Danner *et al.* [30]	rat pancreatic stem cell	EBs formation, spontaneous differentiation (9d)	Oocyte-like cells and tissue-like structures
Lacham-Kaplan *et al.* [6]	mESC-XY	EBs formation, conditioned media from testicular cell (7d)	Ovarian structures (83%) containing putative oocytes
Novak *et al.* (2006) [30]	mESC—XY	Spontaneous differentiation in adherent cultures and through EBs formation (20d)	Ovarian follicles
Kee *et al.* [32]	hESC—XX	EBs formation, growth factors addition (7d)	Oocyte-like cells

Table 1: cont....

Qing *et al.* [33]	mESC—XY	EBs formation, co-cultured with ovarian granulose cells (10d)	Oocyte-like cells
Chen *et al.* [34]	hESC—XX	Spontaneous differentiation in adherent cultures and through EBs formation (21d)	Oocyte-like cells
Salvador *et al.* [35]	mESC—XX (Gdf9-reporter)	EBs formation, culture medium with LIF (45d)	Oocyte-like cells but no follicular structures
Linher *et al.* [36]	porcine skin stem cells	Monolayer, medium with porcine follicular fluid (50d)	Oocyte-like cells (1.4%)
Nicholas *et al.* [37]	mESC-XX(oct4-GFP and stra8-GFP reporter)	EBs formation, culture in media containing a germ cell maturation factor cocktail in the absence of feeder layer support (19d)	Meiotic germ cells, and primary ovarian follicle formation following ovarian niche transplantation under the kidney capsule (21d) (0.023%)
Lavagnolli *et al.* [38]	hybrid cells (ES-SCH): fusion of ESCs and splenocytes from adult female mice	EBs formation, culture medium with RA (8d)	Enlarged cells resembling PGC on the EB periphery

miRNAs and their Functions

miRNAs are a type of small RNAs, which consist of 18-25 nucleotides that originate from endogenous hairpin transcripts called primary miRNAs (pri-miRNA). The pri-miRNAs are processed by the RNase (Drosha) in the nucleus before being exported as ~70-nucleotide pre-miRNAs. In the cytosol, pre-miRNAs are further processed by another RNase (Dicer) into mature double-stranded miRNAs [39]. One of the miRNA duplex strands serves as a guide strand and is incorporated into the Argonaute-containing RISC complex, then bind to the 3' untranslated region (UTR) of the target mRNAs and regulate their translation[40], while the other strand is released and degraded. The eight-nucleotide "seed sequence" in the 5′ terminus of the miRNA guide strand, is considered to be the most critical region for selecting targets[41]. Based on the degree to which the seed region complements the mRNA target, the mRNA transcript is either degraded (with high complementation) or translationally repressed (see Fig. **1**).

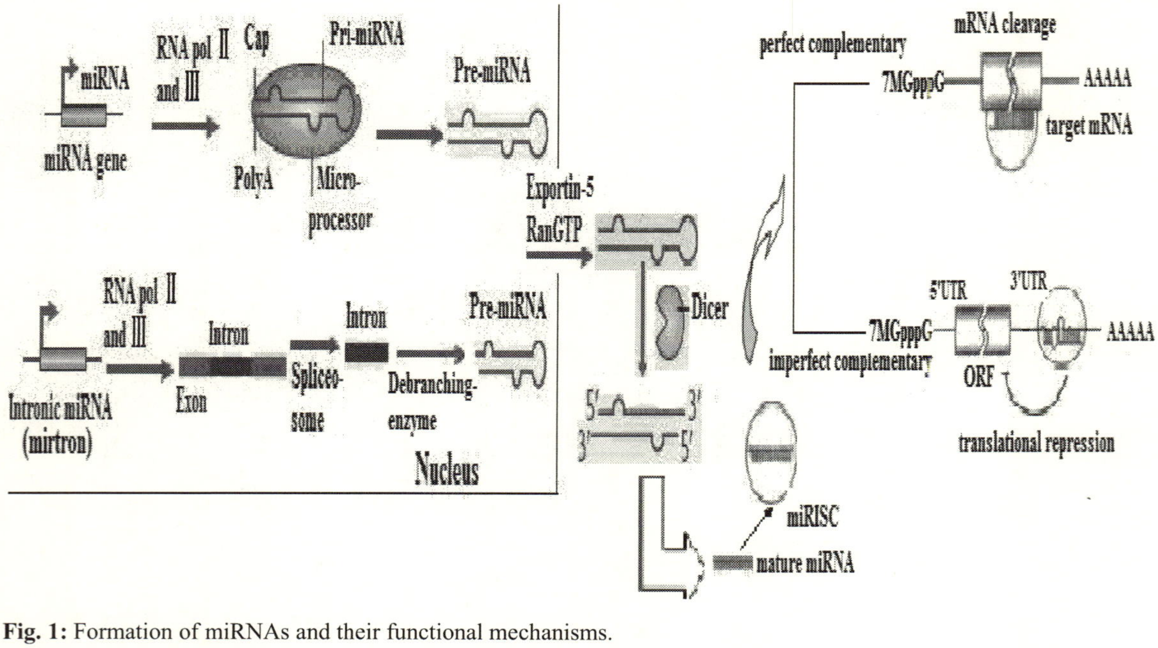

Fig. 1: Formation of miRNAs and their functional mechanisms.

These miRNAs are expressed in a time- and tissue-specific manner, and show a great deal of conservation among a diverse range of species [42]. miRNAs are involved in almost every biological process, including developmental timing, cell differentiation, cell proliferation, cell death, metabolic control, transposon silencing and antiviral defence, and so on (Fig. **2**).

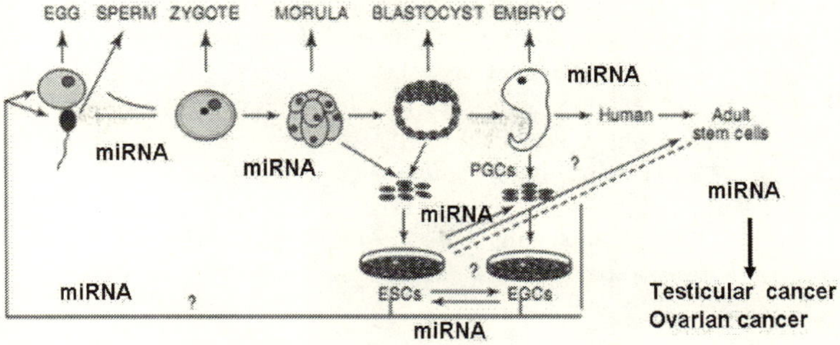

Fig. 2: The roles of miRNAs in animal development and germ cell formation (Adapted from [43]).

miRNAs in Pluripotent Stem Cells

In the past, several genes (including transcription factors such as Oct4, Nanog and Sox2) were cloned and thought to be essential players in stem cell self-renewal and differentiation. However, the exact molecular mechanism still remains a mystery. Recent studies indicate that the expression profiles of miRNAs in stem cells are different from other tissues, this phenomenon plus other emerging evidences suggest that miRNAs may play an essential role in stem cell self-renewal and differentiation.

miRNAs play a key role in the maintenance and differentiation of pluripotent ES cells. As demonstrated by two independent studies [44, 45], Dicer-deficient ES cells, which lack microprocessor activity and hence all canonical miRNAs, exhibited a reduced proliferation rate and an altered cell-cycle profile, with a slight increase in G1 and G0 cells and a corresponding decrease of cells in G2 and M [44, 45]. Furthermore, *in vitro*, when Dicer-null ES cells were subjected to conditions that normally induce ES cell differentiation, they exhibited abnormal activation of multiple markers of differentiation, coupled with the inability to silence expression of pluripotency markers [47]; *in vivo*, Dicer-null ES cells did not contribute to the generation of chimeric mice when injected into blastocysts, and failed to generate teratomas upon subcutaneous injection into nude mice. These results strongly indicated that endogenous miRNAs produced by Dicer are essential for maintaining the normal self-renewal and differentiation of ES cells.

Studies found that there are some ES cell–specific miRNAs, some of which could regulate cell cycle and called the ES cell–specific cell cycle–regulating (ESCC) miRNAs, such as the miR-290, -302 families, they are highly expressed in mouse ES cells [48]. In the developing mouse zygote, the expressions of the miR-290-295 cluster are among the first embryonic miRNAs to be detected [50]. By further study, we found that the miR-290 family suppresses multiple inhibitors of the G1/S transition (p21、Rbl2、Lats2) to enable a short G1 phase in ES cells (see Fig. **3**). Upon differentiation, expression of the miR-290 cluster is rapidly down-regulated, coincident with the elongation of the cell cycle [40, 48, 50].

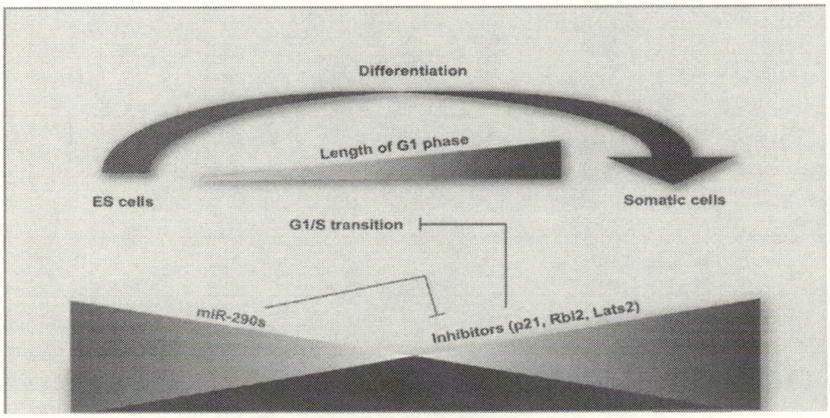

Fig. 3: The mechanism of ESC cell cycling regulation by miR-290s (From [50]).

A second cluster of ES cell-specific miRNAs, the miR-302-367 cluster, is also highly expressed in ES cells with no noticeable expression in somatic cells [51]. miR-302-367 cluster regulates cellular growth, metabolism and transcription. This cluster also has several predicted targets associated with chromatin modification[52]. Both the miR-290-295 and miR-302-367 clusters appear to have regulatory functions that have a direct impact on the pluripotency potential of ES cells. Tightly controlled regulation of the cell cycle, transcription, and epigenetic modification are essential for the maintenance of the pluripotency and proliferation observed in ES cells.

Introduction of Oct4, Sox2 and Klf4 plus miR-291-3p, miR-294 or miR-295 consistently increased the number of Oct4-GFP$^+$ colonies compared with controls transduced with Oct4, Sox2 and Klf4 plus transfection reagent [53].

Since 2006, Yamanaka *et al.* discovered that somatic cells could be reprogrammed into iPS cells [54], the original combination of reprogramming factors (Sox2, Oct4, Klf4, and c-Myc) has been used to generate iPS cells from a wide-range of cell types. Other combinations have also been used to generate iPS cells. However, relatively little is known about the mechanism of cell reprogramming, especially the roles played by miRNAs.

Judson and co-workers infected mouse embryonic fibroblasts with retroviruses that express Sox2, Oct4 and Klf4, and ESC-specific miRNAs by transient transfection, and found that ESC miRNAs increase the generation of mouse iPS cells induced by the combination of Sox2, Oct4, and Klf4. Among different ESC miRNAs, miR-294 exhibited the greatest effect on reprogramming and increased efficiency of iPS cell generation from 0.01–0.05% to 0.4–0.7% [53]. However, when miR-294 was introduced with Sox2, Oct4, Klf4, and c-Myc, it had no effect on reprogramming. Therefore, ESC miRNAs appear to promote Sox2, Oct4, and Klf4 mediated reprogramming by substituting for c-Myc. Studies showed that, miRNAs from the miR-302 cluster have also been shown to promote reprogramming.

Recently, two independent studies have demonstrated that expression of the imprinted Dlk1–Dio3 gene cluster, which codes for ~50 conserved miRNAs, is often silenced in iPS cells [55, 56]. Moreover, treatment of iPS cells with HDAC inhibitors led to activation of the Dlk1–Dio3 miRNA cluster and the generation of iPS cells whose developmental potential appears to be equivalent to that of ES cells [56]. These findings highlight the importance of achieving the appropriate epigenetic status of cells during reprogramming, and further reinforce the critical roles played by miRNAs in the establishment and maintenance of pluripotency.

The following figure shows the interaction of ESC-related miRNAs and its target genes (Fig. **4**).

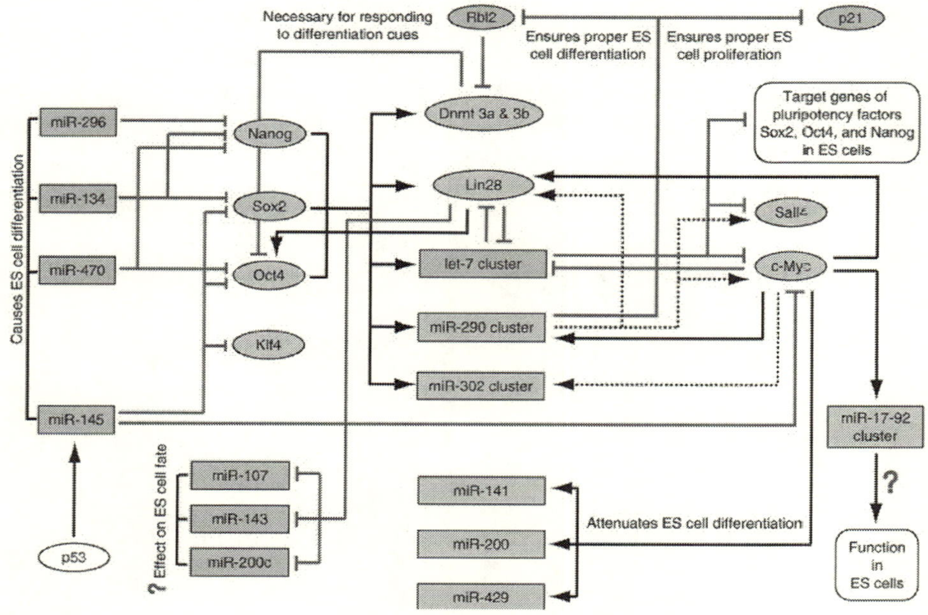

Fig. 4: Interaction between ESC-related miRNAs and its target genes (From [57]).

miRNAs in Germ Cells Derived from Pluripotent Stem Cells

A lot of studies showed that miRNA also plays an important part in germ cell differentiation. To establish techniques reliable for genome-wide miRNA profiling, miRNAs were first cloned from the testis only a few years ago [58, 59]. Shortly afterwards, two studies reported the miRNA profiles of the murine testis using either microarrays [60] or cloning analysis [61], and found some highly-expressed miRNAs in testis. For example, the expression of Mir-34b is much higher in adult testis compared to the pre-puberal testis [58], indicating that Mir-34b plays a potential role in the differentiation of male germ cells. In newborn and adult mouse ovaries, the let-7 cluster of miRNAs was highly expressed, but levels in specific cell types were not clearly examined [62]. This association makes let-7 miRNAs likely to be candidates for involvement in the differentiation of PGCs into more mature gametes.

By microarray and real-time PCR, studies found that miR-17-92, 290-295 clusters were highly expressed in mouse PGCs and spermatogonia. Expression of 4 miRNAs (miR-141, -200a, -200c and -323) decreased gradually in developing PGCs, The reduction was indistinguishable between male and female germ cells; By contrast, expression levels of several miRNAs belonging to let-7 miRNA family (let-7a, d, e, f and g), as well as miR-125a and miR-9 increased in male PGCs but not in female PGCs. Dicer-deleted PGCs and spermatogonia exhibited poor proliferation, which showed that these miRNAs play an important part in PGC development [62]. Transcripts of these clusters were also observed preferentially in ES cells, but notably, they decreased in differentiating cells[64], suggesting the role of the clusters in pluripotency of ES cells. Taking into account the fact that PGCs have the potential to give rise to pluripotent embryonic germ cells (EGCs), it is likely that the expression of the miRNA clusters in PGCs may underlie the plasticity of PGCs to dedifferentiate into pluripotent EGCs.

Since Dicer and Ago2, which play an important part in miRNA maturation, are localized in the chromatoid body, which is a characteristic fibrous structure in the cytoplasm of spermatocyte and spermatid [65], it indicates that miRNAs play a role in late spermatogenesis.

Furthermore, elongating spermatids exhibit abnormal morphology and motility, and consequently, male infertility occurs in the Dicer1-knockout mice, indicating that both Dicer1 and miRNAs play crucial roles in proper differentiation during spermatogenesis [66]. Together, these studies further indicate that normal miRNA biogenesis is required for proper spermatogenesis.

Some specific miRNAs and its' probable role during spermatogenesis are listed as follows (Table **1**)

Table 2. Some specific miRNAs during the development of germ cells.

miRNA	Species	Sex	Function	Target Gene	Reference
miR-17-92 cluster; miR-290-295 cluster	mouse	Male/female	promoting the survival and/or the proliferation of PGCs	/	Hayashi K *et al.* [63]
let-7 miRNA family; miR-125a; miR-9	mouse	male	Male PGC differentiation	suppression of Ras/MAPK pathway	Hayashi K *et al.* [63]
miR-t25; miR-t27	mouse	male	Highly and specifically expressed in spermatocytes and spermatids	/	Ro *et al.* [61]
miR-296; miR-470; miR-134	mouse	male	Lower pluripotency and promote differenciation	CDSs of Nanog, Oct4 and Sox2	Tay *et al.* [67]
miR-34c	mouse	male	Highly and specifically expressed in testis especially in late stage	/	Bouhallier *et al.* [68]
miR-122a	mouse	male	Highly expressed in postmeiotic germ cells	Transition Protein 2 (Tnp2)	Yu *et al.* [69]

Table 2: cont....

miR-17-5p	human	male	Expressed greatly in later stages of germ cell maturation, especially in pachytene spermatocytes	E2F1	Novotny *et al.* [70]
miR-372;miR-373	human	male	Oncogenes participating in the development of human testicular germ cell tumors	LATS2	Voorhoeve *et al.* [71]
miR-430	zebrafish	male	PGC development and somatic cell differentiation	nanos1、TDRD7(both are PGC-specific)	Mishima *et al.* [72]

*:/, no detailed.

Apart from miRNAs, another type of small RNAs has been isolated from germ cells. These are Piwi-interacting RNA (piRNA) [74], or germline small RNA (gsRNA) [74] of approximately 26–30 bp, which are specifically expressed in the testis. They are highly diverse in sequence, and many of them are derived from non-repetitive sequences, especially in mice. piRNAs interact with mouse piwi-family proteins, such as Miwi, Miwi2 and Mili. These piwi-family proteins play an essential role in spermatogenesis. Spermatogenesis was arrested at the spermatid stage in Miwi-deficient mice [75], and as spermatocytes in Miwi2- or Mili-deficient mice [76, 77]. A further research proved that during spermatogenesis, piwi-family proteins are associated with a germ cell-specific perinuclear granule called nuage or chromatoid body [65].

All previous evidences offer such a conclusion that miRNAs and piRNAs play a very role during every stage of spermatogenesis, from PGC migration, proliferation and differentiate into sperms and oocytes. By transfecting germ cell-specific miRNAs, maybe we could promote the efficiency of gamete culture *in vitro* and improve the induction efficiency of ES cells to functional gametes and apply it to human reproduction diseases.

CONCLUSIONS AND PROSPECTS

Tens of thousands of small RNAs have been discovered, and more small RNAs are expected to be discovered owing to rapidly expanding sequencing power. miRNAs and their varied effects on ES cells and germ cells have provided a far better understanding of the molecular mechanisms that regulate the complex gene regulatory networks which control the proliferation and differentiation of pluripotent stem cells.

Obviously, this is an exciting era in RNA biology, but at the same time we are facing multiple challenges. Such as, how many small RNAs finally existed in organisms? What a crucial part they play in species evolution? How the expression of these miRNAs is regulated? What are the cis-elements (promoters and enhancers), the transcription factors, and the epigenetic regulators that sustain extremely high levels of expression of miRNAs during early development and are then silenced in differentiated somatic cells? Recent studies have begun to gain insights into the transcriptional regulation of these miRNAs, at least in ES cells.

The discovery of miRNAs also has inevitably its impact on the field of reproduction. Reproductive biology is slowly entering into a new era, which demands a thorough revision of the current theories, so we need to complement them with the most recent findings from RNA biology. Novel essential roles in the male reproductive function have been revealed by the Dicer-dependent pathway. It is now clear that a global loss of miRNAs, whether in the supporting or having detrimental effects on male fertility, should be taken into consideration when discussing about treatment of male infertility or –on the opposite-, male contraception. Surely, more detailed analysis at the molecular level are required to identify individual miRNAs and its key roles during spermatogenesis. Luckily, the need to generate knockouts for individual miRNAs now seems feasible, this will allow us to identify their target genes and thus understand in depth how the process of spermatogenesis is controlled by the miRNA machinery. Certainly, this means that we would face a lot of challenges: such as, there are just a little miRNAs in different stages of germ cells (PGCs, GSCs, spermatocytes and sperms), and isolation of these germ cells itself is not a easy task; in addition, the heterogeneity of different cells in development greatly hinders our experimental outcomes. so we should focus on single germ cell, combined with miRNAs over-expression or knock down technology to reveal the molecular mechanism of spermatogenesis regulated by miRNAs during further research.

It's believed that, with some modern biological technologies such as microarray analysis, laser capture microscopy, confocal microscopy, we would find more specific germ cell miRNAs and further understand the regulatory mechanisms during the differentiation of spermatogenesis. Combined with RNAi technology, we could apply these technologies to infertility, treatment of germ cell tumor, and male contraception may also be achieved by injecting directly into the testis synthetic siRNAs or shRNAs to knock down the expression of genes that are required for normal spermatogenesis.

REFERENCES

[1] Hanna JH, Saha K and Jaenisch Rudolf. Pluripotency and cellular reprogramming: facts, hypotheses, unresolved issues. Cell, 2010; 143(4): 508-524.

[2] Thomson JA, Itskovitz-Eldore J, Shapiro SS *et al.* Embryonic stem cell line from human blastocysts. Science 1998; 282(5391): 1145-1147.

[3] Shamblott MJ, Axelman J, Wang SP, *et al.* Derivation of pluripotent stem cells from cultured human primordial germ cells. Proc Natl Acad Sci USA,1998; 95(23): 13726-13731.

[4] Clark AT. Establishment and differentiation of human embryonic stem cell derived germ cells. Soc Reprod Fertil Suppl. 2007; 63: 77-86.

[5] Nagano McCa. *In vitro* gamete derivation from pluripotent stem cells: progress and perspective. Biol Reprod 2007; 76(4): 546-551.

[6] Lacham-Kaplan O, Chy H and Trounson A. Testicular cell conditioned medium supports differentiation of embryonic stem cells into ovarian structures containing oocytes. Stem Cells 2006; 24(2): 266-273.

[7] Toyooka Y, Tsunekawa N, Akasu R, *et al.* Embryonic stem cells can form germ cells *in vitro*. Proc. Natl. Acad. Sci. USA. 2003; 100(20): 11457-11462.

[8] Geijsen N, Horoschak M, Kim K, Gribnau J, Eggan K and Daley GQ. Derivation of embryonic germ cells and male gametes from embryonic stem cells. Nature 2004; 427(6970): 148-154.

[9] Nayernia K, Nolte J, Michelmann HW, *et al. In vitro*-differentiated embryonic stem cells give rise to male gametes that can generate offspring mice. Dev. Cell 2006; 11(1): 125-132.

[10] Clark AT, Bodnar MS, Fox M, *et al.* Spontaneous differentiation of germ cells from human embryonic stem cells *in vitro*. Hum Mol Genet 2004; 13(7): 727-739.

[11] Park TS, Galic Z, Conway AE, *et al.* Derivation of primordial germ cells from human embryonic and induced pluripotent stem cells is significantly improved by coculture with human fetal gonadal cells. Stem Cells. 2009; 27(4): 783-795.

[12] Tilgner K, Atkinson S, Golebiewska A, Stojkovic, M, Lako M and Armstrong L. Isolation of primordial germ cells from differentiating human embryonic stem cells. Stem Cells 2008; 26(12): 3075-3085.

[13] West F, Machacek D, Boyd N, Pandiyan K, Robbins K and Stice S. Enrichment and differentiation of human germ-like cells mediated by feeder cells and basic fibroblast growth factor signalling. Stem Cells 2008; 26(11): 2768-2776.

[14] Kee K, Angeles V, Flores M, Nguyen H and Reijo Pera RA. Human DAZL, DAZ and BOULE genes modulate primordial germ cell and haploid gamete formation. Nature 2009; 462(7270): 222-225.

[15] Bucay N, Yebra M, Cirulli V, *et al.* A novel approach for the derivation of putative primordial germ cells and Sertoli cells from human embryonic stem cells. Stem Cells 2009; 27(1): 68-77.

[16] Panula S, Medrano JV, Kee K, *et al.* Human germ cell differentiation from fetal- and adult-derived induced pluripotent stem cells. Hum Mol Genet 2010; 20(4): 752-62.

[17] Dyce P, Shen W, Huynh E, *et al.* Analysis of oocyte-like cells differentiated from porcine fetal skin-derived stem cells. Stem Cells Dev 2010; 20(5): 809-19.

[18] Yu Z, Ji P, Cao J. Dazl promotes germ cell differentiation from embryonic stem cells. J Mol Cell Biol. 2009; 1(2): 93-103.

[19] Aflatoonian B, Ruban L, Jones M, Aflatoonian R, Fazeli A, and Moore HD. *In vitro* post-meiotic germ cell development from human embryonic stem cells. Human Reproduction 2009; 24(12): 3150-3159.

[20] Yamauchi K, Hasegawa K, Chuma S, Nakatsuji N, Suemori H. *In vitro* germ cell differentiation from cynomolgus monkey embryonic stem cells. PLoS One. 2009; 4(4): e5338.

[21] Zohreh M, Mansoureh M, Mehdi FM. Effects of Treatment with Bone Morphogenetic Protein 4 and Co-culture on Expression of Piwil2 Gene in Mouse Differentiated Embryonic Stem Cells. Int J Fertil Steril 2009; 3: 78-83.

[22] Kerkis A, Fonseca SAS, Serafim RC, *et al. In vitro* differentiation of male mouse embryonic stem cells into both presumptive sperm cells and oocytes. Cloning and Stem Cells 2007; 9(4): 535-548.

[23] Lee DR, Kim KS, Yang YH, *et al.* Isolation of male germ stem cell-like cells from testicular tissue of non-obstructive azoospermic patients and differentiation into haploid male germ cells *in vitro*. Hum Reprod. 2006; 21(12): 471-476.

[24] Nayernia K, Li M, Jaroszynski L, *et al.* Stem cell based therapeutical approach of male infertility by teratocarcinoma derived germ cells. Hum Mol Genet. 2004; 13(14): 1451-60.

[25] Hua J and Sidhu KS Coaxing hESC to form oocyte-like structures by co-culture with testicular extract and hormones. Open Stem Cell J Suppl 2011; 3: 34-45

[26] Hübner K, Fuhrmann G, Christenson LK, *et al.* Derivation of oocytes from mouse embryonic stem cells. Science 2003; 300(5623): 1251-1256.

[27] Johnson J, Canning J, Kaneko T, *et al.* Germline stem cells and follicular renewal in the postnatal mammalian ovary. Nature 2004; 428(6979): 145-150.

[28] Johnson, J, Bagley J, Skaznik-Wikiel M, *et al.* Oocyte generation in adult mammalian ovaries by putative germ cells in bone marrow and peripheral blood. Cell 2005; 122(2): 303-315.

[29] Bukovsky A, Svetlikova M, Caudle MR. Oogenesis in cultures derived from adult human ovaries. Reprod Biol Endocrinol 2005; 3: 1-13.

[30] Dyce PW, Wen L, Li J. *In vitro* germline potential of stem cells derived from fetal porcine skin. Nat Cell Biol 2006; 8(4): 384-390.

[31] Danner S, Kajahn J., Geismann C, Klink E and Kruse C. Derivation of oocyte-like cells from a clonal pancreatic stem cell line. Mol Human Reprod 2007; 13(1): 11-20.

[32] Novak I, Lightfoot DA, Wang H, *et al.* Mouse embryonic stem cells form follicle-like ovarian structures but do not progress through meiosis.Stem Cells 2006; 24(8): 1931-1936.

[33] Kee K, Gonsalves JM, Clark AT, and Pera RAR. Bone morphogenetic proteins induce germ cell differentiation from human embryonic stem cells. Stem Cells Dev 2006; 15(6): 831-837.

[34] Qing TT, Shi Y, Qin H, *et al.* Induction of oocyte-like cells from mouse embryonic stem cells by co-culture with ovarian granulosa cells. Differentiation 2007; 75(10): 902-911.

[35] Chen HF, Kuo HC, Chien C-L, *et al.* Derivation, characterization and differentiation of human embryonic stem cells: comparing serum-containing versus serum-free media and evidence of germ cell differentiation. Human Reprod 2007; 22(2): 567-577.

[36] Salvador LM, Silva CP, Kostetskii I, Radice GL, and Strauss JF. The promoter of the oocyte-specific gene, Gdf9, is active in population of cultured mouse embryonic stem cells with an oocyte-like phenotype. Methods 2008; 45(2): 172-181.

[37] Linher K, Dyce P, and Li JL. Primordial germ cell-like cells differentiated *in vitro* from skin-derived stem cells. Plos One 2009; 4(12): e8263.

[38] Nicholas CR, Haston KM, Grewall AK, Longacre TA, Pera RAR. Transplantation directs oocyte maturation from embryonic stem cells and provides a therapeutic strategy for female infertility. Hum Mol Genet 2009; 18(22): 4376-4389.

[39] Lavagnolli TMC, Fonseca SAS, Serafim RC, *et al.* Presumptive germ cells derived from mouse pluripotent somatic cell hybrids. Differentiation 2009; 78(2-3): 124-130.

[40] van den Berg A, Mols J and Han J RISC-target interaction: cleavage and translational suppression. Biochim Biophys Acta 2008; 1779(11): 668-677.

[41] Smalheiser N.R. Regulation of mammalian microRNA processing and function by cellular signalling and subcellular localization. Biochim Bioph Acta 2008; 1779(11): 678-681.

[42] Singh SK, Pal Bhadra M, Girschick HJ and Bhadra U. MicroRNAs--micro in size but macro in function. FEBS J 2008; 275(20): 4929-4944.

[43] Niwa R and Slack FJ. The evolution of animal microRNA function. Curr Opin Genet Dev 2007; 17(2): 145-150.

[44] Hua J, Sidhu K. Recent advances in the derivation of germ cells from the embryonic stem cells. Stem Cells and Development 2008; 17(3): 399-411.

[45] Bernstein E, Kim SY, Carmell MA, Czech MP, Imbalzano AN, Jones SN. Dicer is essential for mouse development. Nat Genet 2003; 35(3): 215-217.

[46] Murchison EP, Partridge JF, Tam OH, Cheloufi S, Hannon GJ. Characterization of Dicer-deficient murine embryonic stem cells. Proc Natl Acad Sci USA 2005; 102(34): 12135-12140.

[47] Kanellopoulou C, Muljo S, Kung A, *et al.* Dicer-deficient mouse embryonic stem cells are defective in differentiation and centromeric silencing. Genes Dev 2006; 19(4): 489-501.

[48] Wang Y, Medvid R, Melton C, Jaenisch R, Blelloch R. DGCR8 is essential for microRNA biogenesis and silencing of embryonic stemcell self-renewal. Nat Genet 2007; 39(3): 380-385.

[49] Wang Y, Baskerville S, Shenoy A, Babiarz JE, Baehner L, Blelloch R. Embryonic stem cell-specific microRNAs regulate the G1-S transition and promote rapid proliferation. Nat Genet 2008; 40(12): 1478-1483.

[50] Tang F, Hajkova P, Barton SC, Lao K, Surani MA. MicroRNA expression profiling of single whole embryonic stem cells. Nucleic Acids Res 2006; 34(2): e9.

[51] Wang Y, Blelloch R. Cell cycle regulation by MicroRNAs in embryonic stem cells. Cancer Res 2009; 69(10): 4093-4096.

[52] Chen C, Ridzon D, Lee CT, Blake J, Sun Y and Strauss WM. Defining embryonic stem cell identity using differentiation- related microRNAs and their potential targets. Mamm. Genome 2007; 18(5): 316-327.

[53] Ren J, Jin P, Wang E, Marincola FM and Stroncek DF. MicroRNA and gene expression patterns in the differentiation of human embryonic stem cells. J Transl Med 2009; 7: 20.

[54] Judson RL, Babiarz JE, Venere M, Blelloch R. Embryonic stem cell-specific microRNAs promote induced pluripotency. Nat Biotechnol 2009; 27(5): 459-461.

[55] Takahashi, K., Yamanaka, S. Induction of pluripotent stem cells from mouse embryonic and adult fibroblast cultures by defined factors. Cell 2006; 126(4): 663-676.

[56] Liu L, Luo GZ, Yang W, *et al.* Activation of the imprinted Dlk1-Dio3 region correlates with pluripotency levels of mouse stem cells. J Biol Chem 2010; 285(25): 19483-19490.

[57] Stadtfeld M, Apostolou E, Akutsu H, *et al.* Aberrant silencing of imprinted genes on chromosome 12qF1 in mouse induced pluripotent stem cells. Nature 2010; 465(7295): 175-181.

[58] Mallanna SK, Rizzino A. Emerging roles of microRNAs in the control of embryonic stem cells and the generation of induced pluripotent stem cells. Dev Biol 2010; 344(1): 16-25.

[59] Barad O, Meiri E, Avniel A, *et al.* MicroRNA expression detected by oligonucleotide microarrays: system establishment and expression profiling in human tissues. Genome Research 2004; 14(12): 2486-2494.

[60] Liu CG, Calin GA, Meloon B, *et al.* An oligonucleotide microchip for genome-wide microRNA profiling in human and mouse tissues. Proc Natl Acad Sci USA 2004; 101(26): 9740-9744.

[61] Yan N, Lu Y, Sun H, Tao D, Zhang S, Liu W, Ma Y. A microarray for microRNA profiling in mouse testis tissues. Reproduction 2007; 134(1): 73-79.

[62] Ro S, Park C, Sanders KM, McCarrey JR, Yan W. Cloning and expression profiling of testis-expressed microRNAs. Dev Biol 2007; 311(2): 592-602.

[63] Zhao H. and Rajkovic A. MicroRNAs and mammalian ovarian development. Semin Reprod Med 2008; 26(6): 461-468.

[64] Hayashi K, Chuva de Sousa Lopes SM, Kaneda M, *et al.* MicroRNA biogenesis is required for mouse primordial germ cell development and spermatogenesis. PLoS One, 2008; 3(3): e1738.

[65] Houbaviy, H.B., Murray, M.F., and Sharp, P.A. Embryonic stem cell-specific MicroRNAs. Dev Cell 2003; 5(2): 351-358.

[66] Kotaja N, Bhattacharyya SN, Jaskiewicz L, *et al.* The chromatoid body of male germ cells: similarity with processing bodies and presence of Dicer and microRNA pathway components. Proc Natl Acad Sci USA 2006; 103(8): 2647-2652.

[67] Maatouk DM, Loveland KL, McManus MT, Moore K & Harfe BD. Dicer1 is required for differentiation of the mouse male germline. Biology of Reproduction 2008,79(4): 696-703.

[68] Tay Y, Zhang J, Thomson AM, Lim B, Rigoutsos I. MicroRNAs to Nanog, Oct4 and Sox2 coding regions modulate embryonic stem cell differentiation. Nature 2008; 455(7216): 1124-1128.

[69] Bouhallier F, Allioli N, Lavial F, *et al.* Role of miR-34c microRNA in the late steps of spermatogenesis. RNA 2010,16(4): 720-731.

[70] Yu Z, Raabe T, Hecht NB. MicroRNA Mirn122a reduces expression of the posttranscriptionally regulated germ cell transition protein 2 (Tnp2) messenger RNA (mRNA) by mRNA cleavage. Biol Reprod 2005; 73(3): 427-433

[71] Novotny GW, Sonne SB, Nielsen JE, *et al.* Translational repression of E2F1 mRNA in carcinoma in situ and normal testis correlates with expression of the miR-17-92 cluster. Cell Death Differ 2007, 14(4): 879-882.
[72] Voorhoeve PM, le Sage C, Schrier M, *et al.* A genetic screen implicates miRNA-372 and miRNA-373 as oncogenes in testicular germ cell tumors. Cell 2006, 124(6): 1 169-1181.
[73] Mishima Y, Giraldez AJ, Takeda Y, *et al.* Differential regulation of germline mRNAs in soma and germ cells by zebrafish miR-430. Curr Biol 2006, 16(21): 2135-2142.
[74] Aravin A, Gaidatzis D, Pfeffer S, *et al.* A novel class of small RNAs bind to MILI protein in mouse testes. Nature 2006; 44297099): 203-207.
[75] Watanabe T, Takeda A, Tsukiyama T, *et al.* Identification and characterization of two novel classes of small RNAs in the mouse germline: retrotransposon-derived siRNAs in oocytes and germline small RNAs in testes. Genes Dev 2006, 20(13): 1732-1743.
[76] Deng W, Lin H. Miwi, a murine homolog of piwi, encodes a cytoplasmic protein essential for spermatogenesis. Dev Cell 2002; 2(6): 819-830.
[77] Carmell MA, Girard A, van de Kant HJ, *et al.* MIWI2 Is Essential for Spermatogenesis and Repression of Transposons in the Mouse Male Germline. Dev Cell 2007; 12(4): 503-514.
[78] Kuramochi-Miyagawa S, Kimura T, Ijiri TW, *et al.* Mili, a mammalian member of piwi family gene, is essential for spermatogenesis. Development 2004; 131(4): 839-849.

CHAPTER 13

Stem Cell Technologies and their Application in Modern Toxicology-Current State of the Art

Martina Klarić, Roberto E.-Waser, Kinga Vojnits and Susanne B.-Hoffmann*

Systems Toxicology unit/ ECVAM (European Centre for the Validation of Alternative Methods), T.P. 580, IHCP, JRC, European Commission, via E. Fermi 2749, 21027 Ispra (VA), Italy

Abstract: Chemicals produced by various industrial sectors such as drugs, cosmetics, agrochemicals and industrial chemicals are obliged to undergo a rigorous safety assessment before their entry into the market. Today the standard information requirements providing the necessary details on the toxicity of the individual compounds are mainly derived from tests performed in laboratory animals. The principal drawbacks of such *in vivo* tests are their length, cost, the remaining uncertainty for humans and the associated ethical concerns. Therefore there is a strong desire of toxicologists to compile the increasing scientific knowledge as well as emerging technologies and methodologies in a framework that drives safety assessments towards a mechanistic understanding of chemically induced toxicity.

An integer part of such modern toxicology are human based *in vitro* tests that are designed to allow a high throughput of substances and to specifically elucidate relevant toxicological mechanisms. The most promising source of human cells are stem cells since these cells can be propagated *in vitro* indefinitely, provide all toxicologically relevant target cells and they do not exhibit unwanted characteristics of cell lines with a carcinogenic origin.

However, the establishment of *in vitro* toxicity tests based on human pluripotent stem cells is challenging since differentiation protocols are often unstable, the cell cultures are not pure and the differentiation is not leading to fully matured cells that are toxicologically relevant. A first step in test development to tackle these challenges is the establishment of high quality standards which will support the reproducibility of test results within a laboratory set up but also between laboratories.

In the last years, many progress have been made to develop test systems particularly in areas of toxicity where human cellular models are not available but strong interspecies variations exists such as developmental toxicity, cardiotoxicity and hepatotoxicity. However, no pluripotent stem cell- based toxicity test has been formally validated yet.

Keywords: Human pluripotent stem cells, toxicity testing, *in vitro* tests, safety assessment.

INTRODUCTION

Every year hundreds of new chemicals need to be assessed for their safety before they can be released on the market. A variety of regulatory frameworks have been established to ensure that new products will not cause harm to consumers. For more than 50 years now regulatory defined standard information requirements have been obtained by toxicity tests that rely largely on experiments in animals [1]. However, since the mid-eighties, ethical concerns and the growing knowledge on *in vitro* and *in silico* testing created a clear tendency to replace those tests with alternative methods [2]. Furthermore, there is an increasing concern on the relevance of animal based data for certain toxicological assessments due to reports demonstrating a high level of interspecies variations compromising the level of safety for the users.

***Address correspondence to Susanne B.-Hoffmann:** *In Vitro* Methods Unit (ECVAM), Institute for Health & Consumer, Protection Joint Research Centre, European Commission *via* E. Fermi, Ispra (VA) Italy; E-mail: susanne.bremer@jrc.ec.europa.eu

Kuldip S. Sidhu (Ed)

The adoption of the 7th Amendment of the Cosmetic Directive paved further the way towards an animal free, mechanistically based safety assessment by requesting a phase out of animal testing for cosmetic substances. It was the first legislation establishing a market ban for cosmetic products that have been tested on animals [3]. As such the 7th Amendment supported the need for a paradigm shift towards a new toxicology that assesses and regulates substances based on their mode of action without using laboratory animals. It can be expected that this legislation will also introduce a breakthrough for new toxicological assessment schemes in other EU chemical legislations such as: REACH [4], CLP [5], the new regulation on plant protection products [6], the legal instrument on biocides [7] (currently under revision) and workers' protection legislation [8], once it can be demonstrated that safety evaluation without using animals can be performed. This new vision was presented by the National Research Council in 2007 and suggested chemical safety assessments that are based on the perturbation of 'toxicity' pathways [9]. The knowledge of cellular response pathways that are significantly perturbed by chemicals (defined as toxicity pathways) is the key element in this new paradigm of toxicity evaluations. Emerging methodologies and technologies in areas of computational biology, high- throughput methods, and stem cell based methods will support the implementation of such a vision. However, initial research activities are necessary to establish such a new framework of safety assessments. The knowledge on the cellular toxicity pathways has to be gathered and compiled in a searchable database, the extent of perturbation that will cause an adverse effect has to be defined, the effect of different life stages as well as the exposure duration triggering biological responses and the effect on human variability will be crucial for the implementation of the vision. Such efforts are currently addressed in an activity called SEURAT (Safety Evaluation Ultimately Replacing Animal Testing) [10]. Nevertheless, the availability of human cell models is the key for the identification of 'toxicity' pathways and their relevance in the chemical space.

Human Cells as Toxicological Test Systems in a New Paradigm of Safety Assessment

The development of human cell models requires a series of standards to assure that the toxicological information provided by the model is meaningful for safety assessments. Such standards include rigorous assessments of the functionality of the cells and the evaluation of their genetic and phenotypic stability, which are required for a normal physiological response to toxicants. Good cell culture practice (GCCP) has been applied in order to increase the quality and reproducibility of any cell system to be used in toxicity testing [11]. These principles include quality and identity controls of reagents, aspects of sterility of the cell cultures in terms of contamination with pathogens or cross contamination with other cells *etc.*

By applying the GCCP criteria when selecting an appropriate human cellular model, it becomes obvious that primary cell cultures of a specific tissue have some constraints: i) the limited availability of donor tissues ii) the variable quality of the derived cells, iii) limitation in the quantity of cells, iv) the lifespan of the cells as well as their tendency to de-differentiate when cells are cultured *in vitro*. This last issue is particularly relevant for hepatocytes, one of the most toxicologically relevant cell types, as they need to maintain the properties and functionality to be targeted by xenobiotics.

Established cell lines are another source of human cells for toxicity testing. Their major advantage is their unlimited proliferation capacity induced by spontaneous or induced transformation. On the other hand, the same genetic modifications might have a great impact on cell functionality, rendering these cells less suitable for specific testing. In addition, several unwanted characteristics such as chemical resistance, have been reported for a number of cell lines with carcinogenic origin. Due to the above mentioned scientific facts, human pluripotent stem cells received more and more attention as an unlimited cell source for toxicity assessments.

In 2007, GCCP principles have been further expanded and discussed in an ECVAM workshop in order to cover the specific characteristics of stem cells and establish a quality framework which allow stem cells based *in vitro* tests to fulfil international test development criteria [12]. These obligations are a prerequisite to use toxicological methods in the world of regulatory toxicology.

Pluripotent Stem Cells

Since detailed descriptions and approaches to pluripotent stem cells are given elsewhere in this eBook, we will focus on issues which have to be considered if pluripotent stem cells are used as cellular models for *in vitro* test development.

Actually it is the combination of two characteristics which render human stem cells as an interesting model for toxicity testing:

1. Their self-renewal can overcome the expansion limitations of conventional somatic cells and allows their application in high throughput applications
2. Their differentiation potential to generate theoretically any toxicologically relevant cell

Despite these advantages and the advances in stem cell biology made in the recent years, the application of stem cell technologies to toxicological questions is still in their infant status. A major challenge is the stabilization of differentiation protocols that can vary significantly from one laboratory to another. Unstable differentiation protocols will have a major impact on the reliability of stem cell based *in vitro* tests in validation studies. Another aspect which urgently needs to be addressed is the relatively low output of target cells available for toxicological testing [13-15]. Optimal cell lines and differentiation protocols have to be established before proper test development is started.

At this time, banking of pluripotent stem cell lines and accurate description of their individual characteristics and differentiation potentials has been provided by different national Stem Cell Banks (UK Stem Cell Bank, US National Stem Cell Bank, WISC Bank or the Spanish Stem Cell Bank) and different Stem Cell Registries (NIH Human Embryonic Stem Cell Registry or the European Human Embryonic Stem Cell Registry) and these information is openly available. However, it is worthwhile to take the technological challenges on since the opportunities of pluripotent stem cells are huge. The availability of cell material representing different human populations, life stages, genetic predispositions, diseases can be used as a unique tool in drug discovery, drug development and better safety assessments

Currently significant efforts in various national and international projects [16, 17] are undertaken to understand key parameters for stem cell based toxicity tests and to define crucial criteria for a successful test.

In this context, it is relevant to distinguish between sophisticated stem cell models which are used as research models *e.g.* to unmask new toxicological pathways *via* emerging 'omics' technologies and such stem cell based tests which can be used in routine applications and in high-throughput formats for safety assessments.

Recent reports showed that human induced pluripotent stem cells (hiPSCs) might be considered as an alternative to human embryonic stem cells (hESCs) for the delivery of toxicologically relevant target cells [18, 19]. Importantly, since hiPSCs derive from various individuals, they allow -for the first time- to address the genetic diversity of human beings in toxicological applications. In particular, hiPSCs which are generated from donors carrying a predisposition for a disease are of special interests. These cellular models might improve the pipeline of drug discovery and drug development [18, 20].

The establishment of quality criteria is part of a successfully developed *in vitro* test that can be considered for formal validation. The criteria that have to be fulfilled are laid down in ECVAM's criteria for prevalidation [21]. In summary Fig. **1** demonstrates a stepwise approach defining scientific and technical requirements which will lead to acceptable levels of test development.

In any case test developers need to take the specific nature of stem cells into account. This includes in particular the characterisation of undifferentiated stem cell populations in culture that are used to derive toxicologically relevant cell types. A panel of markers including acceptable threshold levels is needed to support the standardization of the cellular model which will serve as a test system later on. The starting culture of undifferentiated hESC can spontaneously accumulate mutations in their karyotype rendering them unstable [22-24], also this limitation has to be monitored by an efficient quality control and systematic testing of the integrity of the cells.

In addition, suitable quantitative acceptance criteria verifying the quality of target cell populations are needed by establishing an agreed set of lineage specific markers or readouts proving the functionality of the cells. These markers, as well as their threshold values, should be part of the standard operating procedures (SOPs) of any toxicological stem cell-based test. If the general principles of test development are addressed, the test can enter into the (pre)validation process in which it has to prove its reliability and relevance under various laboratory set ups. These criteria have been internationally accepted and summarized in the OECD Guidance Document 34 [25].

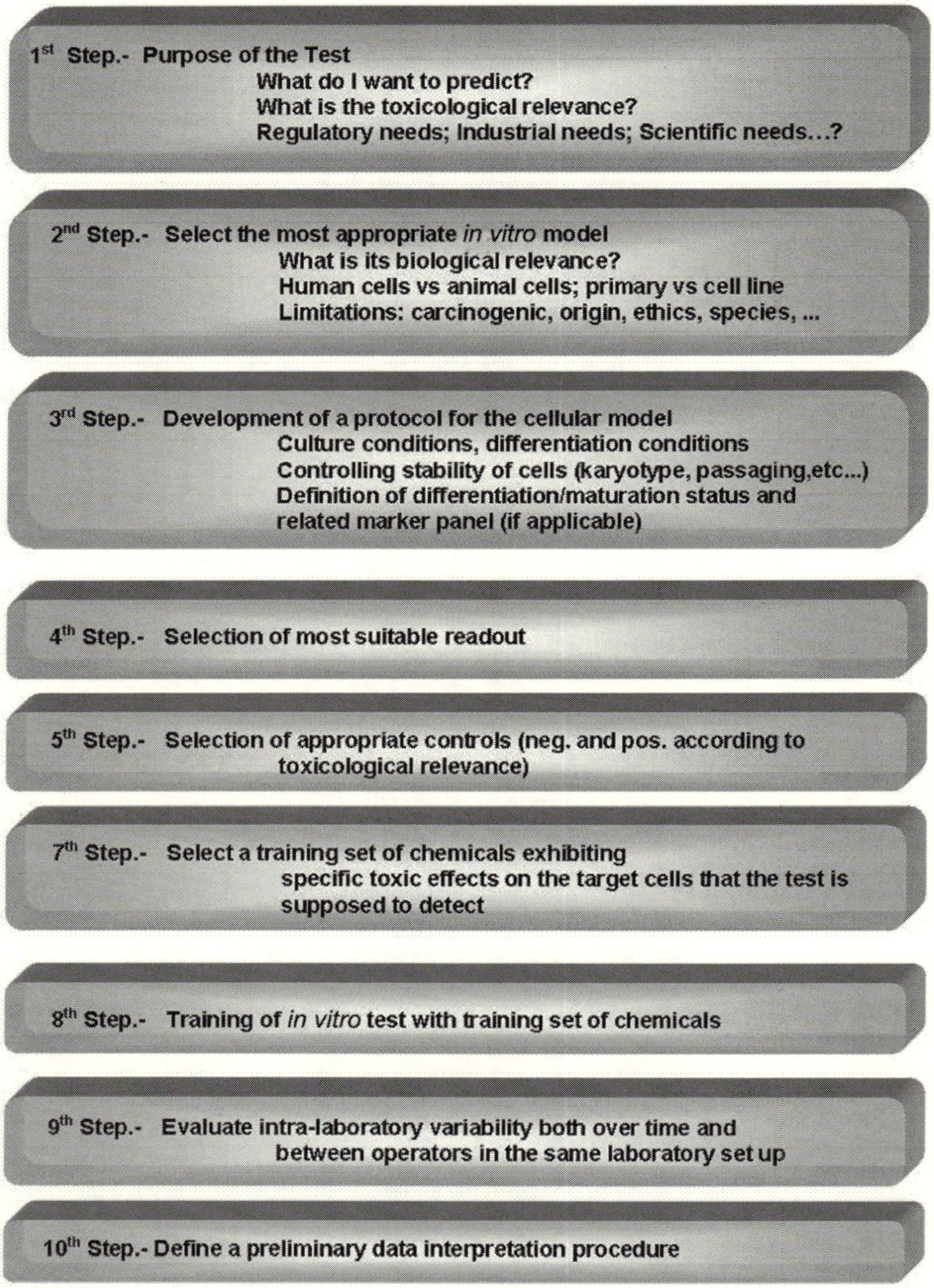

Fig. 1: Scheme of the steps for the development and validation of an *in vitro* toxicity test.

hESC vs hiPSC

Nowadays, it is impossible to define the exact role of hESC or iPSCs in toxicology. Nevertheless, it can be envisaged that both forms of pluripotent stem cells will be the most suitable cell sources for pathway based toxicity testing that is relevant to humans. In addition, for the first time also inter-individual toxicity variabilities, which is often observed in the field of hepatoxicity can be addressed by using hiPSCs that will allow toxicological testing on a variety of cells with different genetic background. This testing of a 'population-in-a-dish' might also be the basis for a personalized toxicological analysis of high risk patients such as cancer patients which often suffer from cardiotoxicity due to the cancer treatment.

Nevertheless, the mutual acceptance of data deriving from hESC based test systems is challenged due to ethical constraints in the various countries [26]. Therefore a stringent analysis on the potential of hiPSCs and hESCs for their suitability in the toxicity assessments is necessary as it was proposed by the European Group on Ethics in their opinion 22 [27]. Currently, huge efforts are ongoing to address this challenges [10, 17]. Table **1** summarizes key aspects that should be analysed if derivatives of stem cells are used in toxicological application.

Table 1. Comparison of characteristics of hESC and hiPSC

Charteristics		**Hesc**	**Hipsc**
Self-Renewal		Yes, unlimited	Yes, unlimited
Differentiation Potential	*In vivo* (Teratomas/Tetraploid Aggregation)	Can form teratomas, but tetraploid aggregation not determined	Can form teratomas, but tetraploid aggregation not determined
	In vitro	Yes, to all three germ layers (germ line under investigation)	Yes, to all three germ layers (not clear germ line)
Pluripotency Markers	Genetic (OCT4, NANOG, DPPA, Etc…)	Yes, only determined once line established	Yes, can be monitored during generation of line
	Surface (SSEA--3, -4, TRA-1-61, -80, ALP etc…)	Yes, only determined once line established	Yes, can be monitored during generation of line
Genetic Stability		Yes, only chromosome aberrations	Partial, if culture conditions appropriate
Phenotype Stability		Yes	Partial, could revert reprogramming
Epigenetics	ESC status	Full	hESC-like, similar but not identical
	Somatic status	None	Partial, presence of somatic memory
Genetic Manipulation		Yes, existing reporter lines	Yes
Ethical issues		Yes, related to status of human embryo	None, just those related to tissue donation
Population Variety	HLA polymorphisms	Limited and possible through banking but not easy	Yes, just using appropriate donors
	Disease Specific	Limited, as preimplantational diagnosis not always allowed	Yes, can use adult sick donors

One major aspect that has to be taken into account is the degree of stemness in hESC and hiPSC cell cultures. hESC serve hereby as the 'golden standard of pluripotency and self-renewal' [28, 29]. hESC are - due to their origin- close to the totipotent zygote and have therefore the broadest differentiation potential together with the most stable expression of pluripotency related markers [30-33]. Furthermore the stemness of hESC has been proven not only at genetic level but also in epigenetic studies, revealing the activity of a complex machinery of epigenetic modifiers that ensures the stable and controlled expression of their phenotype [34-38].

Initial studies demonstrated a high correlation of undifferentiated iPSC and hESCs when comparing i) the endogenous expression levels of pluripotency related genes, ii) defined surface markers and iii) the level of alkaline phosphatase activity, which in fact is one of the first events to take place during reprogramming [39]. However, the analysis of hESC and iPSC at the epigenetic level, and the full assessment of their differentiation potential showed differences between the two forms of pluripotent stem cells [29-33, 40]. Many epigenetic somatic genes which are in control function are still active whereas the triggers of stemness are not fully free to exert their function [29, 41-43]. This results in a rather unstable system which is quite sensitive to modifications in the environment. Thereby oversensitivity of such toxicological systems might be expected. Furthermore, remaining somatic labels, normally referred to as "somatic memory" might have a great influence in the differentiation tendency of iPSC when compared to hESC [44, 45]. As a result iPSC might preferentially differentiate, under spontaneous conditions, into cell lineages related to the donor somatic cell [46-48]. Currently the impact of these findings in toxicological applications is still unclear. Further studies need to be performed to clarify if an iPSC-derived neuron, which might still have some epigenetic triggers of the mesodermal parental cells, will show a different sensitivity than a hESC-derived neuron,. This could be considered as a drawback, but could also turn into a positive issue if the targets of the toxicity assessment with iPSC are cells from the same lineage as the donor somatic cells. In such a scenario, not only the yield of the desired target cell might be higher than the one obtained by hESC but possibly the degree of maturation and functionality might be much closer to the actual somatic cell than what could be achieved by hESC-derived target cells [49].

Increased Safety for Consumers and Patients by Using Mode of Action Based Toxicity Evaluations

In the past, various studies and observations have demonstrated that animal- based toxicity tests cannot detect all possible adverse effects that are relevant to humans. In order to compensate for these inter species variations, risk assessors normally apply a factor of 10 to threshold values such as the no observed adverse effect level or benchmark dose, in order to derive safe levels of exposure in humans for compounds that are assumed to cause threshold toxicities. Nevertheless, this 'pragmatic' approach has some drawbacks as reported by Price and colleagues in a recent study. They have shown that healthy test animals are, on the whole, less sensitive than the human cancer patients when dose is expressed on a body weight basis and in many cases the ratios exceeded the safety factor of 10 associated to inter-species variations. The fractions of the compounds with values above 10 were approximately 3%, 5%,19%, and 37% for the dog, monkey, rat, and mouse proving that animal models are not an accurate prediction of the effect of drugs in humans [50].

The relevance of interspecies variations becomes quite obvious in the sector of pharmaceuticals. Adversities to humans are often discovered in the phase of clinical trials or even when the drug is placed on the market. There have been several cases when a drug has been called back from the market due to unexpected side effects, which raised the public attention on this issue. The most common reasons for a withdrawal are severe hepatotoxicity and adverse effects on the cardiovascular system (Reviewed by mac Donald [51]). Liver toxicity accounted for 29% of drug withdrawals in the last decade while cardiovascular toxicity reached 33% of withdrawals in the same period. The other adverse effects that have been reported are related to renal dysfunctions, carcinogenicity, mutagenicity and others [51].

Drug induced hepatotoxicity causes 50% of acute liver failures in USA [52]. Reasons for this high rates are certainly the low concordance between animal and human toxicity data but also drug induced idiosyncratic liver injuries are a challenge for modern toxicology[52]. Drug induced liver toxicity is one of the most common single adverse effects leading to non- approval of a drug or its withdrawal from the market [53]. Prominent examples are pharmaceuticals such as troglitazone, bromfenac, pemoline ibufenac and benoxaprofen (reviewed by Giri [52]).

Another area of concern is the assessment of cardiotoxicity which often leads to regulatory decision on a drug [54]. In the case of Vioxx (Rofecoxib) that was approved by the FDA in 1999 for the treatment of osteoarthritis and acute pain conditions and was voluntarily called back in 2004 because of concerns about an increased risk of heart attack and stroke associated with long-term, high-dosage use. In the case of a tyrosine- kinase inhibitor sunitinib signs of cardiotoxicity were discovered during the clinical trials where 11% patients given repeating cycles of sunitinib had a cardiovascular event and hypertension was seen in 47% of patients [55]. Although cardiotoxicity tests are performed in the phase of preclinical and clinical studies, some adverse effects remain undetected. A possible explanation is that there is a 20% negative concordance between animal and human cardiotoxicity tests [53].

Prenatal toxicity is one of the human health endpoints for which tests are performed in the phase of preclinical trials. Those tests are obligatory since the well known disaster of thalidomide that in the late 50' and early 60' has caused 7000 children born with severe malformations [56]. However, animal tests for developmental toxicity assessments are poorly concordant with human data (reviewed in [57]).

Another factor of 10 is applied to compensate the toxicities related to the polymorphisms in humans. Often severe side effects are observed only after the approval and entry on the market of the new drug where a large number of patients in different life stages, healthy status and sex are using the drug. Inter- individual differences, such as enzyme polymorphism, can not be predicted in clinical trials where the patients are selected according to predefined criteria.

In particular, the development of *in vitro* models in the 'problematic' toxicological areas with high inters- and intra species variations are progressing well.

Assessing Target Organ Specific Toxicities by Using Stem Cells

Prenatal Toxicity

The capacity of hESCs to differentiate into derivates of all the three germ layers including trophoblast cells [58] makes them an ideal model to study underlying mechanisms of developmental toxicity. Usually tests of developmental toxicity are very complex as they include the stages from the formation of gametes up to the birth, passing through fertilization and the prenatal development. Prenatal toxicity is usually assessed in a rodent and a non rodent species in order to compensate for interspecies variations. These precautionary principles have been applied after the thalidomide disaster in the late 50s and early 60s. Humanized tests that unmask mode of actions of compounds will significantly improve the safety in particular for the unborn child.

A single *in vitro* test using stem cells can not mimic all the stages of the mammalian reproductive cycle. However, in the last years a series of tests have been developed where each test focuses on a different critical window leading to reproductive toxicity. Among them gametogenesis is considered as a highly relevant target for reproductive toxicants. In contrary to the murine ESCs that have been differentiated into early germinal cells and spermatozoa, the differentiation of human pluripotent stem cells into gametes is still in an emergent phase. Low yield of target cells and a lack of final derivatives in the case of male gametes are challenges of test developers working in the field. Protocols for the formation of mature female gametes have not been published yet [14, 59, 60]. However, the importance of the maturity status and functionality of the target cells for toxicity studies still needs to be carefully examined. The mode of action of many developmental toxicants is still unknown. Emerging technologies such as the "omics" technologies will give precious input to toxicologists by defining key markers of developmental toxicity [61, 62].

Despite the progress that has been made in the last years, the methods that are currently in the phase of development still need to be refined, particularly in regard to the amount of target cells as well as the readouts and prediction models. In a recently published draft report on alternative methods for cosmetics testing it was concluded that several years of test development will be still necessary before a test for developmental toxicity based on pluripotent stem cells can enter into the validation process [63].

Neurotoxicity

The interests of regenerative medicine have driven the large expansion of protocols for neural cell differentiation which give a variety of different cell types. At this time, protocols have been established for the differentiation of pluripotent stem cells into neural precursor cells [64] as well as several differentiated cells such as motoneurons [65], glia cells [64], dopaminergic neurons [66, 67]), cerebellar cells [68], *etc.*

iPSCs give also an opportunity to create patient-specific and/or disease specific models for neurotoxicity testing. According to a recent work, iPSCs have been generated from a patient with spinal muscular atrophy and differentiated them towards motoneurons [69].The rapid progress in stabilization of differentiation protocols allowed first successes in the development of *in vitro* tests. Recently, data of two new tests have been published demonstrating the effects of chemicals on dopaminergic neurons [67, 70]. EPA researchers have announced a test system that is using high content imaging analysis to detect chemical effects on the neurite outgrowth opening up opportunities for a medium/high throughput screening system [71]. Additionally, Han and co-workers have developed a screening platform for the toxicity testing of compounds on neural progenitors and differentiated neural cells. Their aim was to select compounds that could selectively kill the progenitors, leaving a population of differentiated cells to be used for therapeutic purposes [70]. The test system could be adapted also for the screening of compounds with neurotoxic effects on the three stages of neuronal differentiation.

Cardiotoxicity

During the last years several protocols have been developed for the differentiation of hESC and hiPSC into cardiomyocytes [72-74]. The two major protocols include coculture with END2 cells or culture in defined conditions in monolayer or in 3D structures named embryoid bodies [73, 75]. Currently, the maximum reported differentiation rates towards the toxicologically relevant ventricular cells are of 20-30% [74, 76]. The differentiation efficiency into nodal cells, atrial cells and ventricular cell types might be improved by using emerging 3D culture conditions.

Such differentiated cells can be used for the assessment of the cardiotoxic potential of a substance by monitoring the gene expression profile, the electrophysiological properties and the release of substances related to cardiac cell stress.

The main purpose of the cardiotoxicity tests that are currently in the phase of test development is the *in vitro* detection of drug-induced cardiomyopathies and arrhythmias. Severe adverse effects such as Torsade de Pointes (TdP) are associated with chemical interactions with various ion channels of the heart [77]. A feasibility study with 12 compounds, having well described effects on the different ion channels, has been described by Braam *et al.* and has demonstrated the biological and toxicological relevance of the developed *in vitro* system by analyzing the extracellular field potentials of hESC-derived cardiomyocytes [13, 78].

Other manifestations of cardiotoxicity, such as various forms of cardiomyopathies, have been recently addressed by Andersson *et al.* In their system, the release of cardiac troponin T and the fatty acid binding protein 3 - two clinically decisive cardiac biomarkers- have been monitored after Doxorubicin treatment by using biosensing methods [72].

Hepatotoxicity

Because of its role as the major metabolizing organ in the human body the liver is subject to many of the adverse effects of chemicals. Besides, hepatotoxicity is one of the major causes of drug withdrawals [79]. However, a reliable model to predict hepatotoxicity is still missing. Primary human hepatocytes have the right characteristics in terms of metabolizing capability that need to be addressed in toxicological studies. The drawbacks of this system is the limited availability of primary cells, the variation of their quality, limited proliferation potential and their rapid dedifferentiation *in vitro* conditions [80]. Additionally, hepatic cell lines are mainly from carcinogenic origin, often exhibit unwanted characteristics and are less functional, which might impact the biological and toxicological relevance of the selected *in vitro* system [81].

The crucial characteristics of an *in vitro* liver model are the presence of a vast array of enzymes and the level of enzymatic activity that should mimic the *in vivo* situation. In the last couple of years, several groups have demonstrated successful differentiation into hepatocyte-like cells, characterized by the expression of relevant markers, which are expressed during liver development *in vivo* [82-84]. However, the proposed differentiation protocols still vary with regard to growth factor combinations driving the differentiation of hESCs into various stages of hepatic development. Even if the majority of publications are based on observations of the hESC lines H1 and H9 (WiCell Research Institute), Song *et al* demonstrated that also two iPSC lines were able to differentiate into hepatocyte-like cells after stimulation with a series of growth factors, to the same extent as H1 [85].

The functionality of hESC-derived hepatocyte-like cells was proven by their capacity to: i) store glycogen, ii) produce urea and albumin and iii) to develop a complete biotransformation system including phase I and II metabolizing enzymes and phase III transporters [86]. A comprehensive overview on the differentiation of embryonic and adult stem cells into hepatic lineages and related markers was provided by Snykers et al [87]. Nevertheless, a definition of which functions and marker panel a hepatocyte culture derived from pluripotent stem cells has to show, to be acceptably used in hepatoxicity testing, has not been defined so far.

Table 2. toxicologically relevant target cells derived from pluripotent stem cells

System/Tissue	Cell Types	Markers	References
Reproductive	Germ cells	VASA, DAZL, DPPA3, TEKT1, acrosin, SCP3, C-KIT	[14, 59, 88]
	Primordial germ cells	VASA, DAZL, C-KIT, DPPA3	[60, 89]
Neural	Neural crest cells		[64]
	Dopaminergic neurons	TH+, beta III tubulin	[67, 70]
	Neurons	bIII-tubulin, MAP2, nestin, neurofilament SMI-312	[71]
	Cerebellar- like cells	Electrophysiological properties, Na, K channels	[68]
	Motoneurons	Electrophysiological properties, AcH release	[65, 90]
	Astrocytes		[91]
	Glial cells		[92]
Circulatory/cardiac	Cardiomyocytes	Electrophysiological properties, beta-adrenergic receptors, IKr channel, Na, Ca and hERG K channels, cTnT release	[13, 73, 74, 78]
	Endothelial cells	CD31 and vascular endothelial cadherin, NO production *in vivo*	[93]
	Vascular cells	Formation of vessels *in vivo*	[94]
Respiratory	Lung epithelium	CC16, NKX2.1, SP-C, Aquaporin 5, FOXJ1, tubulin IV	[95]
	Alveolar epithelial type II	surfactant proteins A, B, C, cystic fibrosis transmembrane conductance receptor	[96]
Digestive	Hepatocyte	albumin+, Glycogen storage, inducible CYP1A2 activity, ICG uptake, Phase I, II and III enzymes	[84, 86, 97-100]
	Islet cells		[101]
Other	Renal progenitors	CD24+, podocaylxin+, GCTM2-, PAX2, LHX1, WT1	[102]
	Ectodermal cells	K18+, p63, K14, lama3	[103]
	Chondrocytes		[104]
	Osteoblasts		[105]
	T cells	CD34+, CD43+, CD45+	[106]

CONCLUSIONS AND PROSPECTS

The recent progress in stem cell technologies for basic research and therapeutic applications also offers unique possibilities in toxicology. Currently regulatory toxicity testing stands at the beginning of a paradigm shift towards a targeted mechanistic and pathway oriented toxicology which will increase the safety of consumers and patients while reducing the number of animals for safety assessments. Major tools of such new framework for safety assessment are human cells which are stable in culture, are available in sufficient amounts and have the physiological characteristics of non carcinogenic cells. However, these technical challenges still need to be solved. Currently only pluripotent stem cells have the potential to fulfil these requirements for a broad range of target cells if existing technical challenges could be overcome. National and international funding bodies have understood the opportunities of stem cells and huge research initiatives are ongoing to tackle the challenges for making stem cell models available for regulatory safety assessments.

REFERENCES

[1] Federal Food, Drug, and Cosmetic Act. US Code title 21.

[2] Council Directive 86/609/EEC of 24 November 1986 on the approximation of laws, regulations and administrative provisions of the Member States regarding the protection of animals used for experimental and other scientific purposes (1986).

[3] Council Directive 76/768 of 27 July 1976 on the approximation of the laws of the Member States relating to cosmetic products ("Cosmetics Directive"). (1976).

[4] Regulation (EC) No 1907/2006 of the European Parliament and of the Council of 18 December 2006 concerning the Registration, Evaluation, Authorisation and Restriction of Chemicals (REACH). (2006).

[5] Regulation (EC) No 1272/2008 of the European Parliament and of the Council of 16 December 2008 on classification, labelling and packaging of substances and mixtures, amending and repealing Directives 67/548/EEC and 1999/45/EC, and amending Regulation (EC) No 1907/2006. (2008).

[6] Regulation (EC) No 1107/2009 of the European Parliament and of the Council of 21 October 2009 concerning the placing of plant protection products on the market and repealing Council Directives 79/117/EEC and 91/414/EEC. (2009).

[7] Directive 98/8/EC of the European Parliament and of the Council of 16 February 1998 concerning the placing of biocidal products on the market. (1998).

[8] Council Directive 98/24/EC of 7 April 1998 on the protection of the health and safety of workers from the risks related to chemical agents at work. (1998).

[9] Krewski D, Andersen ME, Mantus E, Zeise L. Toxicity testing in the 21st century: implications for human health risk assessment. Risk Anal 2009; 29(4): 474-9.

[10] SEURAT. Safety Evaluation Ultimately Replacing Animal Testing. Available from: http: //www.eurtd.com/seurat/kick-off/index.php. [Access date 17.3.2011].

[11] Hartung T, Balls M, Bardouille C, *et al.* Good Cell Culture Practice. ECVAM Good Cell Culture Practice Task Force Report 1. Altern Lab Anim 2002; 30(4): 407-14.

[12] Adler S, Allsopp T, Bremer S, et al. hESC Technology for Toxicology and Drug Development. ECVAM Workshop Report 2007 Available from: http: //ecvam.jrc.ec.europa.eu. [Access date 17.3.2011].

[13] Braam SR, Tertoolen L, van de Stolpe A, Meyer T, Passier R, Mummery CL. Prediction of drug-induced cardiotoxicity using human embryonic stem cell-derived cardiomyocytes. Stem Cell Res; 4(2): 107-16.

[14] Kee K, Angeles VT, Flores M, Nguyen HN, Reijo Pera RA. Human DAZL, DAZ and BOULE genes modulate primordial germ-cell and haploid gamete formation. Nature 2009; 462(7270): 222-5.

[15] Zhao D, Chen S, Cai J. *et al.* Derivation and characterization of hepatic progenitor cells from human embryonic stem cells. PLoS One 2009; 4(7): e6468.

[16] ESNATS. Embryonic stem cell-based Novel Alternative Testing Strategies. Available from: www.esnats.eu. [Access date 17.3.2011].

[17] SCR&Tox. Stem Cells for Relevant Efficient Extended and Normalized Toxicology. Available from: www.scrtox.eu. [Access date 17.3.2011].

[18] Marchetto MC, Carromeu C, Acab A, *et al.* A model for neural development and treatment of Rett syndrome using human induced pluripotent stem cells. Cell; 143(4): 527-39.

[19] Park IH, Zhao R, West JA. *et al.* Reprogramming of human somatic cells to pluripotency with defined factors. Nature 2008; 451(7175): 141-6.

[20] Carvajal-Vergara X, Sevilla A, D'Souza SL. *et al.* Patient-specific induced pluripotent stem-cell-derived models of LEOPARD syndrome. Nature; 465(7299): 808-12.

[21] Curren RDS, J A, Spielmann, H, Liebsch M, Fentem JH and Balls M. The Role of Prevalidation in the Development, Validation and Acceptance of Alternative Methods ATLA 1995; 23: 211-7.

[22] Liu X, Wu H, Loring J. *et al.* Trisomy eight in ES cells is a common potential problem in gene targeting and interferes with germ line transmission. Dev Dyn 1997; 209(1): 85-91.

[23] Longo L, Bygrave A, Grosveld FG, Pandolfi PP. The chromosome make-up of mouse embryonic stem cells is predictive of somatic and germ cell chimaerism. Transgenic Res 1997; 6 (5): 321-8.

[24] Draper JS, Smith K, Gokhale PJ. *et al.* Recurrent gain of chromosomes 17q and 12 in cultured human embryonic stem cells. Nat Biotechnol 2004; 22(1): 53-4.

[25] OECD. Guidance Document on the Validation and International Acceptance of New or Updated Test Methods for Hazard Assessment. Paris: Environment Directorate, OECD; 2005 Contract No.: Document Number|.

[26] Leist M, Bremer S, Brundin P. *et al.* The biological and ethical basis of the use of human embryonic stem cells for *in vitro* test systems or cell therapy. ALTEX 2008; 25(3): 163-90.

[27] The European Group on Ethics in Science and New Technologies to the European Commission. Recommendations on the ethical review of hESC FP7 research projects; 2007 Contract No.: Document Number|.

[28] Tarkowski AK, Ozdzenski W, Czolowska R. Mouse singletons and twins developed from isolated diploid blastomeres supported with tetraploid blastomeres. Int J Dev Biol 2001; 45(3): 591-6.

[29] Wernig M, Meissner A, Foreman R. *et al. In vitro* reprogramming of fibroblasts into a pluripotent ES-cell-like state. Nature 2007; 448(7151): 318-24.

[30] Stewart MH, Bendall SC, Bhatia M. Deconstructing human embryonic stem cell cultures: niche regulation of self-renewal and pluripotency. J Mol Med 2008; 86(8): 875-86.

[31] Brimble SN, Zeng X, Weiler DA. *et al.* Karyotypic stability, genotyping, differentiation, feeder-free maintenance, and gene expression sampling in three human embryonic stem cell lines derived prior to August 9, 2001. Stem Cells Dev 2004; 13(6): 585-97.

[32] Draper JS, Moore HD, Ruban LN, Gokhale PJ, Andrews PW. Culture and characterization of human embryonic stem cells. Stem Cells Dev 2004; 13(4): 325-36.

[33] Sidhu KS, Tuch BE. Derivation of three clones from human embryonic stem cell lines b y FACS sorting and their characterization. Stem Cells Dev 2006; 15(1): 61-9.

[34] Bhattacharya B, Miura T, Brandenberger R, *et al.* Gene expression in human embryonic stem cell lines: unique molecular signature. Blood 2004; 103(8): 2956-64.

[35] Surani MA, Hayashi K, Hajkova P. Genetic and epigenetic regulators of pluripotency. Cell 2007; 128(4): 747-62.

[36] Boyer LA, Plath K, Zeitlinger J. *et al.* Polycomb complexes repress developmental regulators in murine embryonic stem cells. Nature 2006; 441(7091): 349-53.

[37] Bernstein BE, Mikkelsen TS, Xie X. *et al.* A bivalent chromatin structure marks key developmental genes in embryonic stem cells. Cell 2006; 125(2): 315-26.

[38] Farthing CR, Ficz G, Ng RK. *et al.* Global Mapping of DNA Methylation in Mouse Promoters Reveals Epigenetic Reprogramming of Pluripotency Genes. PLoS Genetics 2008; 4(6): e1000116.

[39] Stadtfeld M, Maherali N, Breault DT, Hochedlinger K. Defining molecular cornerstones during fibroblast to iPS cell reprogramming in mouse. Cell Stem Cell 2008 6; 2(3): 230-40.

[40] Hochedlinger K, Plath K. Epigenetic reprogramming and induced pluripotency. Development 2009; 136(4): 509-23.

[41] Yu J, Vodyanik MA, Smuga-Otto K. *et al.* Induced pluripotent stem cell lines derived from human somatic cells. Science 2007; 318(5858): 1917-20.

[42] Lister R, Pelizzola M, Kida YS. *et al.* Hotspots of aberrant epigenomic reprogramming in human induced pluripotent stem cells. Nature 2011; 471(7336): 68-73.

[43] Stadtfeld M, Apostolou E, Akutsu H. *et al.* Aberrant silencing of imprinted genes on chromosome 12qF1 in mouse induced pluripotent stem cells. Nature 2010 13; 465(7295): 175-81.

[44] Sullivan GJ, Bai Y, Fletcher J, Wilmut I. Induced pluripotent stem cells: epigenetic memories and practical implications. Mol Human Reprod 2010; 16(12): 880-5.

[45] Marchetto MC, Yeo GW, Kainohana O, Marsala M, Gage FH, Muotri AR. Transcriptional signature and memory retention of human-induced pluripotent stem cells. PLoS One 2009; 4(9): e7076.

[46] Ji H, Ehrlich LI, Seita J. *et al.* Comprehensive methylome map of lineage commitment from haematopoietic progenitors. Nature 2010; 467(7313): 338-42.

[47] Quattrocelli M, Palazzolo G, Floris G. *et al.* Intrinsic cell memory reinforces myogenic commitment of pericyte-derived iPSCs. The Journal of Pathology 2009; 223(5): 593-603.

[48] Kim K, Doi A, Wen B. *et al.* Epigenetic memory in induced pluripotent stem cells. Nature 2010; 467(7313): 285-90.

[49] Polo JM, Liu S, Figueroa ME. *et al.* Cell type of origin influences the molecular and functional properties of mouse induced pluripotent stem cells. Nat Biotechnol 2010 ; 28(8): 848-55.

[50] Price PS, Keenan RE, Swartout JC. Characterizing interspecies uncertainty using data from studies of anti-neoplastic agents in animals and humans. Toxicol Appl Pharmacol 2008; 233(1): 64-70.

[51] MacDonald JS, Robertson RT. Toxicity testing in the 21st century: a view from the pharmaceutical industry. Toxicol Sci 2009; 110(1): 40-6.

[52] Giri S, Nieber K, Bader A. Hepatotoxicity and hepatic metabolism of available drugs: current problems and possible solutions in preclinical stages. Expert Opin Drug Metab Toxicol 2010; 6(8): 895-917.

[53] Olson H, Betton G, Robinson D. *et al.* Concordance of the toxicity of pharmaceuticals in humans and in animals. Regul Toxicol Pharmacol 2000; 32(1): 56-67.

[54] Lasser KE, Allen PD, Woolhandler SJ, Himmelstein DU, Wolfe SM, Bor DH. Timing of new black box warnings and withdrawals for prescription medications. JAMA 2002; 287(17): 2215-20.

[55] Chu TF, Rupnick MA, Kerkela R. *et al.* Cardiotoxicity associated with tyrosine kinase inhibitor sunitinib. Lancet 2007; 370(9604): 2011-9.

[56] Gilbert SF. Developmental Biology. seventh ed.: Sinauer Associates; 2003.

[57] Stummann TC, Bremer S. The possible impact of human embryonic stem cells on safety pharmacological and toxicological assessments in drug discovery and drug development. Curr Stem Cell Res Ther 2008; 3(2): 118-31.

[58] Thomson JA, Itskovitz-Eldor J, Shapiro SS. *et al.* Embryonic stem cell lines derived from human blastocysts. Science 1998; 282(5391): 1145-7.

[59] Aflatoonian B, Ruban L, Jones M, Aflatoonian R, Fazeli A, Moore HD. *In vitro* post-meiotic germ cell development from human embryonic stem cells. Hum Reprod 2009 ; 24(12): 3150-9.

[60] Tilgner K, Atkinson SP, Golebiewska A, Stojkovic M, Lako M, Armstrong L. Isolation of primordial germ cells from differentiating human embryonic stem cells. Stem Cells 2008; 26(12): 3075-85.

[61] Cezar GG, Quam JA, Smith AM. *et al.* Identification of small molecules from human embryonic stem cells using metabolomics. Stem Cells Dev 2007; 16(6): 869-82.

[62] West PR, Weir AM, Smith AM, Donley EL, Cezar GG. Predicting human developmental toxicity of pharmaceuticals using human embryonic stem cells and metabolomics. Toxicol Appl Pharmacol 2010; 247(1): 18-27.

[63] Spielmann H. The EU Commission's Draft Report on Alternative (Non-animal) Methods for Cosmetics Testing: Current Status and Future Prospects--2010: a missed opportunity. Altern Lab Anim. 2010; 38(5): 339-43.

[64] Lee G, Chambers SM, Tomishima MJ, Studer L. Derivation of neural crest cells from human pluripotent stem cells. Nat Protoc; 5(4): 688-701.

[65] Lee H, Shamy GA, Elkabetz Y. *et al.* Directed differentiation and transplantation of human embryonic stem cell-derived motoneurons. Stem Cells 2007; 25(8): 1931-9.

[66] Iacovitti L, Donaldson AE, Marshall CE, Suon S, Yang M. A protocol for the differentiation of human embryonic stem cells into dopaminergic neurons using only chemically defined human additives: Studies *in vitro* and *in vivo*. Brain Res 2007; 1127(1): 19-25.

[67] Zeng X, Chen J, Deng X. *et al.* An *in vitro* model of human dopaminergic neurons derived from embryonic stem cells: MPP+ toxicity and GDNF neuroprotection. Neuropsychopharmacology 2006; 31(12): 2708-15.

[68] Erceg S, Ronaghi M, Ivan Z. *et al.* Efficient differentiation of human embryonic stem cells into functional cerebellar-like cells. Stem Cells Dev 2010; 19(11): 1745-56.

[69] Ebert AD, Yu J, Rose FF. *et al.* Induced pluripotent stem cells from a spinal muscular atrophy patient. Nature 2009; 457(7227): 277-80.

[70] Han Y, Miller A, Mangada J. *et al.* Identification by automated screening of a small molecule that selectively eliminates neural stem cells derived from hESCs but not dopamine neurons. PLoS One 2009; 4(9): e7155.

[71] Harrill JA, Freudenrich TM, Machacek DW, Stice SL, Mundy WR. Quantitative assessment of neurite outgrowth in human embryonic stem cell-derived hN2 cells using automated high-content image analysis. Neurotoxicology 2010; 31(3): 277-90.

[72] Andersson H, Steel D, Asp J, *et al.* Assaying cardiac biomarkers for toxicity testing using biosensing and cardiomyocytes derived from human embryonic stem cells. J Biotechnol 2010; 150(1): 175-81.

[73] Mummery C, Ward-van Oostwaard D, Doevendans P. *et al.* Differentiation of human embryonic stem cells to cardiomyocytes: role of coculture with visceral endoderm-like cells. Circulation 2003; 107(21): 2733-40.

[74] Zhang J, Wilson GF, Soerens AG. *et al.* Functional cardiomyocytes derived from human induced pluripotent stem cells. Circ Res 2009; 104(4): e30-41.

[75] Kehat I, Kenyagin-Karsenti D, Snir M. *et al.* Human embryonic stem cells can differentiate into myocytes with structural and functional properties of cardiomyocytes. J Clin Invest 2001; 108(3): 407-14.

[76] Mummery CL, Ward D, Passier R. Differentiation of human embryonic stem cells to cardiomyocytes by coculture with endoderm in serum-free medium. Curr Protoc Stem Cell Biol 2007; Chapter 1: Unit 1F 2.

[77] Kannankeril PJ, Roden DM. Drug-induced long QT and torsade de pointes: recent advances. Curr Opin Cardiol 2007; 22(1): 39-43.

[78] Caspi O, Itzhaki I, Kehat I. *et al.* *In vitro* electrophysiological drug testing using human embryonic stem cell derived cardiomyocytes. Stem Cells Dev 2009; 18(1): 161-72.

[79] Goldkind L, Laine L. A systematic review of NSAIDs withdrawn from the market due to hepatotoxicity: lessons learned from the bromfenac experience. Pharmacoepidemiol Drug Saf 2006; 15(4): 213-20.

[80] Elaut G, Henkens T, Papeleu P. *et al.* Molecular mechanisms underlying the dedifferentiation process of isolated hepatocytes and their cultures. Curr Drug Metab 2006; 7(6): 629-60.

[81] Ponten J. Cell biology of precancer. Eur J Cancer 2001; 37 Suppl 8: S97-113.

[82] Agarwal S, Holton KL, Lanza R. Efficient differentiation of functional hepatocytes from human embryonic stem cells. Stem Cells 2008; 26(5): 1117-27.

[83] Baharvand H, Hashemi SM, Kazemi Ashtiani S, Farrokhi A. Differentiation of human embryonic stem cells into hepatocytes in 2D and 3D culture systems *in vitro*. Int J Dev Biol 2006; 50(7): 645-52.

[84] Schwartz RE, Linehan JL, Painschab MS, Hu WS, Verfaillie CM, Kaufman DS. Defined conditions for development of functional hepatic cells from human embryonic stem cells. Stem Cells Dev 2005; 14(6): 643-55.

[85] Song Z, Cai J, Liu Y. *et al.* Efficient generation of hepatocyte-like cells from human induced pluripotent stem cells. Cell Res 2009; 19(11): 1233-42.

[86] Duan Y, Ma X, Zou W. *et al.* Differentiation and characterization of metabolically functioning hepatocytes from human embryonic stem cells. Stem Cells. 2010; 28(4): 674-86.

[87] Snykers S, De Kock J, Rogiers V, Vanhaecke T. *In vitro* differentiation of embryonic and adult stem cells into hepatocytes: state of the art. Stem Cells 2009; 27(3): 577-605.

[88] West FD, Machacek DW, Boyd NL, Pandiyan K, Robbins KR, Stice SL. Enrichment and differentiation of human germ-like cells mediated by feeder cells and basic fibroblast growth factor signalling. Stem Cells 2008; 26(11): 2768-76.

[89] Bucay N, Yebra M, Cirulli V. *et al.* A novel approach for the derivation of putative primordial germ cells and sertoli cells from human embryonic stem cells. Stem Cells 2009; 27(1): 68-77.

[90] Karumbayaram S, Novitch BG, Patterson M. *et al.* Directed differentiation of human-induced pluripotent stem cells generates active motor neurons. Stem Cells 2009 ; 27(4): 806-11.

[91] Zhang SC, Wernig M, Duncan ID, Brustle O, Thomson JA. *In vitro* differentiation of transplantable neural precursors from human embryonic stem cells. Nat Biotechnol 2001; 19(12): 1129-33.

[92] Yuan SH, Martin J, Elia J. *et al.* Cell-surface marker signatures for the isolation of neural stem cells, glia and neurons derived from human pluripotent stem cells. PLoS One 2011; 6(3): e17540.

[93] Kane NM, Meloni M, Spencer HL. *et al.* Derivation of endothelial cells from human embryonic stem cells by directed differentiation: analysis of microRNA and angiogenesis *in vitro* and *in vivo*. Arterioscler Thromb Vasc Biol 2010; 30(7): 1389-97.

[94] Levenberg S, Ferreira LS, Chen-Konak L, Kraehenbuehl TP, Langer R. Isolation, differentiation and characterization of vascular cells derived from human embryonic stem cells. Nat Protoc 2010; 5(6): 1115-26.

[95] Van Haute L, De Block G, Liebaers I, Sermon K, De Rycke M. Generation of lung epithelial-like tissue from human embryonic stem cells. Respir Res 2009; 10: 105.

[96] Wang D, Haviland DL, Burns AR, Zsigmond E, Wetsel RA. A pure population of lung alveolar epithelial type II cells derived from human embryonic stem cells. Proc Natl Acad Sci U S A 2007; 104(11): 4449-54.

[97] Duan Y, Catana A, Meng Y. *et al.* Differentiation and enrichment of hepatocyte-like cells from human embryonic stem cells *in vitro* and *in vivo*. Stem Cells 2007 ; 25(12): 3058-68.

[98] Ek M, Soderdahl T, Kuppers-Munther B. *et al.* Expression of drug metabolizing enzymes in hepatocyte-like cells derived from human embryonic stem cells. Biochem Pharmacol 2007; 74(3): 496-503.

[99] Hay DC, Fletcher J, Payne C. *et al.* Highly efficient differentiation of hESCs to functional hepatic endoderm requires ActivinA and Wnt3a signalling. Proc Natl Acad Sci USA 2008; 105(34): 12301-6.

[100] Rambhatla L, Chiu CP, Kundu P, Peng Y, Carpenter MK. Generation of hepatocyte-like cells from human embryonic stem cells. Cell Transplant 2003; 12(1): 1-11.

[101] Jiang J, Au M, Lu K. *et al.* Generation of insulin-producing islet-like clusters from human embryonic stem cells. Stem Cells 2007; 25(8): 1940-53.

[102] Lin SA, Kolle G, Grimmond SM. *et al.* Subfractionation of differentiating hES cell populations allows the isolation of a mesodermal population enriched for intermediate mesoderm and putative renal progenitors. Stem Cells Dev 2010; 19(10): 1637-48.

[103] Aberdam E, Barak E, Rouleau M. *et al.* A pure population of ectodermal cells derived from human embryonic stem cells. Stem Cells 2008; 26(2): 440-4.

[104] Toh WS, Yang Z, Heng BC, Cao T. Differentiation of human embryonic stem cells toward the chondrogenic lineage. Methods Mol Biol 2007; 407: 333-49.

[105] Bielby RC, Boccaccini AR, Polak JM, Buttery LD. *In vitro* differentiation and *in vivo* mineralization of osteogenic cells derived from human embryonic stem cells. Tissue Eng 2004t; 10(9-10): 1518-25.

[106] Timmermans F, Velghe I, Vanwalleghem L. *et al.* Generation of T cells from human embryonic stem cell-derived hematopoietic zones. J Immunol 2009; 182(11): 6879-88.

CHAPTER 14

Pluripotent Stem Cells for (Neurodegenerative) Disease Modelling

Greg T. Sutherland[1*] and Kuldip S. Sidhu[2]

[1]Discipline of Pathology, Sydney Medical School, University of Sydney, Australia and [2]Stem Cell Lab, Faculty of Medicine, University of New South Wales, NSW Australia

Abstract: The prevalence of late-onset neurodegenerative disorders such as Alzheimer's and Parkinson's disease is increasing alarmingly worldwide but their causes remain largely unknown. In order to study the pathogenesis of these diseases researchers use a variety of models from cell culture through to full organisms. As the complexity of the model system increases so does the potential variables that can confound experimental results and interpretation. Transgenic animal models are the workhorses of disease research but they have been purposefully intra-bred to remove their genetic variability, a major factor in human neurodegenerative disease. Prior to stem cell advances, researchers could obtain differentiated cell types such as fibroblasts from patients and these had reasonable longevity in the laboratory. However it was questionable how much fibroblasts could tell us about the relative susceptibilities of cells in the human brain. Embryonic stem cell lines provided the flexibility to produce all the different cell types of the body including neurons but they were derived from a limited number of individuals. It was the discovery of induced pluripotency by Yamanaka and colleagues that allowed neurons to be produced from a number of patients and unaffected controls. Now individual genetic variability could be incorporated into our experimental paradigms. Looking forward, induced pluripotent stem cells (iPSC) can be perturbed with putative toxins to re-create the natural history of a disease. This *in vitro* phenotype can then be used for the high throughput screening of therapeutic agents. The latter is akin to a "cure in a dish".

Keywords: Embryonic stem cells, induced pluripotent stem cells, neurodegenerative diseases, disease modelling, differentiation, drug secreening.

INTRODUCTION

All medical researchers are trying to find cures that reduce the burden of human suffering. Curing diseases is predicated on understanding causation and pathogenesis. Scientists mostly operate within a specific experimental paradigm or model that complements their background and skills. These models are simplifications of the disease but allow specific hypotheses to be tested in line with the scientific method. The key element is whether these simplifications remain relevant to the disease.

Stem cell technology has evolved in recent years to the point where regenerative medicine is not only feasible but utilised in clinical practice. The ability to cultivate pluripotent cells *in vitro* to replace damaged cells *in vivo* is truly miraculous. Outside regenerative possibilities, a valuable spin-off of stem cell technology has been the development of patient-derived stem cell disease models. Adult stem or somatic cells from a diseased individual can now be manipulated *in vitro* into disease-specific phenotypes. This sets up the possibility of not only recreating the disease history or pathogenesis in a dish, but also exploring disease cures.

The Complexity of Late-Onset Diseases

Diseases often manifest from an interaction between the host and a pathogen. In genetic diseases a gene defect may either increase susceptibility to a pathogen or environmental stress or be sufficiently severe *per se* to cause the disease. In Western society complex diseases are the major health burden. Here the distinction between aetiological agent and host response is blurred with pathogenesis being a complex interplay between environmental stresses and host genetics, often over a prolonged period of time. Cancer,

***Address correspondence to Greg T. Sutherland:** Discipline of Pathology, Sydney Medical School, University of Sydney, Sydney, NSW 2006, Australia; Tel: +61-2-9036 7233; Fax: +61-2-9351 3429; Email: g.sutherland@sydney.edu.au

Kuldip S. Sidhu (Ed)

heart disease, diabetes and Alzheimer's disease (AD) fall into this category with age as their major risk factor. The pathophysiology of ageing is hotly debated, as is its role in these diseases [1]. Some see the ageing effect as a decrease in tissue integrity or resistance to insult, while others see it as having a more interrelated role in the disease; most likely these are just different interpretations of the same scenario. Notwithstanding the biological impact of ageing mechanisms in disease there is also the time period itself. A prolonged disease history allows the effects of subtle risk factors to accumulate sufficiently to manifest as clinical disease. It also means there is sufficient time for many rounds of insult and recovery or compensation before a tissue functionally succumbs to disease. This means that end-stage pathology may bear little resemblance to the early pathogenic changes, an important issue when our understanding of pathogenesis often relies heavily on postmortem tissue samples. In the case of AD, clinical signs manifest due to neuron loss in the temporal lobes but by postmortem the disease has progressed to such an extent that the brain can weigh half its original weight.

Ideally we would like to uncover the primary pathogenic events so we can avert deleterious events early enough in the disease process to retain functionality at levels consistent with a reasonable quality of life. This would require the concurrent discovery of a rational drug target and a pre-symptomatic marker to identify persons at risk before substantial neuronal loss. The term 'rational' is used to describe a drug has been designed to directly attenuate a proven pathogenic mechanism.

In most late-onset diseases it is difficult to determine when the disease process actually began. As all diseases will have a genetic component, individuals are essentially born with a relative susceptibility, although the disease may never manifest in the absence of, or minimal exposure to, environmental factors. The real power of human genetic studies is that once susceptibility factors are identified models can be created. These models can be observed over their natural life or perturbed with putative disease factors. If we regard the susceptible state as a 'ground zero then with sequential exposures we can attempt to mimic the disease history or essentially model time. Assaying such a system at incremental time points would presumably reveal early defects that could serve as targets for both rational therapies and preclinical diagnostic markers.

Rational therapies are of course not the only therapeutic approach and many chapters in this textbook are testament to the exciting prospects of regenerative medicine where damaged cells or tissues can be replaced irrespective of the underlying or precipitating cause. However in this chapter we discuss the use of stem cell technologies to model the disease where the paradigm is very much about repair based on the accurate characterisation of pathogenic events. For most complex diseases where we have only fledgling ideas about their aetiologies this is no mean feat. Human neurodegenerative diseases represent the extreme level of difficulty. Stem cell technology will undoubtedly revolutionise the modelling of Alzheimer's and Parkinson's disease but we need to be cognisant of the specific challenges these diseases impose to fully realise its potential. The first part of this chapter discusses existing disease models to give the reader the necessary background for the subsequent consideration of patient-derived stem cell models and their potential impact on disease research.

Disease Modelling

In diseases there is usually one particular cell type that is most susceptible to the causative agent(s). In type I diabetes it is the β cell of the pancreas, in Parkinson's disease (PD) it is the dopaminergic neuron of the mid-brain nucleus, the substantia nigra. Most modelling strategies to date have used stem cell technology to recreate these cell types *in vitro* from patients and unaffected controls for comparative analysis and manipulation.

It may seem to many readers that 'dopaminergic' neurons in a cell culture dish are quite distinct from the architectural and functional complexity of the human brain. This is of course true but it is also worthwhile considering why we use models and what the competition for such stem cell-based systems is before becoming overly critical.

The statement "most models are wrong, but some are useful", credited to the statistician George Box [2], is very true in the study of neurodegenerative diseases. We model diseases in lieu of observing sufferers directly and although these models are inherently 'wrong' they do allow two key components of the scientific method to be met: a reductionist experimental paradigm where all variables other than the dependent can be tightly controlled and the subsequent ability to perturb the system and observe the consequences.

The importance of the 'Box' dictum is that we need to be aware of how 'useful' our model is before extrapolating the results into relevance for the disease being modelled. In biology, in particular, we are modelling complex and dynamic systems that constantly interact with the environment and these interactions will evolve over the life of the organism. Furthermore our sampling is inherently static – a snapshot of events.

These issues are intensified when we try to model the human brain because our understanding of its functional complexity is at such rudimentary stage.

All models have their advantages and disadvantages. In the simplest models a very tight experimental paradigm can be created: ideally there is one variable and interpretation of the result is relatively straightforward. At the opposite extreme are human clinical trials where potential confounders abound, including genetic variation and differential environmental exposures among subjects, both retrospective and prospective.

Drug candidate toxicity is a major cause of drug attrition in drug development and leads to high costs and few clinical candidates. It was estimated in 2002 that the cost of getting a new drug to the market was US$800 million [3] and took 12-15 years. A more recent study suggests this figure could actually approach US$2 billion [4]. It is only the final stages of drug development where people themselves become part of the research and development process – the so-called phase I-IV clinical trials. The majority of work in bringing a drug to market occurs in model systems, including computerised simulations, cell culture dishes and invertebrate and vertebrate animals. Throughout this process the researcher trades simplicity of experimental paradigm and interpretation against increasing system complexity. For example the single cell brewer's yeast, Saccharomyces cerevisiae, has 6,275 genes that have all been individually knocked out in mutant strains. The transparent nematode, Caenorhabditis elegans has exactly eight dopaminergic neurons that can be easily visualised [5] while the Zebrafish embryo is also transparent making phenotypic abnormalities rapidly recognisable. The International Mouse Phenotyping Consortium (IMPC) predict that US$900 million will be required over the next decade to create a yeast-like gene knockout database for a mammal.

Despite their relative similarity to humans there remain many issues over the suitability of laboratory animals for human disease modelling. These stem from the significant anatomical, physiological, developmental and ageing differences. Only one-third of the mouse studies have translated at the level of human randomised trials, and then only one-tenth of these interventions, were subsequently approved for use in patients [6]. These issues become even more important when we are considering therapeutics for human central nervous system. In diseases such as PD, researchers will typically look to utilise the more human-like primate species but differences still remain and research costs escalate sharply [7]. There are also important ethical issues and societal concerns that need to be considered when contemplating the necessity of primate research [8].

There is a need to find humanised model systems that can perform a large component of preclinical drug discovery economically and patient-derived pluripotent stem cells are emerging as a very cost-effective *in vitro* assay for this purpose.

Genetics and the Modelling of Human Diseases

Genetic diseases can vary from monogenic, caused by a single mutation, through to polygenic where a number of gene variants combine to create the disease phenotype. These combinations might not be just main effects acting in an additive fashion, but include epistatic or gene x gene interactions.

The completion of the Human Genome Project in 2001 brought with it tremendous hope that the pathogenesis of many complex diseases and therefore therapeutic targets were in our immediate grasp [9, 10]. However one of the major challenges of the post-human genome era has been to characterise the nature and extent of variation between humans. Any two humans chosen at random will share 99.5% of their genetic code while their differences will be made up of common variants including single nucleotide polymorphisms, variable repeat regions, small inserts or deletions and larger copy number variations [11, 12]. The Human Genome Project also sparked a technological revolution in high throughput technologies for the characterisation of these variants. An example of their impact is the recent discovery of the major role that copy number variants (CNV) play in CNS diseases including the autism spectrum [13]. Excitingly, next-generation sequencing looks set to accelerate the pace of genetic variation discovery [14] and along with it, the potential sources of disease causation.

While rare polymorphisms, or mutations are responsible for monogenic forms of disease, common genetic variants are postulated to modify the risk for common or sporadic forms (common variant hypothesis). Common variants appear to explain a good proportion of the genetic or heritable component of causation in type II diabetes [15] and macular degeneration [16-18]. In AD the APOE ε4 common variant plays a major role [19] whereas common variants of the genes microtubule associated tau (MAPT) and α-synuclein (PARK 1,4) have a relatively modest effect on modifying the risk for sporadic forms of PD [20, 21]. The opposing idea or 'rare variant hypothesis' assumes far greater genetic heterogeneity in disease causation with rare polymorphisms in many different genes prevalent in the affected population [22]. A very recent paper suggests that, for hypertriglyceridaemia at least, these hypotheses may not be mutually exclusive [23].

However the genetic component of disease rarely exists in isolation with environmental exposures playing some role in causation. This can range from a solely causative peracute toxicological insult through to gene x environmental interactions. The latter describes the scenario when a putative genetic risk factor only becomes important if the individual is exposed to a particular environment hazard. For example, variants of xenobiotic metabolism genes such as the P450 enzymes that confer different detoxification abilities on their subjects appear important in a person's risk of PD [24]. Similarly variants in these enzymes are important in the metabolism of existing drugs such as codeine and prospective pharmaceuticals [25].

Technological advances have afforded us an ever-increasing ability to characterise genetic variation in the human population but also reminded us just how extensive this variation is likely to be. This contrasts sharply with the intrabred strains of mice and rats that serve as the workhorses of human disease modelling and drug development. Here genetic variation has been deliberately removed as a potentially confounding factor.

Patient-derived stem cell models that incorporate the genotype or genome-type of the individual they are derived from represent a very powerful model for disease modelling and drug development.

The Development of Cellular Models of Disease

Disease research often begins in cell culture experiments where the cells have been derived directly from the tissue of interest or more commonly immortalised lines originating from tumours or somatic cells treated with neoplastic agents. Cell culture grew out of organ and tissue culture in the late 19th century and advanced in the mid twentieth century through viral isolation and culture for vaccine development.

The majority of cell lines are monocultures that adhere to the flask surface which may be coated with extracellular matrix proteins to facilitate this process. Co-culturing of different cell phenotypes together is possible while organotypic culture systems that feature a three-dimensional environment are moving cell culture closer to an *in vivo* tissue.

The advantages of immortalised cell lines and cell culture in general are the theoretical unlimited supply of the same phenotype (experimental replication), economics and their ease of use and manipulation. However immortalised lines represent the genotype of a single individual and by definition feature a genetic and physiological anomaly. Furthermore disease modelling will most often involve additional and likely non-physiological toxicological or artificial genetic manipulation (overexpression, knock-in or knock-down). Therefore a major disadvantage of these lines is that the role of human genetic variability in disease cannot be assessed.

In contrast primary cell cultures can be derived from multiple individuals for comparative studies but they have a relatively short lifespan. This is particularly true for mammalian post mitotic brain cells that last only days to weeks in culture. However human brain tissue, whether from surgical excision, biopsy or following a bequest, is a fairly rare resource in medical research. Other cells such as fibroblasts retain an ability to replicate in culture but they too undergo the process of senescence after a certain number of population doublings. The maximum number of divisions which somatic cells undergo *in vitro* is referred to as the Hayflick limit [26] and is generally thought to reflect an incremental shortening of telomere length. The maximum number of doublings for a somatic cell is considered to be around eighty. Cells that can, in the absence of oncogenic transformation, easily exceed this limit are considered capable of extensive proliferation, a criteria first met by embryonic stem cells.

Stem cells are defined by two basic characteristics: *immortality* and *multipotency*. A stem cell can undergo symmetrical division to produce a replica of itself or asymmetrical division to produce a less multipotent descendent, referred to as a transit-amplifying or progenitor cell.

In human embryology, the term pluripotent only applies for cells up to the 16-cell blastocyst stage. From then on lineage determination begins, with dedicated precursors moving towards unipotency or differentiated progeny. This has always been assumed to be a one-way paradigm.

In recent years stem cell biology has greatly expanded our knowledge regarding the extent to which multipotent stem cells are retained and remain functionally active in adult human tissues. Stem cell technology has made it possible to reverse the lineage paradigm by recreating a multipotent, perhaps pluripotent, state from fully differentiated cells [27, 28].

Embryonic Stem Cells

The first successful derivation of human embryonic stem cell (ESC) lines occurred in 1998 [29] and there have now been over one thousand lines produced worldwide [30, 31]. ESC lines are derived from the inner cell mass of pre-implantation embryos, 4-5 days after fertilisation, that have been discarded from *in vitro* fertilisation (IVF) clinics (Fig. **1**). The production of these lines remains an inefficient procedure and excess embryos are initially required. Typically 20 or more embryos will be required to generate one ESC line [32, 33]. ESC lines are propagated on feeder cells such as human fetal fibroblasts. The latter tissue is often sourced from therapeutic pregnancy terminations and hence this entire process must be carried out under strict ethical guidelines. ESC lines are inherently versatile because they are pluripotent and can be propagated indefinitely including through several freeze/thaw cycles (Sidhu, 2010 #116). To be considered pluripotent, cells must display a spontaneous ability to differentiate, be able to be manipulated to form cell lineages derived from all three germ layers and form mixed-cell neoplasms (teratomas) when injected into ectopic sites of immunosuppressed mice [34]. In addition ESCs derived from animal work should be able to generate germline-competent chimaeras when introduced into pre-implantation embryos that are subsequently allowed to develop to full term in foster mothers.

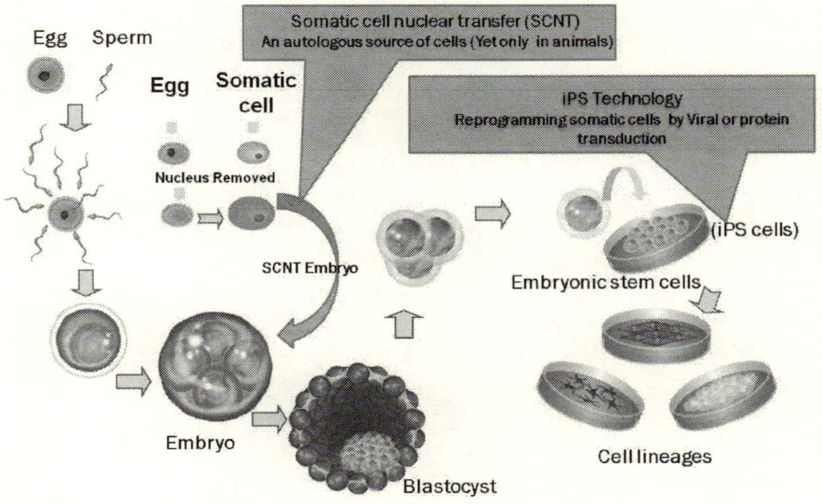

Fig. 1: Stem cell sources for medical research.

The pluripotent human embryonic stem cells (hESCs) are derived after fertilisation from a day 5 embryo called a blastocyst. At this stage it consists of about 200 cells of which 30-40 cells constitute the inner cell mass (ICM) while the others form the trophectoderm that eventually forms the placenta. hESCs are derived by dissecting out the ICM and growing these on feeder cells. hESCs are pluripotent and form all lineages derived from the three germ layers. In animals, an autologous source of such cells have been produced by somatic cell nuclear transfer (SCNT) where an enucleated ovum is combined with a somatic cell nuclei, but this has not yet been achieved in humans. Recently somatic cells have been reprogrammed to a pluripotent state by simply introducing four pluripotent genes, Oct4, Sox2, Klf4, c-Myc either by viral or protein transduction (iPS technology). Such induced pluripotent stem (iPS) cells have also been shown to generate the different lineages derived from the three germ layers but they currently have limitations for future regenerative medicine because of the potential risks associated with these transgenes. Such limitations may not necessarily prevent their immediate use in disease modelling and drug discovery (from Sidhu, 2011#35).

Molecular assays are also used to characterise pluripotency and ESCs typically show expression of the stem cell surface markers (stage-specific embryonic antigens) SSEA-3 and SSEA-4 and tumor recognition antigens TRA-1-60 and TRA-1-81. They should also express the pluripotent gene marker, Nanog [36] and stain for alkaline phosphatase[33]. The expression of SSEA3, SSE4, Nanog and OCT4 (POU5F1) are highly correlated with one another in ESCs [37] although a meta-analysis of microarray studies showed that only OCT4 was reported to be consistently overexpressed when ESCs were compared to differentiated cell types [38]. This suggests that there is a high degree of heterogeneity amongst and within ESC lines and that the truly pluripotent cells represent only a subpopulation within the cultures.

In terms of neurodegenerative diseases ESCs represent a source of all of the brain cells implicated in the disease process including the neurons and glial cells derived from (neuro)ectoderm and microglia derived from mesoderm [39]. The production of neurons from ESCs relies on protocols that incorporate the orchestrated signalling events determined from embryological work. This process incorporates the development of neuroectodermal precursors (neural induction), the subdivision of neural precursors along dorsoventral and rostocaudal axes (patterning) and the differential exposure of these precursors to region-specific morphogenic cocktails to derive neuronal and glial subtypes (recently reviewed [30]).

The neuronal pathway has been regarded as a default lineage as ESCs can spontaneously form neural cells after prolonged culture. This process can be augmented with the introduction of epidermal growth factor (EGF) and basic fibroblast growth factor (bFGF or FGF2) resulting in neurospheres containing a high proportion of neural progenitors [40]. However subsequent studies suggest that these neurospheres probably represented aggregates of less multipotent progenitors [41]. The Zhang *et al.* protocol was based on the development of aggregates of cells called embryoid bodies (EB). Alternatively neural cells can be derived from monolayer ESCs utilising mesodermal feeder cells or derived-media [42]. The withdrawal of FGF2, from protocols that utilise it, precipitates their spontaneous differentiation into mixed neurons and glia.

The production of specific neuronal phenotypes such as motor neurons from ESCs has relied on different combinations of retinoic acid (RA), FGF2 and other factors such as sonic hedgehog (SHH) [43]. The confirmation of the neuronal phenotypes relies on RT-PCR or immunocytochemical demonstration of general (β III tubulin, NeuN) or specific (tyrosine hydroxylase (TH), choline acetyltransferase (ChAT) or homeobox HB9 (MNX1)) neuronal markers as well as functional assays such as the generation of action potentials and their subsequent inhibition by subpopulation-specific voltage-gated ion channel inhibitors [30]. Nevertheless most differentiation protocols still result in heterogenous populations of neurons, glia and progenitor cells. There remain considerable challenges in identifying the cellular and molecular players that regulate the different steps in neurogenesis to improve the efficiency of differentiation from ESCs.

ESCs have been valuable in modelling monogenic diseases where afflicted embryos have been obtained from preimplantation diagnosis such as in Fragile X (FX) syndrome [44]. Furthermore their versatility in culture makes them ideal for basic research into aspects of common disease pathogenesis however ESCs are constrained by their derivation from one or two individuals.

Adult Stem Cells

As mentioned above, there are a number of challenges to the accurate modelling of human neurodegenerative disease. Post-mortem brain tissue is contextually optimal but it is a relatively scarce commodity, analyses are inherently retrospective and there are tissue quality issues that differentially affect RNA-based methodologies [45, 46]. Alternatively researchers can utilise clinically accessible samples such as whole blood serum or cerebrospinal fluid. The major issue here is that the brain lies behind the blood brain barrier and not all potential biomarkers will disseminate from the brain, or be expressed systemically, in sufficient quantities to be detected. In the case of PD the pathways highlighted as disease-specific in transcriptomic analyses of whole blood bear little resemblance to those derived from whole brain samples [47]. Researchers have previously taken advantage of the reasonable proliferative potential of clinically accessible somatic cells such fibroblasts to model neurological diseases like schizophrenia [48, 49] and PD [50] but the contextual relevance of these non-neural cells for understanding neurological and neuropsychiatric disorders appears limited [51].

It has been long known that certain adult tissues such as the bone marrow retain a multipotent cell population and this has been utilised for many years in bone marrow or stem cell transplantation. It has been less appreciated that most tissues actually retain a multipotent population of cells. These 'stem' cells are functionally relevant in high turnover areas such as the skin and intestinal lining but in other tissues such as the human brain, stem cells were always considered quiescent or absent. It was only in 1998 that neurogenesis was shown to continue throughout life in the human hippocampus [52]. These adultborn hippocampal neurons have functional roles [53] as do the neuroblasts that migrate *via* the rostral migratory stream from the subventricular zone to continually repopulate the olfactory bulb [54].

It was also realised that if the stem cells present in clinically accessible tissues such as the skin could be proliferated in culture they might provide a less controversial alternative to ESCs. The important question was how pluripotent these cells would be? Adult stem cell researchers took the optimistic view that these cells would be sufficiently multipotent to differentiate into all the major lineages but insufficient to result in deleterious sequelae like teratoma formation if transplanted [55].

Advances in stem cell technology meant that these adult stem cells could be sourced from the skin [56-58], bone marrow stroma (mesenchymal) [59-61] and olfactory mucosa [55, 62] and manipulated in culture to produce neuronal-like cells. The initial focus of such work was for regenerative medicine where the use of autologous transplantation of patient-derived stem cells was predicted to remove the risk of "graft-host" disease inherent to allografts of hESCs, although the underlying genetic susceptibility to disease would presumably still be present in the transplanted cells. However their alternative use for disease modelling was soon appreciated where cell lines from multiple individuals could be generated facilitating cell culture experiments that could explore the role of genetic variance in complex diseases. There were also obvious modelling applications in drug development as well as potential clinical uses in personalised medicine [63, 64].

An interesting source of stem cells is the olfactory mucosa. This tissue is unique in the mammalian CNS in that all the olfactory receptor neurons (ORN) are continually replaced over the lifetime of the animal. This process relies on a yet to be defined 'stem cell' near the basement membrane of the olfactory epithelium [65]. The axons of these new ORNs are guided by olfactory ensheathing or glial cells through the cribiform plate to synapse with mitral cells within odour-specific glomeruli of the olfactory bulb [66, 67]. This embryological-like environment has tremendous potential for understanding neural biology and as a source of cells for nervous system injuries [68, 69].

Like mammalian neural stem cells from the dentate gyrus or subventricular zone [70], olfactory mucosa-derived stem cells are multipotent and can be propagated in culture to form neurospheres [55]. Cells subsequently grown from these dissociated neurospheres (hONS) have neuronal-like properties [71]. Consistent with previous case-control comparisons [72, 73], hONS cells from schizophrenic patients show alterations in neurogenesis and cell migration pathways while PD patient-derived hONS cells show anomalies in mitochondrial function and oxidative stress [71], a convergent theme from genetic and toxicological studies [24, 74].

The pathogenesis of sporadic forms of chronic diseases like PD is thought to be due to complex gene x environmental interactions and this has been called the 'ecogenetic theory' [24]. The theory refers to common gene (germline) variants conferring differential responses to environmental exposures. However the environment can also modify the pattern of gene expression. Processes such as DNA methylation, RNA editing [75] and chromatin restructuring create an epigenetic signature of lifetime exposures. This is most impressively shown in studies of phenotypic discordance in monozygotic twins [76]. The epigenome of the human brain is particularly responsive to neuronal insults making epigenetic dysregulation a common theme in neurodegenerative disorders [77].

For patient-derived hONS cells it is interesting to consider what the relative impact of the individual's exposure history might be on the antecedent olfactory stem cell. Furthermore would this epigenetic signature be maintained through successive divisions of their descendent cell populations? In the first instance would these stem cells be anymore or less insulated *in vivo* from the environmental stresses that are postulated to cause epigenetic effects in post mitotic neurons? Certainly stem cells with their distinctive transcriptome would appear to be differentially susceptible to exposures that afflict brain cells. The second issue here is how the disease itself or therapies might impact on adult stem cell populations. Critics of stem cell disease models suggest that case-control differences will merely represent the impact of the disease or treatments specific to the cases rather than an underlying genetic susceptibility or physiological difference in the cells. In experimental terms it will be extremely hard to determine whether patient-specific anomalies are 'cause' or 'effect' for adult stem cell derived cultures. The case of resilience of epigenetic effects to cell culture passages is discussed further below.

Work continues in demonstrating the degree of pluripotency inherent to adult stem cell cultures and therefore their overall applicability in medical research, but their overall role in disease modelling may be superseded by a recent development in stem cell technology that has changed the entire landscape of regenerative medicine.

Induced Pluripotent Stem Cells

It was always considered that the transition from pluripotency to a restricted lineage and finally to terminally differentiated cells was a linear and irreversible fate.

In 2006, Yamanaka and colleagues showed in mice that both embryonic and adult fibroblasts could be induced to 'de-differentiate' to a pluripotent state through the iatrogenic expression of four transcription factors, under ESC culture conditions [27]. In quick succession the same group were able to demonstrate that human adult fibroblasts could be induced to undergo the same transformation [28]. These cells were called induced pluripotent stem cells (iPSC) (Fig. **1**).

The pioneering work of Yamanaka and colleagues in iPSCs was driven by both the ethical difficulties and the problem of tissue rejection following transplantation with ESCs [27]. Briefly their experimental paradigm featured an assay system in mouse embryonic fibroblasts (MEF) cultures that incorporated an antibiotic (G418)-sensitivity cassette in the promoter region of the pluripotent state-expressed, but dispensible gene, *Fbx15*. They initially selected 24 transcription factors based on the maintenance of long-term pluripotency in embryos and ECSs. The full combination of these 24 factors induced pluripotency when introduced into the MEFs by retroviral transduction. Takanashi *et al.* then withdrew the factors one by one to ascertain a core requirement for pluripotency. They found that only Oct3/4, Sox2, c-Myc, and Klf4 were indispensable while surprisingly nanog was not required. Their iPSCs satisfied the functional definitions of pluripotency including germline transmission. However they did note that only a small portion of cells that expressed the four factors actually became iPS cells.

Similar to mouse iPSCs and distinct from the parent fibroblasts the promoter regions of pluripotent-associated genes, such as *OCT3/4*, *REX1*, and *NANOG* were highly unmethylated in human iPSCs, their telomerase activity was high and embryoid bodies could be differentiated into cells immunopositive for mesoderm, ectoderm and endoderm markers. Using ESC-derived protocols Yamanaka and colleagues could differentiate their iPSCs into dopaminergic neurons [42] and cardiac myocytes [78] or at least phenocopies that expressed the characteristic markers of these cells. Importantly the iPSCs satisfied the functional definition of teratoma formation in immunocompromised mice. Differentiated cell models of amyotrophic lateral sclerosis [79], Huntington's disease [80], PD [81] Friedreich's ataxia [82] and spinal muscular atrophy have been developed from iPSCs and in the case of the latter, cured *in vitro* [83]. iPSC lines have also been developed from a number of other monogenic disorders [84]. Interestingly the triplet repeat length in the Huntington's disease iPSC cells and differentiated neurons remained constant [80] but the repeat region Friedreich's ataxia patient's fibroblasts lengthened during the induction of pluripotency and as iPSC cells grew in culture [82]. Ku *et al.* found that that silencing of the DNA repair gene MSH2 stalled the expansions and suggested that misplaced attempts at repair were likely to cause the expansions themselves. It may be important that the expansion in Huntington's disease is in the coding region of the Huntingtin protein while the Freidreich's ataxia repeat region is the first intron of the *FRDA* gene.

In contrast to adult stem cells, iPSC technology was originally thought to erode epigenetic signatures, meaning cell memory would be lost, both in the iPSCs themselves and their differentiated progeny [85]. This 'ground zero/naïve state' appears akin to the embryological status of a susceptible individual. In the modelling of neurodegenerative diseases most researchers have looked to differentiate iPSCs to susceptible phenotype such as dopaminergic neurons in PD [81, 86]. We might expect that the putative differences in naïve iPSCs from patients compared to unaffected individuals to manifest themselves prior to or following this differentiation processes. However in cell lines derived from sporadic cases differences may only become apparent when cell lines are subjected to incremental perturbation and recovery. In essence we are trying to recreate the natural history of exposures for an affected individual or their disease history. This process could provide valuable insight into the progressive pathophysiology of that disease, allowing the potential discovery preclinical biomarkers, and providing a continuous supply of afflicted cell types for drug screening and discovery.

However a couple of recent reports suggest that the iPSC process does not necessarily revert cells to this ground state. In Fragile X syndrome, mental retardation results from the silencing of the fragile X mental retardation 1 (*FMR1*) gene. A CAG expansion in the *FMR1* promoter region is synonymous with hypermethylation. The gene is active in FX-ESCs cells but is silenced upon differentiation in teratomas. However induction of pluripotency by the four "Yamanaka" factors failed to reactivate *FMR1* expression [87]. This appears to be due to recalcitrant histone modifications and is not methylation-dependent. A similar scenario has also been reported in other iPSCs concerning the imprinted genes, *H19* [88] and iPSCs from patients with the imprinting disorders Angelman syndrome (AS) and Prader-Willi syndrome (PWS) [89, 90]. These examples may just reflect sites that become highly modified with differentiation [90], a scenario that augurs well for disease modelling of monogenic diseases but also serves a more general caution that a 'ground zero' state for any loci cannot be automatically assumed with current iPSC methodologies.

In their work Yamanaka and colleagues performed techniques to molecularly characterise iPSC with ESC markers and suggested that the two cell types were equivalent in terms of their pluripotency [28]. This has been supported by other studies using embryoid body and teratoma formation assays [84, 91, 92]. A recent transcriptomic study confirmed the similarities between ESCs and iPSCs but also suggested that there are sufficient differences, including miRNA expression, to regard iPSCs as unique subtype of pluripotent cell [93]. The possible physiological consequences of these differences are yet to be determined.

One issue with iPSCs remains the efficiency of the induction process. Like ESCs the propagation of iPSCs generally relies on feeder cells. Recent work in our laboratory suggests that feeder-free conditions up-regulate DNA replication and cell cycle genes lacking in feeder-derived iPS cells and increase their similarities to human ESCs (unpublished work). We have also identified potential biomarkers for fully reprogrammed iPS cells and several for partially reprogrammed iPS cells (POTEG, MX2).

The other major issue acknowledged by Takahashi et al. in their original studies is the probable incompatibility between viral transduction methodologies and subsequent therapeutic applications. The integration of pluripotent genes, particularly c-myc, or perhaps other viral components, into transplanted cells carries too higher risk of oncogenesis. Pluripotency has been induced by three factors excluding c-myc [86, 94] but the oncogenic potential of the other factors, real or perceived, remains. Alternatively the induction of pluripotency, in the absence of genomic integration, can be conferred by transfection of recombinant factors but this is likely to be too inefficient for therapeutic applications [95]. One potential strategy is the induction of pluripotency using Cre-recombinase excisable viruses [81]. It is reported that the transcriptomic analyses of factor-free iPSCs suggested that they were actually more closely related to ESCs than those carrying the transgenes. It has also been suggested that small molecules could enhance efficiency or even replace transcription factors during somatic cell reprogramming [96]. To date screening of small molecules libraries has identified substitutes for Klf4 in the reprogramming process [97].

One of the potential uses of patient-derived cell lines is the idea of curing diseases *in vitro*. Here molecular or functional anomalies would be rectified to levels seen in iPSCs or differentiated descendents from control individuals through targeted perturbations or high throughput screening of pharmacologically active compounds [98]. Patient derived cell models can potentially overcome the issues with animal models and decrease the natural history of human diseases from years to months. PD [81], familial dysautonomia [99], and spinal muscular atrophy models have been developed and in the case of the latter, cured *in vitro* [83]. Soldner *et al.* [81] induced pluripotency in fibroblasts derived from PD patients and controls and subsequently differentiated both into dopaminergic neurons but they did not examine the cells for case-control differences. They went on to suggest that it might still be necessary to further accelerate PD-pathology related phenotypes *in vitro* with neurotoxins such as MPTP, or the overexpression of PD-related (*PARK*) genes such as *SNCA* or *LRKK2 (encoding* α-synuclein and leucine-rich repeat kinase 2 respectively).

iPSC derived from affected individuals provide a unique opportunity to gain insights into the neurobiology of neurodegenerative diseases. Disease modelling with iPSCs is likely be a three part process: first the derivation of disease-specific iPSC, secondly to differentiate these cells effectively to cell lineage of choice like neurons, oligodendrocytes and astrocytes that hopefully express disease phenotype and thirdly to use these derived lineage-specific cells for high through put screening to model the disease and functional analyses if used for therapeutics down the track [35] (Fig. **2**). In PD this is likely to be tyrosine hydroxylase-positive dopaminergic neurons [81] or in Huntington's disease, the dopamine- and cyclic AMP-regulated phosphoprotein (DARPP-32)-positive striatal neurons [80]. It is still important to realise that these are *in vitro* phenotypes, whose identification (in the current absence of transcriptomic data) is based on only a few markers. Interestingly in diseases with a neurodevelopmental component, and this does not exclude neurodegenerative diseases, aspects of dysfunction can be modelled through this differentiation process [99] rather than solely relying on differentiated ('end-point') phenotypes.

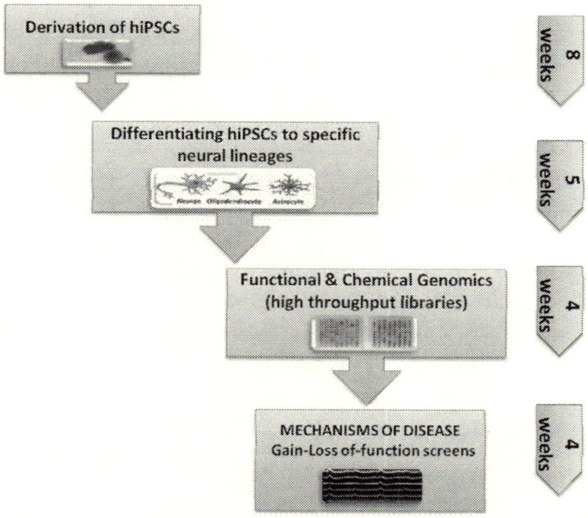

Fig. 2: Modelling Neurodegenerative Diseases *in vitro*.

It is interesting to consider the numbers of cell lines that might be needed to find statistically significant differences between affected and unaffected individuals. In the case of monogenic diseases such as spinal muscular atrophy differences are obvious from comparisons with single unaffected cell lines [83]. However in the case of sporadic forms of neurodegenerative disease it may require multiple individuals to detect the predicted subtle differences in pathophysiology. This subtlety is based on the late-onset of these diseases and the contextual disparity between the cell lines and the brain itself. However the detection of differences will also depend on the variation within both the control and patient groups. Previous experiments with hONS cell lines suggest that disease-specific differences can be detected with around 20 patient and 20 unaffected cell lines [71].

The use of human induced pluripotent stem cells (iPSCs) as cellular models for neurodegenerative diseases could be a three step phenomenon *i.e.* derivation, differentiation and functional annotation of hiPSCs and their derived neural progenies by high throughput assays. A prospective time line is indicated for the individual components (from Sidhu, 2011#35).

CONCLUSIONS AND PROSPECTS

Most of the chapters in this eBook discuss the implications of stem cell technology in regenerative medicine. In this chapter we have attempted to outline the revolutionary impact that stem cell technologies will also have on disease modelling. It is now possible to differentiate patient-derived stem cells into disease-susceptible phenotypes and compare them with those from unaffected individuals. Pathogenic mechanisms can be teased out by perturbing control lines to match levels seen in the patients' cells or alternatively by treating patients' cells to revert particular indices to those of controls. The latter is akin to a "cure in a dish". Outside its role in basic science iPSC technology will be an important driver of personalised medicine. Prior to treatment, a patient's iPS cells or differentiated progeny can be used to tailor a particular drug type and dose to their genetic and cellular profile. Here iPSCs might be differentiated into hepatocytes as these cells are likely to express the relevant xenobiotic metabolism machinery. Similarly cells derived from patients could be used to predict, in a dish, the toxicity of a given drug prior to treatment.

Patient-derived cell lines potentially have a major role to play in preclinical drug development. As described in the introduction to this chapter the vast majority of 'lead' candidates don't translate into clinical applications. This includes drugs that have shown impressive cures in mammalian model systems. This is a massive expense for the pharmaceutical industry and creates an opportunity cost for other diseases. Humanised models could impact positively and early in the drug discovery pipeline from disease models for high throughput screening of compound libraries to drug safety screens [63, 64]. This has the added bonus of reducing the animal dependency of the current drug development pipeline.

However a number of issues need to be resolved before the full potential of these cells can be exploited in clinical practice including the erosion of epigenetic signatures during reprogramming process, the efficiency of the induction and differentiation process and the resultant phenotypic heterogeneity within the culture flask.

The identification of a dysfunctional pathway in patients suffering from complex diseases such as AD and PD is the primary requirement for rational therapeutic drug development. Stem cell models promise to accelerate the pace of pathogenic discovery in complex diseases but they remain fairly basic systems. As with all cell culture models they are proliferative in nature and commit the majority of their energy to mitosis and adaptation to their plastic environment. This is a far cry from the post mitotic neuron in the forebrain nuclei or substantia nigra of an AD or PD patient. Having said this, disease clues could come from this type of environment as these cell culture conditions are a stress and mitosis is the status quo in the human brain until early adulthood. We also know that enclaves of neurogenesis remain throughout life in humans and that this process can be stimulated in pathogenic conditions [100].

Nevertheless it seems most likely that stem cell models will be able to answer pieces of the puzzle in these diseases. Optimistically we must hope that these simple models can accurately represent these pieces but of course there is no guarantee that these individual parts and their manipulation will retain relevance for the disease. Our best defence against expensive blind allies in disease research is collaboration and validation across multiple model systems.

REFERENCES

[1] Kern A and Behl C. The unsolved relationship of brain aging and late-ons*et al*zheimer disease. Biochimica et Biophysica Acta 2009; 1790(10): 1124-1132.

[2] Box GEP and Draper NR. Empirical Model-Building and Response Surfaces. 1987: John Wiley, USA.

[3] DiMasi JA, Hansen RW and Grabowski HG.The price of innovation: new estimates of drug development costs. J Health Econ 2003; 22(2): 151-85.

[4] Adams CP and Brantner VV. Estimating the cost of new drug development: is it really 802 million dollars? Health Aff (Millwood) 2006; **25**(2): 420-8.

[5] Sulston J, Dew M and Brenner S. Dopaminergic neurons in the nematode Caenorhabditis elegans. J Comp Neurol 1975; 163(2): 215-26.

[6] Hackam DG and Redelmeier DA.Translation of research evidence from animals to humans. JAMA 2006; 296(14): 1731-2.

[7] Emborg ME. Nonhuman primate models of Parkinson's disease. ILAR J 2007; 48(4): 339-55.

[8] Busko CA. Green light signals caution for primate research. Lancet 2009; 373(9679): 1920.

[9] Lander ES, Linton LM, Birren B, *et al.* Initial sequencing and analysis of the human genome. Nature 2001; 409(6822): 860-921.

[10] Venter JC, Adams MD, Myers EW, *et al.* The sequence of the human genome. Science 2001; 291(5507): 1304-51.

[11] Levy S, Sutton G, Ng PC, *et al.* The Diploid Genome Sequence of an Individual Human. PLoS Biol 2007; 5(10): e254.

[12] Kidd JM, Cooper GM, Donahue WF, *et al.* Mapping and sequencing of structural variation from eight human genomes. Nature 2008; 453(7191): 56-64.

[13] Pinto D, Pagnamenta AT, Klei L, *et al.* Functional impact of global rare copy number variation in autism spectrum disorders Nature 2010; 466(7304): 368-72.

[14] Oetting WS. Impact of next generation sequencing: the 2009 Human Genome Variation Society Scientific Meeting. Hum Mutat 2010; 31(4): 500-3.

[15] Wellcome Trust Case Control Consortium. Genome-wide association study of 14,000 cases of seven common diseases and 3,000 shared controls. Nature 2007; 447(7145): 661-78.

[16] Klein RJ, Zeiss C, Chew EY, *et al.* Complement factor H polymorphism in age-related macular degeneration. Science 2005; 308(5720): 385-9.

[17] Edwards AO, Ritter R 3rd, Abel KJ, *et al.* Complement factor H polymorphism and age-related macular degeneration. Science 2005; 308(5720): 421-4.

[18] Haines JL, Hauser MA, Schmidt S, *et al.* Complement factor H variant increases the risk of age-related macular degeneration. Science 2005; 308(5720): 419-21.

[19] Bertram L, Lange C, Mullin K, *et al.* Genome-wide association analysis reveals putative Alzheimer's disease susceptibility loci in addition to APOE. Am J Hum Genet 2008; 83(5): 623-32.

[20] Satake W, Nakabayashi Y, Mizuta I, *et al.* Genome-wide association study identifies common variants at four loci as genetic risk factors for Parkinson's disease. Nat Genet 2009; 41(12): 1303-7.

[21] Simón-Sánchez J, Schulte C, Bras JM, *et al.* Genome-wide association study reveals genetic risk underlying Parkinson's disease. Nat Genet 2009; 41(12): 1308-12.

[22] Pritchard, J.K., Are rare variants responsible for susceptibility to complex diseases? Am J Hum Genet 2001; 69(1): 124-37.

[23] Johansen CT, Wang J, Lanktree MB, *et al.* Excess of rare variants in genes identified by genome-wide association study of hypertriglyceridemia. Nat Genet 2010; 42(8): 684-7.

[24] Mellick G.D. CYP450, genetics and Parkinson's disease: gene x environment interactions hold the key. J Neural Transm Suppl 2006; (70): 159-65.

[25] Cascorbi I. Genetic basis of toxic reactions to drugs and chemicals. Toxicol Lett 2006; 162(1): 16-28.

[26] Hayflick L and Moorhead PS.The serial cultivation of human diploid cell strains. Exp Cell Res 1961; 25: 585-621.

[27] Takahashi K and Yamanaka S. Induction of pluripotent stem cells from mouse embryonic and adult fibroblast cultures by defined factors. Cell 2006; 126(4): 663-76.

[28] Takahashi K, Tanabe K, Ohnuki M, *et al.* Induction of pluripotent stem cells from adult human fibroblasts by defined factors. Cell 2007; 131(5): 861-72.

[29] Thomson JA, Itskovitz-Eldor J, Shapiro SS, *et al.* Embryonic stem cell lines derived from human blastocysts. Science 1998; 282(5391): 1145-7.

[30] Sidhu, K.S., Motoneurons from Human Embryonic Stem Cells: Present Status and Future Strategies for the use in Regenerative Medicine, in K Maiese, Ed. Neurovascular Medicine Pursuing Cellular Longevity for Healthy Ageing, 2009, Oxford University Press: Oxford. pp. 580.

[31] Löser P, Schirm J, Guhr A, Wobus AM, Kurtz A. Human embryonic stem cell lines and their use in international research. Stem Cells 2010; 28(2): 240-6.

[32] Sidhu KS, Ryan JP, Lees JG, Tuch BE. Derivation of a new human embryonic stem cell line, Endeavour-2, and its characterization. *In Vitro* Cell Dev Biol Anim 2010; 46(3-4): 269-75.

[33] Sidhu KS, Ryan JP and Tuch BE. Derivation of a new human embryonic stem cell line, endeavour-1, and its clonal propagation. Stem Cells Dev 2008; 17(1): 41-51.

[34] Sidhu KS and Tuch BE. Derivation of three clones from human embryonic stem cell lines by FACS sorting and their characterization. Stem Cells Dev 2006; 15(1): 61-9.

[35] Sidhu KS. Induced pluripotent stem cells as slippery slope for neurodegenerative disease modelling? The open stem cell journal Suppl. Issue 2011; 3: 46-51.

[36] Silva J, Nichols J, Theunissen TW, *et al.* Nanog is the gateway to the pluripotent ground state. Cell 2009; 138(4): 722-37.

[37] Adewumi O, Aflatoonian B, Ahrlund-Richter L, *et al.* Characterization of human embryonic stem cell lines by the International Stem Cell Initiative. Nat Biotechnol 2007; 25(7): 803-16.

[38] Assou S, Le Carrour T, Tondeur S, *et al.* A meta-analysis of human embryonic stem cells transcriptome integrated into a web-based expression atlas. Stem Cells 2007; 25(4): 961-73.

[39] Lobsiger CS and Cleveland DW. Glial cells as intrinsic components of non-cell-autonomous neurodegenerative disease. Nat Neurosci 2007; 10(11): 1355-60.

[40] Reubinoff BE, Itsykson P, Turetsky T, *et al.* Neural progenitors from human embryonic stem cells. Nat Biotechnol, 2001. 19(12): 1134-40.

[41] Zhang SC, Wernig M, Duncan ID, Brüstle O, Thomson JA, *et al. In vitro* differentiation of transplantable neural precursors from human embryonic stem cells. Nat Biotechnol 2001; 19(12): 1129-33.

[42] Kawasaki H, Mizuseki K, Nishikawa S, *et al.* Induction of midbrain dopaminergic neurons from ES cells by stromal cell-derived inducing activity. Neuron 2000; 28(1): 31-40.

[43] Lim UM, Sidhu KS and Tuch BE. Derivation of Motor Neurons from three Clonal Human Embryonic Stem Cell Lines. Curr Neurovasc Res 2006; 3(4): 281-8.

[44] Eiges R, Urbach A, Malcov M, *et al.* Developmental study of fragile X syndrome using human embryonic stem cells derived from preimplantation genetically diagnosed embryos. Cell Stem Cell 2007; **1**(5): 568-77.

[45] Atz M, Walsh D, Cartagena P, *et al.* Methodological considerations for gene expression profiling of human brain. J Neurosci Methods 2007; 163(2): 295-309.

[46] Monoranu CM, Apfelbacher M, Grünblatt E, *et al.* pH measurement as quality control on human post mortem brain tissue: a study of the BrainNet Europe consortium. Neuropathol Appl Neurobiol 2009; **35**(3): 329-37.

[47] Sutherland GT, Matigian NA, Chalk AM, *et al.* A cross-study transcriptional analysis of Parkinson's disease. PLoS ONE 2009; 4(3): e4955.

[48] Mahadik SP and Mukherjee S. Cultured skin fibroblasts as a cell model for investigating schizophrenia. J Psychiatr Res 1996; 30(6): 421-39.

[49] Miyamae Y, Nakamura Y, Kashiwagi Y, Tanaka T, Kudo T, Takeda M. Altered adhesion efficiency and fibronectin content in fibroblasts from schizophrenic patients. Psychiatry Clin Neurosci 1998; 52(3): 345-52.

[50] Hoepken HH, Gispert S, Azizov M, *et al.* Parkinson patient fibroblasts show increased alpha-synuclein expression. Exp Neurol 2008; 212(2): 307-13.

[51] Matigian NA, McCurdy RD, Féron F, *et al.* Fibroblast and lymphoblast gene expression profiles in schizophrenia: are non-neural cells informative? PLoS ONE 2008; 3(6): e2412.

[52] Eriksson PS, Perfilieva E, Björk-Eriksson T, *et al.* Neurogenesis in the adult human hippocampus. Nat Med 1998; 4(11): 1313-7.

[53] Deng W, Aimone JB, and Gage FH. New neurons and new memories: how does adult hippocampal neurogenesis affect learning and memory? Nat Rev Neurosci 2010; 11(5): 339-50.

[54] Curtis MA, Monzo HJ and Faull RL. The rostral migratory stream and olfactory system: smell, disease and slippery cells. Prog Brain Res 2009; 175: . 33-42.

[55] Murrell W, Féron F, Wetzig A, *et al.* Multipotent stem cells from adult olfactory mucosa. Dev Dyn 2005; 233(2): 496-515.

[56] Amoh Y, Kanoh M, Niiyama S, *et al.* Human hair follicle pluripotent stem (hfPS) cells promote regeneration of peripheral-nerve injury: an advantageous alternative to ES and iPS cells. J Cell Biochem 2009; 107(5): 1016-20.

[57] Joannides A, Gaughwin P, Schwiening C, *et al.* Efficient generation of neural precursors from adult human skin: astrocytes promote neurogenesis from skin-derived stem cells. Lancet 2004; 364(9429): 172-8.

[58] Valenzuela MJ, Dean SK, Sachdev P, Tuch BE, Sidhu KS. Neural precursors from canine skin: a new direction for testing autologous cell replacement in the brain. Stem Cells Dev 2008; 17(6): 1087-94.

[59] Pittenger MF and Martin BJ. Mesenchymal stem cells and their potential as cardiac therapeutics. Circ Res 2004; 95(1): 9-20.

[60] Wislet-Gendebien S, Bruyère F, Hans G, Leprince P, Moonen G, Rogister B. Nestin-positive mesenchymal stem cells favour the astroglial lineage in neural progenitors and stem cells by releasing active BMP4. BMC Neurosci 2004; 5: 33.

[61] Wang G, Bunnell BA, Painter RG, *et al.* Adult stem cells from bone marrow stroma differentiate into airway epithelial cells: potential therapy for cystic fibrosis. Proc Natl Acad Sci U S A 2005; 102(1): 186-91.

[62] Roisen FJ, Klueber KM, Lu CL, *et al.* Adult human olfactory stem cells. Brain Res 2001; 890(1): 11-22.

[63] Gunaseeli I, Doss MX, Antzelevitch C, Hescheler J, Sachinidis A. Induced Pluripotent Stem Cells as a Model for Accelerated Patient- and Disease-specific Drug Discovery. Curr Med Chem 2010; 17(8): 759-66.

[64] Kiskinis E and Eggan K. Progress toward the clinical application of patient-specific pluripotent stem cells. J Clin Invest 2010; 120(1): 51-9.

[65] Leung CT, Coulombe PA and Reed RR. Contribution of olfactory neural stem cells to tissue maintenance and regeneration. Nat Neurosci 2007; 10(6): 720-6.

[66] Buck L and Axel R. A novel multigene family may encode odorant receptors: a molecular basis for odor recognition. Cell 1991; 65(1): 175-87.

[67] Breer H. The sense of smell: reception of flavors. Ann N Y Acad Sci 2008; 1126: 1-6.

[68] Franssen EH, de Bree FM and Verhaagen J. Olfactory ensheathing glia: their contribution to primary olfactory nervous system regeneration and their regenerative potential following transplantation into the injured spinal cord. Brain Res Rev 2007; 56(1): 236-58.

[69] Lindsay SL, Riddell,JS and Barnett SC. Olfactory mucosa for transplant-mediated repair: a complex tissue for a complex injury? Glia 2010; 58(2): 125-34.

[70] Reynolds BA, Tetzlaff W. and Weiss S. A multipotent EGF-responsive striatal embryonic progenitor cell produces neurons and astrocytes. J Neurosci 1992; 12(11): 4565-74.

[71] Matigian N, Abrahamsen G, Sutharsan R, *et al.* Disease-specific, neurosphere-derived cells as models for brain disorders. Dis Model Mech 2010; 3(11-12): 785-98.

[72] Corfas G, Roy K. and Buxbaum J.D. Neuregulin 1-erbB signalling and the molecular/cellular basis of schizophrenia. Nat Neurosci 2004; 7(6): 575-80.

[73] Lovestone S, Killick R, Di Forti M, Murray R. Schizophrenia as a GSK-3 dysregulation disorder. Trends Neurosci 2007; 30(4): 142-9.

[74] Matsuda N, Sato S, Shiba K, *et al.* PINK1 stabilized by mitochondrial depolarization recruits Parkin to damaged mitochondria and activates latent Parkin for mitophagy. J Cell Biol 2010; 189(2): 211-21.

[75] Mattick JS, Amaral PP, Dinger ME, Mercer TR, Mehler MF. RNA regulation of epigenetic processes. Bioessays 2009; 31(1): 51-9.

[76] Fraga MF, Ballestar E, Paz MF, *et al.* Epigenetic differences arise during the lifetime of monozygotic twins. Proc Natl Acad Sci U S A 2005; 102(30): 10604-9.

[77] Abel T and Zukin RS. Epigenetic targets of HDAC inhibition in neurodegenerative and psychiatric disorders. Curr Opin Pharmacol 2008; 8(1): 57-64.

[78] Laflamme MA, Chen KY, Naumova AV, *et al.* Cardiomyocytes derived from human embryonic stem cells in pro-survival factors enhance function of infarcted rat hearts. Nat Biotechnol 2007; 25(9): 1015-24.

[79] Dimos JT, Rodolfa KT, Niakan KK, *et al.* Induced pluripotent stem cells generated from patients with ALS can be differentiated into motor neurons. Science 2008; 321(5893): 1218-21.

[80] Zhang N, An MC, Montoro D, Ellerby LM. Characterization of Human Huntington's Disease Cell Model from Induced Pluripotent Stem Cells. PLoS Curr 2010; 2: RRN1193.

[81] Soldner F, Hockemeyer D, Beard C, *et al.* Parkinson's disease patient-derived induced pluripotent stem cells free of viral reprogramming factors. Cell 2009; 136(5): 964-77.

[82] Ku S, Soragni E, Campau E, *et al.* Friedreich's ataxia induced pluripotent stem cells model intergenerational GAATTC triplet repeat instability. Cell Stem Cell 2010; 7(5): 631-7.

[83] Ebert AD, Yu J, Rose FF Jr, *et al.* Induced pluripotent stem cells from a spinal muscular atrophy patient. Nature 2009; 457(7227): 277-80.

[84] Park IH, Arora N, Huo H, *et al.* Disease-Specific Induced Pluripotent Stem Cells. Cell 2008; 134(5): 877-86.

[85] Meissner A. Epigenetic modifications in pluripotent and differentiated cells. Nat Biotechnol 2010; 28(10): 1079-88.

[86] Wernig M, Zhao JP, Pruszak J, *et al.* Neurons derived from reprogrammed fibroblasts functionally integrate into the fetal brain and improve symptoms of rats with Parkinson's disease. Proc Natl Acad Sci U S A 2008; 105(15): 5856-61.

[87] Urbach A, Bar-Nur O, Daley GQ, Benvenisty N. Differential modelling of fragile X syndrome by human embryonic stem cells and induced pluripotent stem cells. Cell Stem Cell 2010; 6(5): 407-11.

[88] Pick M, Stelzer Y, Bar-Nur O, Mayshar Y, Eden A, Benvenisty N. Clone- and gene-specific aberrations of parental imprinting in human induced pluripotent stem cells. Stem Cells 2009; 27(11): 2686-90.

[89] Chamberlain SJ, Chen PF, Ng KY, *et al.* Induced pluripotent stem cell models of the genomic imprinting disorders Angelman and Prader-Willi syndromes. Proc Natl Acad Sci U S A 2010; 107(41): 17668-73.

[90] Yang J, Cai J, Zhang Y, *et al.* Induced pluripotent stem cells can be used to model the genomic imprinting disorder Prader-Willi syndrome. J Biol Chem 2010; 285(51): 40303-11.

[91] Lowry WE, Richter L, Yachechko R, *et al.* Generation of human induced pluripotent stem cells from dermal fibroblasts. Proc Natl Acad Sci U S A 2008; 105(8): . 2883-8.

[92] Yu J, Vodyanik MA, Smuga-Otto K, *et al.* Induced pluripotent stem cell lines derived from human somatic cells. Science 2007; 318(5858): 1917-20.

[93] Chin MH, Mason MJ, Xie W, *et al.* Induced pluripotent stem cells and embryonic stem cells are distinguished by gene expression signatures. Cell Stem Cell 2009; 5(1): 111-23.

[94] Nakagawa M, Koyanagi M, Tanabe K, *et al.* Generation of induced pluripotent stem cells without Myc from mouse and human fibroblasts. Nat Biotechnol 2008; 26(1): 101-6.

[95] Okita K, Nakagawa M, Hyenjong H, Ichisaka T, Yamanaka S.Generation of mouse induced pluripotent stem cells without viral vectors. Science 2008; 322(5903): 949-53.

[96] Li W and Ding S. Small molecules that modulate embryonic stem cell fate and somatic cell reprogramming. Trends Pharmacol Sci 2010; 31(1): 36-45.

[97] Lyssiotis CA, Foreman RK, Staerk J, *et al.* Reprogramming of murine fibroblasts to induced pluripotent stem cells with chemical complementation of Klf4. Proc Natl Acad Sci U S A 2009; 106(22): 8912-7.

[98] Ebert AD and Svendsen CN. Human stem cells and drug screening: opportunities and challenges. Nat Rev Drug Discov 2010; 9(5): 367-72.

[99] Lee G, Papapetrou EP, Kim H, *et al.* Modelling pathogenesis and treatment of familial dysautonomia using patient-specific iPSCs. Nature 2009; 461(7262): 402-6.

[100] Curtis MA, Eriksson PS and Faull RL. Progenitor cells and adult neurogenesis in neurodegenerative diseases and injuries of the basal ganglia. Clin Exp Pharmacol Physiol 2007; 34(5-6): 528-32.

CHAPTER 15

Stem Cell Therapy in Alzheimer's Disease Models: Neurogenesis Versus Trophic Support

Jürgen Götz* **and Sven Büttner**

Alzheimer's and Parkinson's Disease Laboratory, Brain and Mind Research Institute, University of Sydney, Australia

Abstract: With an ageing population, Alzheimer's disease (AD) represents a serious social and economic threat to most societies. While AD's key histopathological features, the amyloid-β plaques and the tau tangles, have been described more than a century and the first disease-causing mutations in familial cases of AD already two decades ago, there is still no cure for this debilitating disease. Current treatment is limited to various acetylcholine esterase (AChE) inhibitors and the NMDA receptor antagonist, memantine, yet neither of these strategies halts the degenerative process that characterizes AD [1]. More recently, strategies have been developed that target both amyloid-β and tau, but none of these has reached the patient [1]. Here we review what we have learned from animal models about the pathogenesis of AD and treatment options. We further review current attempts to use stem cells to *restore the function of degenerating neurons or to replace them.*

Keywords: Alzheimer, amyloid-beta, frontotemporal dementia, tau, therapy, transgenic,

THE DEFINING CHARACTERISTICS OF ALZHEIMER'S DISEASE

The AD brain is histopathologically characterized by overt amyloid-β (Aβ)-containing plaques and tau-containing neurofibrillary lesions [2]. The 39-42 amino acid Aβ peptide in the plaques is fibrillar as is tau, a microtubule-associated protein tau, that becomes hyperphosphorylated in the disease process and eventually aggregates to form neurofibrillary tangles (NFTs), neuropil threads, and abnormal neurites that are associated with some amyloid-β plaques [3]. Tau has been generally perceived as a neuronal protein, with a mainly axonal localization, but very recently we challenged this view by demonstrating an essential dendritic function in mediating Aβ toxicity [4, 5]. In addition to plaques and NFTs, the AD brain is also characterized by massive neuronal cell and synapse loss [6].

Tau pathology, in the absence of overt Aβ plaques, characterizes a series of diseases that are collectively known as tauopathies. This includes FTLD-T (frontotemporal lobar degeneration with tau inclusions), with an inherited form known as FTDP-17 (frontotemporal dementia with Parkinsonism linked to chromosome 17). There is a second, major subset of FTD characterized by tau-negative yet ubiquitin-positive lesions, again in the absence of overt plaques, that has been termed FTLD-U or FTDU-17 [7]. In this subset, the transcription and splicing factor TDP-43 (TAR DNA-binding protein 43) has been identified as aggregated protein. Similar to tau, TDP-43 in the aggregates is hyperphosphorylated [8]. The aggregates are also found in sporadic forms of ALS (amyotrophic lateral sclerosis), and interestingly, can also co-occur with the typical lesions in AD [9].

The majority of AD cases are sporadic, there are however familial (FAD) cases, with autosomal dominant mutations being identified in a total of three genes: *APP*, *presenilin 1 (PSEN1)* as well as *presenilin 2 (PSEN2)* [10]. It is the amyloid precursor protein, APP, from which the Aβ peptide is derived by proteolytic cleavage, and the presenilins are components of a multimeric complex that generates Aβ. In addition to the

***Address correspondence to Jürgen Götz:** Alzheimer's and Parkinson's Disease Laboratory, Brain and Mind Research Institute, University of Sydney, 100 Mallett St, Camperdown, NSW 2050, Australia; Tel: +61-2-9351 0789 ; Fax: +61-2-9114-4200; E-mail: juergen.goetz@sydney.edu.au

Kuldip S. Sidhu (Ed)

three FAD genes, *apolipoprotein E (APOE)* has been confirmed as a susceptibility gene in sporadic AD (SAD) [10]. More recently, additional risk factor genes have been identified, *CLU* encoding clusterin, *PICALM* the phosphatidylinositol-binding clathrin assembly protein and CR1 the complement component (3b/4b) receptor 1 [11, 12]. SAD and FAD cases cannot be discriminated on the basis of histological or clinical features, what is different however is the earlier age of onset in SAD.

Compared to AD, FTLD is a more heterogeneous dementing disorder, with variants that are, *e.g.*, characterized by language or behavioural problems. It was only several years after the first mutations had been identified in the *APP* and *PSEN* genes in AD, that the first mutations were identified in a subset of FTLD, FTDP-17, where they were found in exonic and intronic sequences of the *MAPT* gene that encodes tau [7]. Another subset of familial FTD is characterized by mutations in the *PGRN* gene that encodes the pleiotropic protein progranulin, and in the *VCP* gene *that* encodes valosin-containing protein; these cases are characterized by inclusions of TDP-43 [8]. Finally, mutations in *CHMP2B* encoding chromatin modifying protein 2B lead to FTLD, in the absence of either tau or TDP-43 inclusions [7].

ANIMAL MODELS OF ALZHEIMER'S DISEASE

Traditionally, the mouse has been the major species used in AD and FTD research, but increasingly, other species such as the roundworm *C. elegans*, the fruit fly *D. melanogaster*, or the zebrafish *D. rerio* are employed, reflecting a general trend in neurodegeneration research [13, 14]. In rodents, classical transgenesis [15] and knockout approaches [16] helped in establishing a host of mouse strains that model different aspects of AD and FTLD [17]. Representative models with a robust APP/Aβ pathology are APP mutant strains such as PDAPP, J20, APP23 or Tg2576 [18-21], and P301L FTLD tau mutant strains with NFT formation, such as JNPL3 or pR5 [22, 23]. The histopathology in these strains is associated with behavioural impairment [17].

While these models investigate memory impairment as the major clinical feature characterizing AD, another clinical feature, Parkinsonism, that characterizes a significant subset of FTLD cases, has also been modelled in mice. One strategy employed the generation of several mini-gene tau constructs to establish a series of transgenic mouse strains. One particular strain reproduced tau accumulation in astrocytes and neurons including these of the nigro-striatal pathway [24]. Here, the rate-limiting enzyme in dopamine synthesis, tyrosine hydroxylase (TH), was found to accumulate in so-called varicosities (axonal swellings). Furthermore, the mice showed memory and motor impairment [24]. A second model developed by us was based on the identification of the K369I mutation of tau in a patient with Pick's disease, a form of FTLD [25]. As known for Pick's disease, the tau lesions in the brain of this particular K369I patient showed a remarkable feature when analysed histologically: silver impregnations revealed that the tau lesions were Bielschowsky-positive and Gallyas-negative; moreover, tau in these lesions was phosphorylated at many epitopes but not 12E8 (Ser262/Ser356). We succeeded in reproducing this distinct Pick pathology in the K3 mouse strain that expresses K369I mutant tau [26]. The transgenic mice also modeled early-onset Parkinsonism (resting tremor, bradykinesia, postural instability and gait anomalies). They showed an increased cataleptic response to haloperidol and an early, but not late response to L-Dopa, indicating that the dopaminergic system is impaired in these mice [26].

In recent years, impaired axonal transport has been identified as a central pathomechanism in AD [27-32]. Conceptually, impaired axonal transport is linked to oxidative stress and mitochondrial dysfunction: mitochondria need to be efficiently transported along the long processes for neurons to function; furthermore they are a major source of reactive oxygen species [33, 34]. We found that in the K3 mice, there was a selectively impaired axonal transport of distinct cargos including mitochondria and TH-containing vesicles [26]. At a molecular level, we found that this is due to a component of the kinesin motor machinery, the adapter protein Jip1, being trapped by elevated phospho-tau, preventing it from executing its physiological function. A pathological interaction between tau and JIP1 was also revealed in AD and not, control brain, highlighting the validity of transgenic animal models in dissecting pathomechanisms in AD [35]. As tau only traps JIP1 when it is hyperphosphorylated, this study presents inhibition of abnormal hyperphosphorylation of tau as a promising therapeutic strategy in AD [36].

TREATMENT STRATEGIES – TARGETING AMYLOID-BETA AND TAU

The current treatment of AD is symptomatic. The approved drugs fall into two categories: there are acetylcholine esterase inhibitors that stabilize the neurotransmitter acetylcholine in the synaptic cleft of cholinergic neurons, and memantine, a partial NMDA receptor [1]. None of these drugs halts the neurodegeneration that characterizes AD, and hence, strategies are needed that target the underlying cause of the disease. Of the compounds that are currently in various stages of clinical trials many target Aβ, and a few, tau [1].

Many of the newer therapies have their foundation in animal work. This includes the vaccination approach, pioneered by researchers of the biotech company Elan. Using both an active and passive immunization approach, efficacy was shown in removing existing Aβ plaques and preventing their formation; likewise, age-dependent learning deficits could be reduced or prevented [37, 38]. Driven by the success in animals, an immunization trial was initiated in humans. First, good safety and tolerability data were obtained from different species. Then, a phase I clinical trial was initiated with the AN-1792 vaccine, a pre-aggregated synthetic Aβ42 preparation, and the adjuvant QS-21. First, a single-dose phase I study with 24 patients and, subsequently, a multiple-dose phase I study with 70 patients, were reported with good safety and tolerability data [39]. It was encouraging to find that approximately 25% of the patients produced Aβ-specific antibodies. In October 2001, a phase IIa clinical trial was started in a multicenter placebo-controlled double-blind design and patients were randomized to receive AN1792/QS-21 or placebo (in a ratio of 4:1) by intramuscular injection [40]. Unfortunately, the trial had to be halted prematurely in January 2002 when four patients who had received the vaccine developed signs of subacute meningoencephalitis [41]. Until then, 298 of 372 patients had received the vaccine. Of these, 18 patients (*i.e.*, 6% of the AN1792/QS-21-treated group) developed signs of meningoencephalitis. Upon unblinding, they turned out to belong to the *verum* (AN1792/QS-21-treated) group [40]. In a cohort of 30 patients from one of the centres, additional data were obtained. It could be demonstrated that 24 of the 30 patients developed specific antibodies against Aβ which did not cross-react with human full-length APP, with the specific immune reaction against Aβ being stable even after one year [42]. Moreover, this immune reaction showed a significant correlation with slowed cognitive decline. These beneficial clinical effects could also be demonstrated in two of three patients who had experienced immunization-related aseptic meningoencephalitis [43]. Together, these results indicated that vaccination against Aβ might be an interesting, causal treatment for AD, provided that a safe treatment modality can be introduced. Since these early days when vaccination was established as a possible treatment avenue, many modifications have been tested in mice, using a variety of routes of administration, changes to the peptide and adjuvans, a vaccination strategy using DNA, or even coupling of Aβ to retro-particles [44]. A subset of these modifications made it into the clinics with both active and passive strategies being currently pursued [1].

Regarding tau, one would assume that it would be difficult to achieve efficacy, as different from Aβ, tau is an intracellular, non-secreted protein and hence, at least at first sight, clearing by the immune system would seem to be a highly unlikely process to occur. However, the first immunization studies targeting the PHF1 phospho-epitope of tau (containing the phospho-sites Ser396/Ser404) revealed efficacy: aggregated tau levels were reduced, there was a shift from the insoluble to the soluble pool of tau, and progression of an NFT-related motor phenotype was delayed [45]. It remains to be seen in independent studies whether a tau-targeted strategy will be effective. The finding that tau knock-out mice, at least at a young age, show no overt pathological phenotype, is encouraging [46-48].

Life style factors have been addressed by several studies. One of the early, important studies addressed the impact of environment enrichment in transgenic mice with an Aβ plaque pathology. Housing in large cages and the availability of running wheels, colored tunnels, toys and chewable material to play with caused significant reductions in Aβ levels and plaque load [49]. Feeding of other plaque-forming transgenic mice with the green tea active compound EGCG had similar effects: again, Aβ levels and plaque load were reduced [50, 51]. When plaque-forming mice were fed with omega-3 unsaturated fatty acids, this also reduced Aβ levels and plaque load indicating a crucial role for environmental factors in the progression of AD [52].

Both tau and APP are phospho-proteins, with phosphorylation being regulated by a balanced interplay of kinases and phosphatases. Therefore, numerous molecules which interfere with the function of these enzymes have been tested *in vitro* and in animals [53]. Very recently, an inhibitor of the kinase GSK3β, AR-534, showed efficacy in a tau transgenic zebrafish model, massively reducing tau hyperphosphorylation [54]. Another strategy is the targeting of phosphatases. Here, the multimeric enzyme PP2A is particularly attractive, as instead of modulating general catalytic activity it may be possible to activate specifically the one of its many regulatory subunits that confers specificity for tau as a substrate [55, 56]. PP2A is highly expressed in brain, making it a prime target. We identified sodium selenate as a compound that reduces tau phosphorylation both *in vitro* and *in vivo* [57]. Importantly, chronic oral treatment of two independent tau transgenic mouse strains with NFT pathology, P301L mutant pR5 and K369I mutant K3 mice, reduced tau hyperphosphorylation and completely abrogated NFT formation. Furthermore, treatment improved contextual memory and motor performance, and prevented neurodegeneration. As hyperphosphorylation of tau precedes NFT formation, the effect of selenate on tau phosphorylation was assessed in more detail. We found that selenate stabilized PP2A-tau complexes. Moreover, there was an absence of therapeutic effects in sodium selenate-treated tau transgenic mice that co-express a dominant-negative mutant form of PP2A, suggesting a mediating role for PP2A [57]. Sodium selenate mitigated tau pathology in several AD models, thus making it a promising lead compound for tau-targeted treatments of AD and related dementias. Together, this presents a strategy that modulates either kinases or phosphatases as an attractive therapeutic approach [36].

CONCLUSIONS AND PROSPECTS

Different from Parkinson's disease with a mainly focal degeneration of dopaminergic neurons in the substantia nigra pars compacta, AD is characterized by widespread neuronal degeneration raising the question as to whether a stem cell therapy would be a suitable avenue. One could argue that disease is initiated focally, as in the entorhinal cortex and the hippocampus, NFTs appear first [58]. Hence, a stem cell strategy might be suitable even for AD, provided the intervention would occur early enough before the pathology has spread throughout the brain.

Several aspects have to be considered: How does a diseased brain react to a stem cell transplantation approach? Would the stem cells have to be modified in order to survive and be functional in such an environment? And also, would a stem cell approach work because it replaces dying neurons or because of neurotrophic support by the grafted cells. Several studies addressed these questions in animal models and *in vitro*.

When human neural stem cells (HNSCs) were exposed to high concentrations of secreted amyloid-precursor protein *in vitro*, they differentiated mainly into astrocytes, suggesting that pathological alterations in APP processing during neurodegenerative conditions such as AD may prevent a proper neuronal differentiation of HNSCs [59]. Thus, a successful neuro-placement therapy for AD may require regulating APP expression to favorable levels to enhance neuronal differentiation of HNSCs. Both aspects were achieved using phenserine, a recently developed cholinesterase inhibitor [59]. In another study, neurogenesis of young adult subventricular zone precursors was shown to be stimulated by Aβ both *in vitro* and in plaque-forming mice. The neurogenic effect of Aβ42 was found to require expression of the p75 neurotrophin receptor by the precursor cells. However, precursors from 12-month-old plaque-forming mice failed to respond to Aβ42 suggesting that overstimulation of the neurotrophin receptor-positive progenitors during early life might result in depletion of the stem cell pool and thus a more rapid decline in basal neurogenesis [60]. Similarly, when wild-type neuronal precursor cells (NPCs) were co-cultured with primary microglia from mice expressing PSEN1 variants, they exhibited impaired proliferation and neuronal lineage commitment, argueing for non-cell-autonomous mechanisms that govern FAD-linked impairments in adult hippocampal neurogenesis [61]. In plaque-forming APP23 mice, environmental enrichment more so than physical exercise was shown to contribute to a 'neurogenic reserve' (analogous to 'cognitive reserve'), in the absence of changes to Aβ plaque load [62].

Transplantation of bone marrow-derived mesenchymal stem cells (BM-MSCs) has been suggested as a potential therapeutic approach in AD. To evaluate this strategy, BM-MSC transplantation was performed in plaque-forming mice, significantly reducing Aβ deposition. Interestingly, these effects were associated with a restoration of defective microglial function, as evidenced by increased Aβ-degrading factors, decreased inflammatory responses, and elevation of activated microglial markers. Furthermore, the treated mice showed decreased tau hyper-phosphorylation and improved cognitive function [63].

Neural stem cell (NSC) transplantation represents other strategy. When aged so-called triple transgenic mice (3xTg-AD) with a widespread and established Aβ plaque and NFT pathology were transplanted with hippocampal NSCs, this rescued the spatial learning and memory deficits, without altering the Aβ or tau pathology. The mechanism underlying the improved cognition turned out to involve an enhancement of hippocampal synaptic density, mediated by brain-derived neurotrophic factor (BDNF). In additional experiments, such as by depleting the NSCs of BDNF, and by using recombinant BDNF it was shown that it is this neurotrophin that is at the core of the beneficial effects of the NSC transplantation [64].

The future will tell which approach will help if not in providing a cure for AD at least by modifying the disease and the age of onset. Whether any of these approaches involve stem cell therapies remains to be seen.

REFERENCES

[1] Mangialasche F, Solomon A, Winblad B, Mecocci P, Kivipelto M. Alzheimer's disease: clinical trials and drug development. Lancet Neurol 2010; 9(7): 702-16.

[2] Gotz J, Ittner LM, Schonrock N. Alzheimer's disease and frontotemporal dementia: prospects of a tailored therapy? Med J Aust 2006; 185(7): 381-4.

[3] Chen F, David D, Ferrari A, Gotz J. Posttranslational modifications of tau - Role in human tauopathies and modelling in transgenic animals. Curr Drug Targets 2004; 5(6): 503-15.

[4] Ittner LM, Ke YD, Delerue F, *et al.* Dendritic function of tau mediates amyloid-beta toxicity in Alzheimer's disease mouse models. Cell 2010; 142(3): 387-97.

[5] Ittner LM, Gotz J. Amyloid-beta and tau - a toxic pas de deux in Alzheimer's disease. Nat Rev Neurosci 2011; 12(2): 65-72.

[6] Selkoe DJ. Alzheimer's disease is a synaptic failure. Science 2002; 298(5594): 789-91.

[7] Cruts M, Van Broeckhoven C. Loss of progranulin function in frontotemporal lobar degeneration. Trends Genet 2008.

[8] Neumann M, Sampathu DM, Kwong LK, *et al.* Ubiquitinated TDP-43 in frontotemporal lobar degeneration and amyotrophic lateral sclerosis. Science 2006; 314(5796): 130-3.

[9] Chen-Plotkin AS, Lee VM, Trojanowski JQ. TAR DNA-binding protein 43 in neurodegenerative disease. Nat Rev Neurol 2010; 6(4): 211-20.

[10] Bertram L, Tanzi RE. Thirty years of Alzheimer's disease genetics: the implications of systematic meta-analyses. Nat Rev Neurosci 2008; 9(10): 768-78.

[11] Lambert JC, Heath S, Even G, *et al.* Genome-wide association study identifies variants at CLU and CR1 associated with Alzheimer's disease. Nat Genet 2009; 41(10): 1094-9.

[12] Harold D, Abraham R, Hollingworth P, *et al.* Genome-wide association study identifies variants at CLU and PICALM associated with Alzheimer's disease. Nat Genet 2009; 41(10): 1088-93.

[13] Gotz J, Deters N, Doldissen A, *et al.* A Decade of Tau Transgenic Animal Models and Beyond Brain Pathol 2007; 17: 91-103.

[14] Pienaar IS, Gotz J, Feany MB. Parkinson's disease: Insights from non-traditional model organisms. Prog Neurobiol 2010.

[15] Ittner LM, Gotz J. Pronuclear injection for the generation of transgenic mice. Nat Protoc 2007; 2(5): 1206-15.

[16] Wu S, Ying G, Wu Q, Capecchi MR. A protocol for constructing gene targeting vectors: generating knockout mice for the cadherin family and beyond. Nat Protoc 2008; 3(6): 1056-76.

[17] Gotz J, Ittner LM. Animal models of Alzheimer's disease and frontotemporal dementia. Nat Rev Neurosci 2008; 9(7): 532-44.

[18] Games D, Adams D, Alessandrini R, *et al.* Alzheimer-type neuropathology in transgenic mice overexpressing V717F beta-amyloid precursor protein [see comments]. Nature 1995; 373(6514): 523-7.

[19] Hsiao K, Chapman P, Nilsen S, *et al.* Correlative memory deficits, Abeta elevation, and amyloid plaques in transgenic mice [see comments]. Science 1996; 274(5284): 99-102.

[20] Sturchler-Pierrat C, Abramowski D, Duke M, *et al.* Two amyloid precursor protein transgenic mouse models with Alzheimer disease-like pathology. Proc Natl Acad Sci U S A 1997; 94(24): 13287-92.

[21] Mucke L, Masliah E, Yu GQ, *et al.* High-level neuronal expression of abeta 1-42 in wild-type human amyloid protein precursor transgenic mice: synaptotoxicity without plaque formation. J Neurosci 2000; 20(11): 4050-8.

[22] Lewis J, McGowan E, Rockwood J, *et al.* Neurofibrillary tangles, amyotrophy and progressive motor disturbance in mice expressing mutant (P301L) tau protein. Nat Genet 2000; 25(4): 402-5.

[23] Gotz J, Chen F, Barmettler R, Nitsch RM. Tau Filament Formation in Transgenic Mice Expressing P301L Tau. J Biol Chem 2001; 276(1): 529-34.

[24] Dawson HN, Cantillana V, Chen L, Vitek MP. The tau N279K exon 10 splicing mutation recapitulates frontotemporal dementia and parkinsonism linked to chromosome 17 tauopathy in a mouse model. J Neurosci 2007; 27(34): 9155-68.

[25] Neumann M, Schulz-Schaeffer W, Crowther RA, *et al.* Pick's disease associated with the novel Tau gene mutation K369I. Ann Neurol 2001; 50(4): 503-13.

[26] Ittner LM, Fath T, Ke YD, *et al.* Parkinsonism and impaired axonal transport in a mouse model of frontotemporal dementia. Proc Natl Acad Sci U S A 2008; 105(41): 15997-6002.

[27] Stokin GB, Lillo C, Falzone TL, *et al.* Axonopathy and transport deficits early in the pathogenesis of Alzheimer's disease. Science 2005; 307(5713): 1282-8.

[28] Gotz J, Ittner LM, Kins S. Do axonal defects in tau and amyloid precursor protein transgenic animals model axonopathy in Alzheimer's disease? J Neurochem 2006; 98(4): 993-1006.

[29] Magnani E, Fan J, Gasparini L, *et al.* Interaction of tau protein with the dynactin complex. Embo J 2007; 26(21): 4546-54.

[30] Thies E, Mandelkow EM. Missorting of tau in neurons causes degeneration of synapses that can be rescued by the kinase MARK2/Par-1. J Neurosci 2007; 27(11): 2896-907.

[31] Dixit R, Ross JL, Goldman YE, Holzbaur EL. Differential regulation of dynein and kinesin motor proteins by tau. Science 2008; 319(5866): 1086-9.

[32] Pigino G, Morfini G, Atagi Y, *et al.* Disruption of fast axonal transport is a pathogenic mechanism for intraneuronal amyloid beta. Proc Natl Acad Sci U S A 2009; 106(14): 5907-12.

[33] Su B, Wang X, Nunomura A, *et al.* Oxidative stress signalling in Alzheimer's disease. Curr Alzheimer Res 2008; 5(6): 525-32.

[34] Tatsuta T, Langer T. Quality control of mitochondria: protection against neurodegeneration and ageing. Embo J 2008; 27(2): 306-14.

[35] Ittner LM, Ke YD, Gotz J. Phosphorylated Tau Interacts with c-Jun N-terminal Kinase-interacting Protein 1 (JIP1) in Alzheimer Disease. J Biol Chem 2009; 284(31): 20909-16.

[36] Iqbal K, Grundke-Iqbal I. Alzheimer neurofibrillary degeneration: significance, etiopathogenesis, therapeutics and prevention. J Cell Mol Med 2008; 12(1): 38-55.

[37] Schenk D, Barbour R, Dunn W, *et al.* Immunization with amyloid-beta attenuates Alzheimer-disease-like pathology in the PDAPP mouse. Nature 1999; 400(6740): 173-7.

[38] Bard F, Cannon C, Barbour R, *et al.* Peripherally administered antibodies against amyloid beta-peptide enter the central nervous system and reduce pathology in a mouse model of Alzheimer disease. Nat Med 2000; 6(8): 916-9.

[39] Schenk D. Amyloid-beta immunotherapy for Alzheimer's disease: the end of the beginning. Nat Rev Neurosci 2002; 3(10): 824-8.

[40] Orgogozo JM, Gilman S, Dartigues JF, *et al.* Subacute meningoencephalitis in a subset of patients with AD after Abeta42 immunization. Neurology 2003; 61(1): 46-54.

[41] Senior K. Dosing in phase II trial of Alzheimer's vaccine suspended. Lancet Neurol 2002; 1(1): 3.

[42] Hock C, Konietzko U, Papassotiropoulos A, *et al.* Generation of antibodies specific for beta-amyloid by vaccination of patients with Alzheimer disease. Nat Med 2002; 8(11): 1270-5.

[43] Hock C, Konietzko U, Streffer JR, *et al.* Antibodies against beta-Amyloid Slow Cognitive Decline in Alzheimer's Disease. Neuron 2003; 38(4): 547-54.

[44] Bach P, Tschape JA, Kopietz F, *et al.* Vaccination with Abeta-displaying virus-like particles reduces soluble and insoluble cerebral Abeta and lowers plaque burden in APP transgenic mice. J Immunol 2009; 182(12): 7613-24.

[45] Asuni AA, Boutajangout A, Quartermain D, Sigurdsson EM. Immunotherapy targeting pathological tau conformers in a tangle mouse model reduces brain pathology with associated functional improvements. J Neurosci 2007; 27(34): 9115-29.

[46] Harada A, Oguchi K, Okabe S, *et al.* Altered microtubule organization in small-calibre axons of mice lacking tau protein. Nature 1994; 369(6480): 488-91.

[47] Dawson HN, Ferreira A, Eyster MV, *et al.* Inhibition of neuronal maturation in primary hippocampal neurons from tau deficient mice. J Cell Sci 2001; 114(Pt 6): 1179-87.

[48] Tucker KL, Meyer M, Barde YA. Neurotrophins are required for nerve growth during development. Nat Neurosci 2001; 4(1): 29-37.

[49] Lazarov O, Robinson J, Tang YP, *et al.* Environmental enrichment reduces Abeta levels and amyloid deposition in transgenic mice. Cell 2005; 120(5): 701-13.

[50] Obregon DF, Rezai-Zadeh K, Bai Y, *et al.* ADAM10 activation is required for green tea (-)-epigallocatechin-3-gallate-induced alpha-secretase cleavage of amyloid precursor protein. J Biol Chem 2006; 281(24): 16419-27.

[51] Rezai-Zadeh K, Shytle D, Sun N, *et al.* Green tea epigallocatechin-3-gallate (EGCG) modulates amyloid precursor protein cleavage and reduces cerebral amyloidosis in Alzheimer transgenic mice. J Neurosci 2005; 25(38): 8807-14.

[52] Lim GP, Calon F, Morihara T, *et al.* A diet enriched with the omega-3 fatty acid docosahexaenoic acid reduces amyloid burden in an aged Alzheimer mouse model. J Neurosci 2005; 25(12): 3032-40.

[53] Pei JJ, Sjogren M, Winblad B. Neurofibrillary degeneration in Alzheimer's disease: from molecular mechanisms to identification of drug targets. Curr Opin Psychiatry 2008; 21(6): 555-61.

[54] Paquet D, Bhat R, Sydow A, *et al.* A zebrafish model of tauopathy allows *in vivo* imaging of neuronal cell death and drug evaluation. J Clin Invest 2009; 119(5): 1382-95.

[55] Kins S, Crameri A, Evans DR, *et al.* Reduced PP2A activity induces hyperphosphorylation and altered compartmentalization of tau in transgenic mice. J Biol Chem 2001; 276(41): 38193-200

[56] Schild A, Ittner LM, Gotz J. Altered phosphorylation of cytoskeletal proteins in mutant protein phosphatase 2A transgenic mice. Biochem Biophys Res Commun 2006; 343(4): 1171-8.

[57] van Eersel J, Ke YD, Liu X, *et al.* Sodium selenate mitigates tau pathology, neurodegeneration, and functional deficits in Alzheimer's disease models. Proc Natl Acad Sci U S A 2010; 107(31): 13888-93.

[58] Braak H, Braak E. Neuropathological stageing of Alzheimer-related changes. Acta Neuropathol (Berl) 1991; 82(4): 239-59.

[59] Marutle A, Ohmitsu M, Nilbratt M, *et al.* Modulation of human neural stem cell differentiation in Alzheimer (APP23) transgenic mice by phenserine. Proc Natl Acad Sci U S A 2007; 104(30): 12506-11.

[60] Sotthibundhu A, Li QX, Thangnipon W, Coulson EJ. Abeta(1-42) stimulates adult SVZ neurogenesis through the p75 neurotrophin receptor. Neurobiol Aging 2009; 30(12): 1975-85.

[61] Choi SH, Veeraraghavalu K, Lazarov O, *et al.* Non-cell-autonomous effects of presenilin 1 variants on enrichment-mediated hippocampal progenitor cell proliferation and differentiation. Neuron 2008; 59(4): 568-80.

[62] Mirochnic S, Wolf S, Staufenbiel M, Kempermann G. Age effects on the regulation of adult hippocampal neurogenesis by physical activity and environmental enrichment in the APP23 mouse model of Alzheimer disease. Hippocampus 2009; 19(10): 1008-18.

[63] Lee JK, Jin HK, Endo S, *et al.* Intracerebral transplantation of bone marrow-derived mesenchymal stem cells reduces amyloid-beta deposition and rescues memory deficits in Alzheimer's disease mice by modulation of immune responses. Stem Cells 2010; 28(2): 329-43.

[64] Blurton-Jones M, Kitazawa M, Martinez-Coria H, *et al.* Neural stem cells improve cognition *via* BDNF in a transgenic model of Alzheimer disease. Proc Natl Acad Sci U S A 2009; 106(32): 13594-9.

CHAPTER 16

The Prospects and Challenges of Cell-Based Therapies

Kuldip S. Sidhu*

Stem Cell Laboratory, Faculty of Medicine, University of New South Wales, Australia

Abstract: With the emergence of induced pluripotent stem cell (iPSC) technology, the somatic cells can be reprogrammed to pluripotent state, and thus the new paradigm to generate patient-specific pluripotent stem cells for disease pathogenesis and cell therapy. These iPSCs produced from a variety of somatic cell sources are found to be very similar if not identical to embryonic stem cells. Currently the most efficient way to produce such cells is by viral transduction with a combination of transcriptional factors and as such that renders these cells unfit for therapeutic purposes. However, recently there is further development in methods for generating iPSCs with minimal or no genetic modifications *via* excisable lentiviral and transposon vectors or through repeated application of transient plasmid, episomal, and adenovirus vectors and very recently the use of small molecules, synthetic mRNA and microRNAs. However, as it is becoming evident now that the cell type of origin influences the molecular and functional properties of derived iPSC. The indications that reprogramming may erase the cell memory also raises the question if the disease phenotype may not be correctly represented or also erased in iPSC unless coaxed by further perturbation *in vitro* culture conditions. Similarly other tissue-derived stem cells and hESC-derived lineages offer an unprecedented opportunity in biomedical research, cell therapy and regenerative medicine. However, to harness the full potential of these technologies, a number of issues need to be resolved pertaining to their safety, stability, culture variability, and better ways to direct a specific reprogramming process including lineage specifications. Both hESC and iPSC fields offer phenomenal opportunities to study the basic molecular mechanistic of human development, their use in disease modelling, regenerative medicine and drug discovery. A steady progress in transitional and translational research particularly in various disease animal models and very recently some human trials with these cells are very encouraging. This chapter briefly summarises (not review) the recent developments in stem cell therapy, pitfalls and development.

Keywords: Cell therapy, stem cells, induced pluripotency, transplantation, graft rejection, regulation.

INTRODUCTION

To realize the promise of novel stem cell-based therapies for many debilitating human diseases, like diabetes, Alzheimer's, Parkinson's, Strokes, Cardiac to name a few, scientists must be able to manipulate stem cells so that they possess the necessary characteristics for successful differentiation, transplantation, and engraftment. The following is a list of steps in successful cell-based treatments that scientists will have to learn to control to bring such treatments to the clinic. To be useful for transplant purposes, stem cells must be reproducibly made to (i) proliferate extensively and differentiate into the desired cell type(s), (ii) survive and integrate into the surrounding tissue after transplant, (iii) function appropriately for the duration of the recipient's life, (iv) avoid harming the recipient in any way and to avoid the problem of immune rejection.

i) Proliferate Extensively and Differentiate into the Desired Cell Type(s)

The advantage of using iPSCs and hESCs for that purpose is they both self renew, proliferate extensively in culture, are immortal and can be preserved indefinitely and are pluripotent. They can produce all the 240 types of the cells found in human body and as such can have tremendous therapeutic potentials. On the contrary, the tissue-derived adult stem cells can be multipotent and have limited proliferation capacity. The argument that adult stem cells that are ethically more acceptable than embryonic stem cells and the former

***Address correspondence to Kuldip S. Sidhu:** Director, Stem Cell Laboratory, Faculty of Medicine, The University of New South Wales, Level 2, Clinical Sciences Building, Prince of Wales Hospital, Randwick, NSW 2031, Australia; Tel: +61 2 9382 2622; Fax: +61 2 9382 2622; E-mail: k.sidhu@unsw.edu.au can also be immortalized by genetic manipulations that may add to their therapeutic limitations. They still have a limited differentiation potential that is restricted to the site of their isolation. Nevertheless a number of preclinical trials with such adult stem cells are in progress with some reaching to the stage of preclinical safety trials in humans *i.e.* immortalized human fetal neural cells for strokes [1]. However, there could be a practical limitation in supplying such cells in sufficient numbers required for transplantation *i.e.* it is estimated that for successful reversal of type-1 diabetes with cell therapy, each patient would require about 3-4 millions insulin producing cells and that currently with adult islet transplantation is met by tissue from at least three pancreas donors. The number of people suffering with Type-1 diabetes is escalating worldwide and with limited number of donors, this strategy with adult stem cells will simply not stand up because of the supply and demand issue. This may also hold good for many other like neurodegenerative diseases *i.e.* Alzheimer's and Parkinson's where specific neurons are affected and lost and need replacement. Pluripotent stem cells proliferate but also know to acquire culture based epigenetic changes and mutations that may be deleterious for their use in cell therapy. This has been more prevalent in iPSCs than in hESCs as brought out in a series of papers recently [2-6]. It was demonstrated previously that that during cultivation *in vitro*, human embryonic stem cells (ESCs) also can become aneuploid and acquire an abnormal number of chromosomes [7]. These studies were performed using whole genome profiling of the DNA methylomes of multiple human ES cell, iPSC and somatic progenitor lines, encompassing reprogramming performed in different laboratories, using different iPSC-inducing technologies and cells derived from distinct germ layers. They demonstrated that although on a global scale ES cell and iPSC methylomes are very similar, every iPSC line shows significant reprogramming variability compared to both ES cells and other iPSCs, including both somatic 'memory' and iPSC-specific differential DNA methylation. Further, all iPSC lines share numerous non-randomly distributed megabase-scale regions that are aberrantly methylated in the non-CG context, associated with alterations in CG methylation, histone modifications and gene expression. Also they have shown that differentially methylated regions in iPSCs are transmitted to differentiated cells at a high frequency. According to Pera [8] despite extensive evaluation of recurrent genetic change in a vast number of cell lines, we are only slightly closer to identifying which particular genes within the larger chromosomal regions that are commonly subject to duplication in iPSCs and ESCs might be under selection. Years of cytogenetic studies of germ-cell tumors have also identified large genomic regions that are commonly over-represented in these cancers, but the identification of the specific genes involved in the transformation of these pluripotent cells has remained elusive [7]. With regard to evaluating the safety of ESCs and iPSCs, a key issue is the biological significance of the changes that these studies. Clearly, aneuploid cell lines would not be used in therapy (although they might be useful for research into the basis of genetic disorders associated with anomalies in chromosome number or other genetic abnormalities). Cell lines bearing mutations of established functional consequence in oncogenes or tumor suppressors, or in genes associated with Mendelian disorders (those usually due to a single gene), could equally not be used therapeutically. However, the many sub chromosomal changes, point mutations that are not obviously associated with known disease-related genetic abnormalities pose challenges to interpretation. This is because it is unclear how best to assess the effects of new genetic lesions on the growth, differentiation, tumorigenicity and functionality of pluripotent stem cells or their differentiated progeny. High-throughput functional genomics will probably be required to answer these questions. Pluripotent cells themselves will provide the most promising platform for such studies. Once these issues are sorted out, mass production of these cells in clinical grade for transplantation purposes will be required by using appropriate state-of-the- art bioreactors. This field is evolving but slowly at the present moment. As also explained in Chapter 1, the potential to generate clinically relevant numbers and differentiate to any cell types in the body make pluripotent stem cells as alternative source of cells for regenerative medicine for various degenerative diseases such as heart failure, type 1 diabetes, stroke, haematological disorders, neurodegenerative disorders, spinal cord injury, osteoarthritis and kidney failure. Cell fate specifications are controlled by both internal and external cues. It is challenging to control the differentiation and to direct the development along specific pathways.

Although ES cells are known to differentiate to different lineages but complete organogenesis is limiting under *in vitro* conditions unless these cells are injected into mice, thus elucidating full developmental potential of these cells [9]. However, if ES cells are injected into severe combined immunodeficient mice they form teratomas that include differentiated tissues *i.e.* smooth muscle, striated muscle, bone, cartilage, gut, respiratory epithelium, squamous epithelium, hair, neural epithelium, and ganglia all derived from three germ layers [10].

Human ES cell lines exhibit more consistant developmental patterns compared to that found with human EC lines *i.e.* the human EC cell line NTERA2 c1.D1 when form teratomas in immunodeficient mice generally contain simple tubular structures like primitive gut, neural rosettes, and neuropile [11]. In order to fully understand the mechanism of induction and modelling that occur during organogenesis, the future studies should involve in depth exploring all the tissues extracted from teratomas and/or inducing morphogenesis *in vivo* by using modulators, growth factors including, co-culturing with known tissues, cell sorting based on reporters and gene cloning with homeobox transcription factors [12-20].

A significant progress has been made in defining signals that determine stem cell fate during differentiation and reporter lines have been produced for enrichment of specific cell types using high throughput assays. It is a growing field. As emphasized in chapter 8, several groups now have developed protocols for the relatively pure isolation of oligodendrocytes. These cells are now under clinical trials in patients after Food and Drug Administration (FDA) final approval. Similarly, another recent FDA approval has been provided for clinical trials with hESC-derived retina pigment epithelial cells for patients with macular degeneration and we expect more such trials in future.

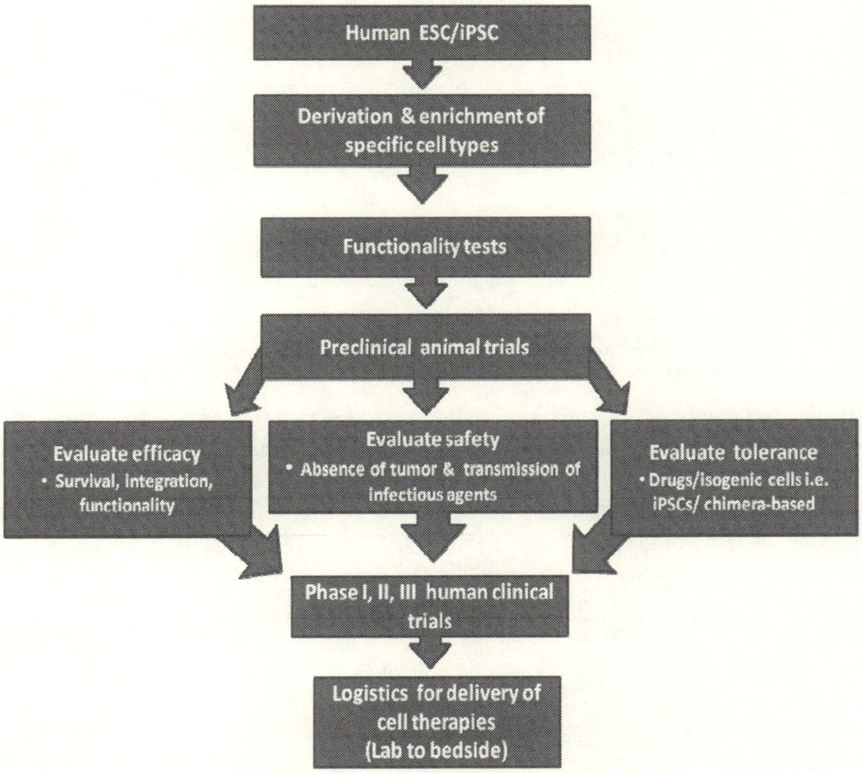

Fig. 1: Milestones for developing stem cell-based cell therapies-lab to bedside.

(ii) Survive and Integrate into the Surrounding Tissue after Transplant

With the development of stem cell technology it is now possible to generate almost all type(s) of cells in our body and consequently, the therapeutic potential of such technology could be enormous.

Diseases that include destruction and/or dysfunction of a specific cell type, such as pancreatic islet cells in diabetes mellitus, or destruction of brain dopaminergic neurons in Parkinson's disease could be treated with cells derived from hESC/iPSC. As also described in Chapter 8 and 15 of this eBook, a number of recent studies with known animal models demonstrated improvements when transplanted with an appropriate cell derivative from ES cells or fetal cells in such diseases like diabetes, Parkinson's disease, spinal cord injury, Purkinje cell degeneration, liver failure, heart failure, Duchenne's muscular dystrophy, and osteogenesis imperfecta [16, 21-28]. However, issues still remain about the safety and graft rejection with these transplantations. Isolation of a pure population of desired cells from ES cells and eliminating pluripotent stem cells in future will offer respite from undesirable side effects such as teratomas fromation (Fig. **1**).

A potential alternative to avoid immunological rejection after nonautologous transplantation of stem cells is the use of reprogrammed adult cell (iPSC) technology, which means derivation of patient-specific and pluripotent cells derived from adult somatic cells. These cells have been generated from mouse and human somatic cells by overexpression of several defined factors [29-36]. Recently, generation of iPSCs from human NSCs with a single transcription factor, OCT4, or using direct delivery of recombinant proteins has been described [37, 38]. iPSCs have identical patterns in gene expression, chromatin methylation, and embryoid body and viable chimera formation [39] as ESCs. They are capable of differentiation toward all cell types, including neurons, glia, NPCs, and motoneurons [40, 41]. Furthermore, the derivation of iPSCs using nonviral methods [42] or by chemicals and small molecules [43] including protein iPSCs [44] makes this cell replacement strategy very attractive. Nevertheless, these cell types share similar disadvantages as other cell sources: teratoma formation, aberrant reprogramming, and the presence of transgenes in iPSC populations are the most concerning obstacles, which should be addressed before their clinical application [45].

(iii) Function Appropriately for the Duration of the Recipient's Life

It is critical to demonstrate that the transplanted cells function in a physiologic manner; *i.e.*, pancreatic islet cells display a normal glucose-responsive insulin secretion. Earlier studies with cardiomyocites and and dopaminergic neurons derived from murine ES cells do display normal physiologic functioning *in vitro* and *in vivo* [16, 46, 47]. However, mixed population of cells containing both the differentiated and undifferentiated populations may be present [48]. It is generally considered that cell lineage derived from ES cells are immature and efficacy of such cells must be demonstrated in animal models before any human trials (Fig. **1**). Some trials are being carried out for diabetes and Parkinson's disease in large animals like rhesus monkeys [49, 50].

(iv) Avoid Harming the Recipient in Any Way and to Avoid the Problem of Immune Rejection

It is pertinent that future studies demonstrate cell integration into host system in a physiological manner and such cells are reasonably modified to prevent rejection(cf therapeutic cloning)[51]. Other possibility of combating such rejection is by homologous recombination to "knock-out" major histocompatibility complex (MHC) class I and class II molecules in mouse ES cells including some measures to protect against natural killer cells-mediated destruction [52-54]. Alternatively, engineering ES cells such that Fas-ligand could be inserted and B7 antigens or CD40-ligand, could be deleted [55, 56].

Therapeutic cloning by nuclear transfer technology as a mean to produce patient specific cell lines may provide a more precise means to prevent rejection of transplanted cells.

However, this technique is very challenging and so far not successful in humans. Inducing tolerance by chimerism is an another approach that is becoming successful in small animals [16, 17, 21, 57-59]

CONCLUSIONS AND PROSPECTS

Human ES cells/iPSC are pluripotent cells, and are capable of indefinite self-renewal and differentiate into all cell types. Therefore the therapeutic potentials with these cells is high and thus represents a prime research interest to successfully treat a variety of diseases (*e.g.*, Parkinson's disease, diabetes, and heart failure), however, many barriers remain in the way of successful clinical trials. [60]. There are potential advantages and disadvantages of both embryonic stem cells as well as induced pluripotent stem cells. Recent work suggests that it might be possible to reprogram adult somatic cells into mature neurons without the intermediate step of induced pluripotent stem cells – a way towards autologous cell therapy. A number of issues need to be addressed prior to clinical trials investigating the efficacy of stem cell therapies in animal models. Further research into safety, optimal cell doze, cell delivery method, and techniques for *in vivo* cell tracking need to be completed to ensure the safety of potential trials. Additional studies to delineate the functionality of stem cell-derived lineages both *in vitro* and *in vivo* will aid in the planning of quality-controlled clinical studies.

REFERENCES

[1] Mack GS. ReNeuron and StemCells get green light for neural stem cell trials. Nat Biotechnol 2011 29(2): 95-7.

[2] Hussein SM, Batada NN, Vuoristo S. Copy number variation and selection during reprogramming to pluripotency. Nature 2011 471(7336): 58-62.

[3] Gore A, Li Z, Fung HL. Somatic coding mutations in human induced pluripotent stem cells. Nature 2011; 471(7336): 46-7.

[4] Lister R, Pelizzola M, Kida YS. Hotspots of aberrant epigenomic reprogramming in human induced pluripotent stem cells. Nature 2011; 471(7336): 46-7.

[5] Mayshar Y, Ben-David U, Lavon N. Identification and classification of chromosomal aberrations in human induced pluripotent stem cells. Cell Stem Cell 2011; 7(4): 521-31.

[6] Laurent LC, Ulitsky I, Slavin I. Dynamic changes in the copy number of pluripotency and cell proliferation genes in human ESCs and iPSCs during reprogramming and time in culture. Cell Stem Cell 2011; 8(1): 106-18.

[7] Baker DE, Harrison NJ, Maltby E. Adaptation to culture of human embryonic stem cells and oncogenesis *in vivo*. Nat Biotechnol. 2007; 25(2): 207-15.

[8] Pera, MF. The dark side of induced pluripotency. Nature 2011; 471(7336), 46-47.

[9] Hochedlinger K, Jaenisch R. Nuclear reprogramming and pluripotency. Nature. 2006; 441(7097): 1061-7.

[10] Kucia M, Reca R, Campbell FR. *et al.* A population of very small embryonic-like (VSEL) CXCR4(+)SSEA-1(+)Oct-4+ stem cells identified in adult bone marrow. Leukemia 2006; 20(5): 857-69.

[11] Sohur US, Emsley JG, Mitchell BD, Macklis JD. Adult neurogenesis and cellular brain repair with neural progenitors, precursors and stem cells. Philos Trans R Soc Lond B Biol Sci 2006; 361(1473): 1477-97.

[12] Schuldiner M, Yanuka O, Itskovitz-Eldor J *et al.* Effects of eight growth factors on the differentiation of cells derived from human embryonic stem cells. Proc Natl Acad Sci USA 2000; 97(21): 11307-11312.

[13] Bonner-Weir S, Taneja M, Weir GC *et al. In vitro* cultivation of human islets from expanded ductal tissue. Proc Natl Acad Sci USA 2000; 97(14): 7999-8004.

[14] Levinson-Dushnik M, Benvenisty N. Involvement of hepatocyte nuclear factor 3 in endoderm differentiation of embryonic stem cells. Mol Cell Biol 1997; 17(7): 3817-3822.

[15] Ferber S, Halkin A, Cohen H *et al.* Pancreatic and duodenal homeobox gene 1 induces expression of insulin genes in liver and ameliorates streptzotocin-induced hyperglycemia. Nat Med 2000; 6(5): 568-572.

[16] Soria B, Roche E, Berna G *et al.* Insulin-secreting cells derived from embryonic stem cells normalize glycemia in streptozotocin-induced diabetic mice. Diabetes 2000; 49(2): 157-162.

[17] Klug MG, Soonpaa MH, Koh GY *et al.* Genetically selected cardiomyocytes from differentiating embryonic stem cells form stable intracardiac grafts. J Clin Invest 1996; 98(1): 216-224.

[18] Li M, Pevny L, Lovell-Badge R *et al.* Generation of purified neural precursors from embryonic stem cells by lineage selection. Curr Biol 1998; 8(17): 971-974.

[19] Li A, Simmons PJ, Kaur P. Identification and isolation of candidate human keratinocyte stem cells based on cell surface phenotype. Proc Natl Acad Sci USA 1998; 95(7): 3902-3907.

[20] Weissman IL. Stem cells(1): units of development, units of regeneration, and units in evolution. Cell 2000; 100: 157-168.

[21] McDonald JW, Liu XZ, Qu Y *et al.* Transplanted embryonic stem cells survive, differentiate and promote recovery in injured rat spinal cord. Nat Med 1999; 5(12): 1410-1412.

[22] Horwitz EM, Prockop DJ, Fitzpatrick LA *et al.* Transplantability and therapeutic effects of bone marrow-derived mesenchymal cells in children with osteogenesis imperfecta. Nat Med 1999; 5(3): 309-313.

[23] Li RK, Weisel RD, Mickle DA *et al.* Autologous porcine heart cell transplantation improved heart function after a myocardial infarction. J Thorac Cardiovasc Surg 2000; 119(1): 62-68.

[24] Studer L, Tabar V, Mckay RDG. Transplantation of expanded mesencephalic precursors leads to recovery in parkinsonian rats. Nat Neurosci 1998; 1(4): 290-295.

[25] Brustle O, Choudhary K, Karram K *et al.* Chimeric brains generated by intraventricular transplantation of fetal human brain cells into embryonic rats. Nat Biotechnol 1998; 16(11): 1040-1044.

[26] Zhang W, Lee WH, Triarhou LC. Grafted cerebellar cells in a mouse model of hereditary ataxia express IGF-I system genes and partially restore behavioral function. Nat Med 1996; 2(1): 65-71.

[27] Lilja H, Arkadopoulos N, Blanc P *et al.* Fetal rat hepatocytes: isolation, characterization, and transplantation in the Nagase analbuminemic rats. Transplantation 1997; 64(9): 1240-1248.

[28] Kobayashi N, Miyazaki M, Fukaya K *et al.* Transplantation of highly differentiated immortalized human hepatocytes to treat acute liver failure. Transplantation 2000; 69(2): 202-207.

[29] Kim JB, Sebastiano V, Wu G *et al.* Oct4-induced pluripotency in adult neural stem cells. Cell 2009; 136(3): 411-419.

[30] Takahashi K, Yamanaka S. Induction of pluripotent stem cells from mouse embryonic and adult fibroblast cultures by defined factors. Cell 2006; 126(4): 663-676.

[31] Park IH, Zhao R, West JA *et al.* Reprogramming of human somatic cells to pluripotency with defined factors. Nature 2008; 451(7175): 141-146.

[32] Jaenisch R, Young R. Stem cells, the molecular circuitry of pluripotency and nuclear reprogramming. Cell 2008; 132(4): 567-582.

[33] Takahashi K, Tanabe K, Ohnuki M *et al.* Induction of pluripotent stem cells from adult human fibroblasts by defined factors. Cell 2007; 131(5): 861-872.

[34] Yu J, Vodyanik MA, Smuga-Otto K *et al.* Induced pluripotent stem cell lines derived from human somatic cells. Science 2007; 318(5858): 1917-1920.

[35] Lowry WE, Richter L, Yachechko R *et al.* Generation of human induced pluripotent stem cells from dermal fibroblasts. Proc Natl Acad Sci U S A 2008; 105(8): 2883-2888.

[36] Nakagawa M, Koyanagi M, Tanabe K *et al.* Generation of induced pluripotent stem cells without Myc from mouse and human fibroblasts. Nat Biotechnol 2008; 26(1): 101-106.

[37] Kim JB, Greber B, Arauzo-Bravo MJ *et al.* Direct reprogramming of human neural stem cells by OCT4. Nature 2009; 461(7264): 649-643.

[38] Kim D, Kim CH, Moon JI *et al.* Generation of human induced pluripotent stem cells by direct delivery of reprogramming proteins. Cell Stem Cell 2009; 4(6): 472-476.

[39] Amabile G, Meissner A. Induced pluripotent stem cells: current progress and potential for regenerative medicine. Trends Mol Med 2009; 15(2): 59-68.

[40] Dimos JT, Rodolfa KT, Niakan KK *et al.* Induced pluripotent stem cells generated from patients with ALS can be differentiated into motor neurons. Science 2008; 321(5893): 1218-1221.

[41] Wernig M, Zhao JP, Pruszak J *et al.* Neurons derived from reprogrammed fibroblasts functionally integrate into the fetal brain and improve symptoms of rats with Parkinson's disease. Proc Natl Acad Sci U S A 2008; 105(15): 5856-5861.

[42] Kaji K, Norrby K, Paca A *et al.* Virus-free induction of pluripotency and subsequent excision of reprogramming factors. Nature 2009; 458(7239): 771-775.

[43] Huangfu D, Osafune K, Maehr R *et al.* Induction of pluripotent stem cells from primary human fibroblasts with only Oct4 and Sox2. Nat Biotechnol 2008; 26(11): 1269-1275.

[44] Zhou H, Wu S, Joo JY *et al.* Generation of induced pluripotent stem cells using recombinant proteins. Cell Stem Cell 2009; 4(5): 381-384.

[45] Yamanaka S. A fresh look at iPS cells. Cell 2009; 137(1): 13-17.

[46] Bain G, Kitchens D, Yao M *et al.* Embryonic stem cells express neuronal properties *in vitro*. Dev Biol 1995; 168(2): 342-357.

[47] Wobus AM, Wallukat G, Hescheler J. Pluripotent mouse embryonic stem cells are able to differentiate into cardiomycocytes expressing chronotropic responses to adrenergic andcholinergic agents and Ca2+ channel blockers. Differentiation 1991; 48(3): 173-182.

[48] Okabe S, Forsberg-Nilsson K, Spiro AC *et al.* Development of neuronal precursor cells and functional postmitotic neurons from embryonic stem cells *in vitro*. Mech Dev 1996; 59(1): 89-102.

[49] Burns RS, Chiueh CC, Markey SP *et al.* A primate model of parkinsonism: selective destruction of dopaminergic neurons in the pars compacta of the substantia nigra by N-methyl-4-phenyl-1,2,3,6-tetrahydropyridine. Proc Natl Acad Sci USA 1983; 80(14): 4546-4550.

[50] Jones CW, Reynolds WA, Hoganson GE. Streptozotocin diabetes in the monkey: plasma levels of glucose, insulin, glucagon, and somatostatin, with corresponding morphometric analysis of islet endocrine cells. Diabetes 1980; 29(7): 536-546.

[51] Kaufman DS, Odorico JS, Thomson JC. Transplantation therapies from human embryonic stem cells—circumventing immune rejection. e-biomed 2000; 1: 11-15: available at http: //www.liber/pub.com/EBI/defaultstatic.asp.

[52] Grusby MJ, Auchincloss H Jr, Lee R *et al.* Mice lacking major histocompatibility complex class I and class II molecules. Proc Natl Acad Sci USA 1993; 90(9): 3913-3917.

[53] Westphal CH, Leder P. Transposon-generated 'knock-out' and 'knock-in' gene-targeting constructs for use in mice. Curr Biol 1997; 7(7): 530-533.

[54] Hardy RR, Malissen B. Lymphocyte development. The (knock) ins and outs of lymphoid development [Editorial] Curr Opin Immunol 1998; 10(2): 155-157.

[55] Harlan DM, Kirk AD. The future of organ and tissue transplantation: can T-cell costimulatory pathway modifiers revolutionize the prevention of graft rejection? [Review] JAMA 1999; 282(11): 1076-1082.

[56] Walker PR, Saas P, Dietrich PY. Role of Fas ligand (CD95L) in immune escape: the tumor cell strikes back. [Review] J Immunol 1997; 158(10): 4521-4524.

[57] Thomson JA, Kalishman J, Golos TG *et al.* Isolation of a primate embryonic stem cell line. Proc Natl Acad Sci USA 1995; 92(17): 7844-7848.

[58] Thomson JA, Itskovitz-Eldor J, Shapiro SS *et al.* Embryonic stem cell lines derived from human blastocysts. Science 1998; 282(5391): 1145-1147.

[59] Brustle O, Spiro AC, Karram K *et al. In vitro*-generated neural precursors participate in mammalian brain development. Proc Natl Acad Sci USA 1997; 94(26): 14809-14814.

[60] Odorico, JS, Kaufman, DS, and Thomson JA Multilineage Differentiation from Human Embryonic Stem Cell Lines. Stem Cells 2001; 19(3): 193-204.

Subject Index

Made in United States
Orlando, FL
10 November 2026

90522520R00146